The Facts On File
DICTIONARY
of
ECOLOGY
and
THE ENVIRONMENT

The Facts On File

DICTIONARY
of
ECOLOGY
and
THE ENVIRONMENT

Edited by
Jill Bailey

☑®
Facts On File, Inc.

The Facts On File Dictionary of Ecology and The Environment

Facts On File, Inc.
132 West 31st Street
New York NY 10001

Library of Congress Cataloging-in-Publication Data

The Facts on File dictionary of ecology and the environment / edited by Jill Bailey.
    p. cm.
Includes bibliographical references (p.).
  ISBN 0-8160-4922-X (hc: alk. paper).
  1. Ecology—Dictionaries. 2. Environmental sciences—Dictionaries. 1. Title: Dictionary of ecology and the environment. II. Bailey, Jill. III. Facts On File, Inc.

  QH540.4.F35 2003
  577'.03—dc22   2003060353

Facts On File books are available at special discounts when purchased in bulk quantities for businesses, associations, institutions, or sales promotions. Please call our Special Sales Department in New York at (212) 967-8800 or (800) 322-8755.

You can find Facts On File on the World Wide Web at
http://www.factsonfile.com

Compiled and typeset by Market House Books Ltd, Aylesbury, UK

Printed in the United States of America

  MP   10 9 8 7 6 5 4 3 2 1

This book is printed on acid-free paper

CONTENTS

PREFACE

This dictionary is one of a series covering the terminology and concepts used in important branches of science. *The Facts On File Dictionary of Ecology and The Environment* is planned as an additional source of information for students taking Advanced Placement (AP) Science courses in high schools. It will also be helpful to older students taking introductory college courses.

This volume covers the topics important for an understanding of the basic principles of ecology - the scientific study of the relationships between organisms and their natural environment. It also deals with the wider subject of how human populations interact with and affect the environment as a whole. Environmental science is an area of study involving a number of different disciplines besides biology – chemistry, geology, meteorology, human geography, etc. There is particular interest in the effects of pollution and on conservation and the management of habitats. The definitions are intended to be clear and informative and, where possible, we have provided helpful diagrams and examples. The book also has a selection of short biographical entries for people who have made important contributions to ecology and environmental science. The appendixes contain a short list of useful webpages and an informative bibliography.

The book will be a helpful additional source of information for anyone studying the AP Environmental Science course, and also to students of AP Biology. However, we have not restricted the content to this syllabus. Ecology and the environment are important to everyone, and we have tried to cover these subjects in an interesting and informative way.

ACKNOWLEDGMENTS

Consultant editor

Stewart Thompson B.Sc., Ph.D.

Contributors

John Clark B.Sc.
Robert Hine B.Sc.
Eve Daintith B.Sc.

abaptation   The process by which evolutionary forces acting on ancestral forms have helped determine present-day adaptations of organisms to their environment.

abiotic   Nonliving. Abiotic factors are the physical and chemical aspects of an organism's environment, such as light, temperature, water, oxygen, and carbon dioxide. They include climatic, EDAPHIC and PHYSIOGRAPHIC factors. *Compare* biotic.

abscission   The controlled separation of a part of an organism from the rest. For example, the shedding of leaves in the fall and the dropping of fruits is triggered by changes in the balance of plant growth substances (hormones).

absolute humidity   *See* humidity.

absorbed dose   The amount of energy absorbed by a tissue or other substance from incident radiation. It is measured in grays (Gy): one gray = the transfer of one joule of energy to one kilogram of material.

absorption   1. The retention of radiant energy by an object, e.g. by the pigments of a photosynthetic organism.
2. The uptake of small nutrient molecules into the body of an organism. In animals this takes place after digestion of larger food molecules. In plants it includes the uptake of water and solutes.
3. *See* adsorption.

abundance   A measure of the number of individual organisms in an area. An assessment of FREQUENCY (the relative abundance of each species present in an area), gives a quick subjective estimate. COVER and abundance may be combined for a subjective community description (*see* Braun-Blanquet scale). Quantitative measures of abundance include DENSITY (the number of individuals in a given area), cover, or the frequency of occurrence of a species in randomly placed QUADRATS.

abyssal zone   The ocean-floor environment between 4000 and 6000 meters in depth. It is characterized by extremely high water pressure, low temperatures and nutrient levels, and an almost total absence of light. *See also* benthic zone.

acclimation   *See* acclimatization.

acclimatization   A reversible change in the physiology or morphology of an organism in response to changes in its environment. The term is generally used in the context of the natural environment. The term *acclimation* tends to be used in a laboratory context.

acid   A compound that acts as a proton donor in aqueous solution, i.e. it releases hydrogen ions or protons, giving a pH less than 7. Acids turn litmus paper from blue to red, and react with alkalis (bases) to yield neutral salts. *See* pH. *Compare* alkali.

acid rain   Rain with a very low pH (below pH 5.6 and often below pH 4), due to pollution from oxides of nitrogen and sulfur released by the burning of fossil fuels or other industrial emissions. These pollutants combine with water in the atmosphere, forming nitric and sulfuric acids, a process that may be catalyzed by other pollutants such as ammonia, hydrogen peroxide, or ozone. Acid rain lowers the pH of soil, lakes, and rivers. The sulfuric acid reacts with minerals in the soil to produce ammonium sulfate, which liberates toxic ions of aluminum and certain heavy metals, which in turn inhibit metabolic reactions, cause the leaching out of certain plant nutrients from the soil, and inhibit the uptake of others. Acid rain also has a direct effect on plant life, breaking down lipids and damaging membranes in leaves. When these ions leach into waterways, they damage the gills of fish and contaminate drinking-water supplies.

acid soil A soil with a pH less than 7.0. Soils with pH below 5 are said to be 'very acid', e.g. the surface horizons of some BROWN EARTH soils. High soil acidity may be due to LEACHING of soluble salts, such as calcium carbonates, in areas of high rainfall. Also, in such climates very little clay (derived from the weathering of the bedrock) moves to the upper horizons of the soil because rain is continually percolating downward, so there are few cations to exchange with the hydrogen ions. Alternatively, acidity may be caused by the accumulation of HUMIC ACIDS in waterlogged soils. ACID RAIN has led to the acidification of many soils. In cold regions acid soils result from the accumulation of organic acids due to the slow decomposition of leaf litter in coniferous forests.

acre A unit of area. 1 British statute acre = 0.4047 hectare, or 4840 square yards. In former times, different parts of the British Isles, and even different forests, had different standards for the acre, all larger than the current statute acre.

acrotelm The upper layer of blanket PEAT that forms in areas where the rainfall is so high that decomposition of plant remains is seriously retarded. The acrotelm is above the average level of the WATER TABLE, so is saturated only from time to time. As a result, air can enter it, triggering active decomposition of the peat, and the breakdown of the plant remains into humic acids. When not saturated, water can flow readily through the acrotelm, but heavy surface RUNOFF may cause peat erosion if the peat gets very dry and loose. The layer between the acrotelm and the impermeable mineral subsoil below is called the *catotelm*. It is permanently saturated, with no water flow and very little oxygen. The peat here is very dense and remains undecomposed. The boundary between acrotelm and catotelm varies with the level of the water table, which determines whether the peat is breaking down or building up. *See* blanket bog.

Actinobacteria (actinomycetes; ray fungi) A phylum of EUBACTERIA that includes rodlike coryneform bacteria and branching filamentous true actinobacteria. They produce thick-walled resistant resting spores called actinospores. Most actinobacteria are aerobic SAPROBES, but some species are PATHOGENS of humans, other animals, and plants. Some are involved in lichenlike symbioses. *See* lichens.

actinomycetes *See* Actinobacteria.

activated carbon Charcoal that has been heated in a vacuum to drive off absorbed gas. Activated carbon can absorb large quantities of gases, and is used to remove solvent vapors and clarify liquids; it is also used in gas masks.

activated sludge process A SEWAGE treatment process during which sludge mixed with sewage is aerated by means of mechanical agitation or compressed air (*see* aerator). The aeration encourages the growth of microorganisms that break down the sludge into a suspension of organic solids called *activated sludge*, and absorb organic compounds from the sewage (*primary effluent*). After several hours, the mixture is allowed to flow into a further settling tank, the *clarifier*, where the activated sludge settles out and the clear water above is disinfected and discharged as *secondary effluent*. Some of the sludge is recirculated into the aerator, and mixed with more raw sewage. The remainder of the sludge is treated and disposed of.

active transport The transport of molecules or ions across a cell membrane against a concentration gradient, with the expenditure of energy. The mechanism involves a carrier protein that spans the membrane and transfers substances in or out of the cell by changing shape.

acylglycerols   *See* lipid.

adaptation   1. A genetically determined characteristic, such as a structure, physiological process or behavior, that makes an organism better suited to its environment, thereby increasing its survival chances. NATURAL SELECTION will act on the underlying GENOTYPE: individuals possessing favorable adaptations will leave more surviving offspring in successive generations, so the adaptation will gradually spread through the population.
2. Changes in form and/or behavior of an organism during its lifetime in response to environmental stimuli. For example, the development of sun and shade leaves on the same tree. *See also* acclimatization.
3. Changes in the excitability of a sense organ in response to continuous stimulation. For example, continued repetition of a stimulus may lead to a decrease in responsiveness.

adaptive function   A mathematical function that combines phenotypic FITNESS in different environments to give a measure of the overall fitness of a PHENOTYPE in a heterogeneous environment. Where organisms encounter patches of habitat of type i in direct proportion to their frequency (p_i), the overall fitness of a phenotype (W) is the average of its fitnesses in each habitat type weighted by their frequencies:
$$W = p_1 W_1 + p_2 W_2$$
If a graph of fitness in each of two environments is plotted, the adaptive function forms a line of equal fitness whose position on the graph is determined by the relative frequencies of the two environments. The slope of the adaptive function will vary from place to place as the relative frequencies of the different types of HABITAT PATCHES vary.

The above equation is appropriate in spatially different environments. When ap-plied to temporally varying environments, where habitat changes with time rather than space, the adaptive function is most accurately represented by a geometric mean, since fitness in this case is expressed as changes in numbers of individuals with each phenotype from one generation to the next. The average fitness of a phenotype is then:
$$W = W_1 p_1 W_2 p_2$$
See patch dynamics; patchiness.

adaptive radiation   The process by which new species and varieties adapted to specialized modes of life arise from a common ancestor. This may happen when a species is introduced into a new environment, when competitors are eliminated by a catastrophe, as happens after mass extinctions, or populations become isolated from competition by forces such as mountain-building, inundation, or continental drift. Examples include the evolution of lemurs from primitive primates on the island of Madagascar, and the diversification of primitive mammals into terrestrial, arboreal, and aquatic forms during the Tertiary period.

adenosine triphosphate   *See* mitochondrion; respiration.

adiabatic   Without gain or loss of heat, i.e. without exchange of heat with the surroundings.

adiabatic cooling and heating   The cooling of rising air associated with its expansion as atmospheric pressure declines with increasing altitude. For example, in rising columns of air, the volume of air increases as the pressure of the atmosphere above it decreases with increasing altitude. This cooling effect is usually about 1°C per 100 m for dry air, 0.5°C per 100 m for saturated air (because moisture condenses out of the air, releasing latent heat). This temperature change affects the rate at which the air rises relative to adjacent bodies of air, the stability of the air column, and cloud formation. The *law of adiabatic expansion or compression* states that a gas will cool if allowed to expand freely from a higher pressure to a lower pressure without the transfer of external energy to the gas. Similarly, it will heat if compressed from a lower to a higher pressure in the absence of energy transfer from the gas.

ADP *See* respiration.

adsorption The adhesion of molecules to the surface of solids or liquids. In adsorption, the molecules remain on the surface. In ABSORPTION, the molecules permeate into the bulk of the material.

advection The horizontal movement of a mass of air or liquid (e.g. air or water currents), or the transport of substances (e.g. pollutants) by such movement. Transport of energy (such as heat) in this way is called *advective energy*.

adventitious roots Roots that do not arise in the normal position on the plant. For example, they may arise from the stem, as in the rooting of runners and bulbs, or from the leaf, as in certain begonias, which can propagate from leaf cuttings.

aeolian *See* eolian.

aerator A tank, usually of concrete, through which air is passed during the treatment of sewage by the ACTIVATED SLUDGE PROCESS. Compressed air may be passed through porous diffusers at the base of the aerator, in which case the bubbling of the air mixes and oxygenates the mixture, or from the atmosphere, mixed into the sewage using mechanical propellers at the surface.

aerenchyma A kind of tissue found in many aquatic plants, such as water lilies (*Nymphaea* spp), which contains many large intercellular air spaces that aid the diffusion of oxygen to the roots and may also increase buoyancy.

aerial root A root that arises above the level of the soil. Most aerial roots are adventitious roots, arising from stem tissue, rather than true lateral roots. Aerial roots are typical of EPIPHYTES, and may hang down into the air to absorb moisture from it, as in some orchids; the epidermal layers of these roots develops a sheath of empty cells, the *velamen*, that acts rather like a sponge. The *prop roots* or STILT ROOTS of plants such as maize and some palms, which arise above the soil and grow down into it, provide extra support, as do BUTTRESS ROOTS. Breathing roots, or *pneumatophores*, found in some MANGROVES and other plants of waterlogged areas, are rich in lenticels and serve to get oxygen into the root. Some of these are true aerial roots, arising above the ground, whereas others are lateral roots that grow up into the air.

aerobe An organism that uses aerobic RESPIRATION and can live and grow only in the presence of free oxygen. *Compare* anaerobe.

aerobic 1. Describing an organism that requires oxygen for respiration.
2. Describing an environment that contains free oxygen.

aerobic respiration *See* respiration.

aerodynamic method A technique used to estimate potential EVAPOTRANSPIRATION from a surface (for example, a lake or an area of vegetation) that takes into account both the supply of heat energy and the transportation of water vapor away from the evaporating surface by air movements.

aerosol 1. A dispersion of liquid or solid particles in a gas, e.g. mist, smoke.
2. A pressurized container with a spray nozzle use to dispense aerosols such as deodorants and insecticides. The propellent gases used in many such aerosols are potent POLLUTANTS.

aesthetic injury level The level of visible injury to plants or other commodities at which it is deemed desirable to apply pesticides.

afforestation The establishment of a forest on land not previously forested. This may be by natural colonization by trees or by deliberate planting.

age class   The individuals in a population of a particular age.

age distribution   The number of individuals of each age in a population. When the birth rate and survival rates for each age remain unchanged for a long period of time, the population acquires a *stable age distribution* and the population will grow or decline at a constant rate per head. When the population is not changing in size, the stable age distribution is called the *stationary age distribution*. Such a population will usually have more older and fewer younger individuals than a growing population.

Agent Orange   A defoliant used in the Vietnam War. Composed of esters of trichlorophenoxyacetic acids, it was heavily contaminated with PCDDs (polychlorinated dibenzodioxins) and PCDFs (polychlorinated dibenzofurans). These have a harmful effect on human reproduction, development, and immunity to disease. They are also suspected to be carcinogenic. Agent Orange is a persistent POLLUTANT, especially in the marine environment. *See also* dioxin.

age structure   The relative proportions of a population in different age classes. Age structure affects the rate at which a population grows; a population with only a small proportion of individuals of reproductive age will grow relatively slowly. The growth rate of the population is the sum of the growth rates of the individuals in each age class weighted by the proportion of each age class in the population. The smaller the range of age in each age class, the more accurate the growth rate calculated in this way.

aggregate   1. A rock consisting of mineral or rock fragments.
2. A clump of soil particles, ranging from microscopic granules to small crumbs, which forms the basic structural unit of SOIL.

aggregated dispersion   *See* dispersion.

aggregate size distribution   The proportions of surface soil aggregates of different sizes in a given soil. The aggregate size distribution is affected by the application of fertilizers, the soil tillage system, and by wind erosion. It affects soil properties such as aeration, drainage, and water retention. *See* soil erosion; soil structure; soil texture.

aggregate species   A group of species so closely related that it is very difficult to distinguish between them in the field. Such species usually arise through HYBRIDIZATION between closely related species. An example is the blackberry (*Rubus fruticosus* agg.).

aggregation of risk   A distribution of risk of attack from predators in which in some habitat patches the prey are at greater or lesser risk of predation than would be expected by chance alone. The term is usually applied to the risk of a host species being attacked by a PARASITOID. *See also* aggregative response.

aggregative response   The response of a predator to prey density in which it spends more time in habitat patches with higher densities of prey, leading to higher predator densities in these patches, or in patches with lower densities of prey, leading to lower densities of predators in these patches. *See also* aggregation of risk.

aggression   A behavior pattern that intimidates or injures another organism, except for the purpose of predation. Aggression ranges from fighting over dominance in social animals to the singing of a bird in defense of its territory.

aggressive mimicry   *See* mimicry.

agonistic   Describing behavior that aids survival, including AGGRESSION and defensive and avoidance behaviors. Such behavior as defense of territory, for example, may help an animal obtain sufficient food or access to mates, both vital to survival. In its purest sense, agonistic behavior improves the chances of survival of the individual's genes, so it can also include certain kinds of cooperative behavior between related individuals. *See* territoriality.

agricultural potential   *See* land-use capability.

agricultural potential maps   *See* map.

Agricultural Revolution   The change some 10–12 000 years ago from a nomadic pastoral way of life to a more settled lifestyle based on the cultivation of crops, especially cereals. This began in the Middle East and western Asia, and in the Nile valley. It led to the growth of towns and cities, and to the development of more complex, less portable tools and technologies, more complex societies, and the more rapid evolution of human culture.

agroforestry   An integrated system of farming in which herbaceous crops and tree crops are cultivated simultaneously on the same patch of land.

agronomy   A branch of agriculture that deals with the theory and practice of crop production and the management of soils.

air   The mixture of gases that forms the ATMOSPHERE.

air pollution   *See* pollution.

air temperature   The temperature of the atmosphere at a specific altitude in a specific location, or the temperature of air in a confined space such as a room or a reaction vessel. On a large scale, the NASA GISS surface temperature is a measure of the changing global surface temperature derived from data from many meteorological stations around the world as well as (more recently) satellite data for sea-surface temperature. *See also* global warming; greenhouse effect.

alarm responses   Animal responses to danger that serve to warn other animals. They may be visual, e.g. the flash of the

white underside of the tails of rabbits or white-tailed deer as they run for cover. Auditory signals such as the alarm calls of birds are often recognized by many different species and function even in dense forest. Injured animals may release alarm chemicals; for example, minnows and aquatic snails release chemicals into the water that alert other members of the same species.

alary polymorphism   *See* polymorphism.

albedo   The fraction of incident light that is reflected by a surface or body, such as a cloud or planet. Albedo is an important parameter in calculations of energy balance.

aldrin   A chlorinated hydrocarbon used as an insecticide. Its use is being discontinued because it is toxic to warm-blooded animals through its effects on the central nervous system.

algae   A diverse group of photosynthetic eukaryotes, most of which are not highly differentiated into tissues and organs when compared with lower plants. They lack distinguishable roots, stems, or leaves and there is no true vascular system. The key features distinguishing them from plants (kingdom PLANTAE) is the lack of a layer of sterile cells surrounding the reproductive organs, and lack of an embryo stage during development. The algae range from single-celled organisms to large seaweeds over 50 m long showing distinct tissue differentiation. They are placed in the kingdom PROTOCTISTA.

algal bloom   *See* bloom.

alkali   A substance that dissolves in water to give a solution of pH greater than 7. Alkalis are usually soluble hydroxides or compounds that release hydroxyl (OH⁻) ions in solution. They turn litmus paper from red to blue and react with ACIDS to yield neutral salts. *See* pH.

alkaline soil   A soil with a pH greater

than 7.0. Soils with pH above 9 are said to be 'strongly alkaline'. Such soils are usually rich in calcium ions, and are often derived from rocks such as limestone or chalk, which are mostly composed of calcium carbonate.

alkylmercury fungicides Mercury-based organic compounds once widely used as FUNGICIDES, especially on seeds, paint, and paper. They are skin irritants, and can also affect the central nervous system, sometimes with fatal results. Mercury from the degradation of such fungicides can accumulate in fish including edible species such as salmon, swordfish, and tuna. The use of alkylmercury fungicides has been prohibited in the United States since the early 1970s. *See also* bioaccumulation.

allele (allelomorph) One of several alternative forms of a particular gene. Different alleles often have different effects on the phenotype. Where more than one allele is present in an individual, the phenotypic outcome depends on which alleles are recessive or dominant: a *dominant allele* is expressed regardless of which other alleles are present, whereas a *recessive allele* is expressed only in the absence of dominant alleles. In a few cases, more than one allele may be dominant, and the phenotypic outcome is a blend of the expression of the two alleles. For example, *codominant alleles* for white and red flower color respectively may, when present together, give rise to pink flowers. *See* gene; phenotype.

allelochemical *See* allelopathy.

allelopathy The release by a plant of a chemical (*allelochemical*) that poisons or inhibits the growth of nearby plants, so reducing COMPETITION.

Allen's rule The generalization that mammals from cold climates have shorter extremities (ears and limbs) that otherwise similar mammals from warmer climates. *See also* Bergmann's rule.

alliance *See* association.

allochthonous Describing materials transported into a community or ECOSYSTEM from outside, for example minerals and organic matter carried by streams, lakes, and oceans but derived from adjacent terrestrial ecosystems.

allogenic succession *See* succession.

allopatric speciation *See* speciation.

allopolyploid *See* polyploidy.

alluvial Found in or derived from river or stream deposits. For example, an *alluvial fan* is a fan of debris dropped by a stream or river where it experiences a change of gradient and flow rate slows, so the water can no long carry so large a load. *Alluvial soil* is formed on river floodplains and deltas where new sediment is deposited on the surface at each successive flooding. The repeated addition of nutrients in this way can lead to very fertile soils.

alpha decay The spontaneous emission from a radioactive isotope of an alpha particle – a positively charged particle consisting of the nucleus of a helium atom, comprising 2 neutrons and 2 protons.

alpine tundra *See* tundra.

alternation of generations The occurrence of alternating haploid and diploid individuals in the life cycle of an organism. In bryophytes, vascular plants, many algae, and some fungi a haploid gamete-producing sexual phase, the *gametophyte*, alternates with a diploid spore-producing asexual phase, the *sporophyte*. In bryophytes the dominant generation is the gametophyte, which forms the main plant body. In vascular plants the sporophyte is dominant. Sporophyte and gametophyte may or may not be morphologically similar, depending on the species. Among animals, many invertebrates, such as jellyfish and flatworms, show alternation of sexual and asexual generations, but there are no distinct haploid and diploid generations.

alternative energy Energy derived from renewable sources, such as wind and wave power, solar energy (*see* solar power), hydroelectric power, geothermal energy, hydrogen-based FUEL CELLS, biomass energy, and wood energy. These technologies do not draw on finite resources and are generally perceived to be nonpolluting.

ambient Present on all sides; e.g. the ambient temperature is the temperature of the surrounding air.

amensalism An interaction between two individuals or species in which one is adversely affected by the other. An example is the production of antibiotics by certain molds to kill or inhibit the growth of nearby bacteria. Such associations between microorganisms are termed *antibiosis*. *Compare* commensalism; mutualism.

amino acids Organic acids that contain an acidic carboxylate group (–COOH) and a basic amino group (–NH$_2$) in their molecules. Proteins are formed from chains of amino acids and there are twenty of these that occur naturally. In adult humans ten of the twenty amino acids can be synthesized by the body itself. Since these are not required in the diet they are known as *nonessential amino acids*. The remaining ten cannot be synthesized by the body and have to be supplied in the diet. They are known as *essential amino acids*. Various other amino acids fulfill important roles in metabolic processes other than as constituents of proteins. For example, ornithine and citrulline are intermediates in the production of urea.

ammonia NH$_3$. A colorless pungent gas, highly soluble in water and alcohol. It is highly reactive in solution, reacting with most acids to form salts, and with metals to form nitrides. Ammonia is a by-product of the breakdown of proteins and amino acids in the body. Highly toxic to cells, it is excreted directly by fish and many amphibians, dissolving in the surrounding water. Terrestrial animals convert it to the less toxic urea (CO(NH$_2$)$_2$) during deamination in the liver. Ammonia is formed naturally during the decomposition of proteins, purines, and urea by bacteria. In the soil, nitrifying bacteria (*see* nitrification) such a *Nitrosomonas* convert ammonia to nitrites, an essential stage in the NITROGEN CYCLE. *See also* Haber process.

Amphibia The class of vertebrates that contains the most primitive terrestrial tetrapods – the frogs, toads, newts, and salamanders. Amphibians have four pentadactyl limbs, a moist skin without scales, and a middle-ear apparatus for detecting airborne sounds. They are cold-blooded (exothermic) and the adults have lungs and live on land but their skin, also used in respiration, is thin and moist and body fluids are easily lost, therefore they are confined to damp places. In reproduction, fertilization is external and so they must return to water to breed. The eggs are covered with jelly and the aquatic larvae have gills for respiration and undergo metamorphosis to the adult.

amphidromous *See* diadromous.

Amphipoda (amphipods) An order of small crustaceans (*see* Crustacea), many with flattened shrimplike bodies and long hairy antennae. They breathe by means of gills, and are found in lakes, rivers, oceans, muddy parts of the ocean floor, sandy beaches, caves, and moist tropical island habitats. Amphipods include sand fleas (sand hoppers) and scuds. Most species are herbivores or scavengers, but some have piercing and sucking mouthparts and are external parasites of larger invertebrates such as cnidarians and sponges. They are important sources of food for larger invertebrates, fish, seabirds, shore birds, small whales, seals, and sea lions.

anabolism The metabolic synthesis of complex molecules from simpler ones. *Compare* catabolism.

anadromous *See* diadromous.

anaerobe An organism that is able to live in the absence of oxygen. *Obligate anaerobes* are unable to live if free oxygen

anion

is present. *Facultative anaerobes* are able to live in the presence of free oxygen but can survive in anaerobic conditions by using anaerobic RESPIRATION. Of these, some (e.g. denitrifying bacteria) never use oxygen for respiration, whereas others use it if it is available (e.g. yeasts). *Compare* aerobe.

anaerobic respiration   *See* respiration.

analysis of variance (ANOVA)   A statistical analysis that tests the significance of differences between experimental samples. In its strict sense, the term is applied only to tests for differences between more than two variables. Observed differences between the samples are compared with the values that would obtain if the differences between the samples had arisen purely by chance (the NULL HYPOTHESIS). While T-TESTS, correlation analysis, and regression analysis can be applied to two samples, other tests are needed where more than two variables are involved. *Frequency distributions* – plots of the number of samples with each given value – can be tested by the CHI-SQUARED TEST or by multiple correlation or multiple regression analysis.

anatomy   *See* morphology.

anchorage   A means of securing an organism to a substrate, e.g. the holdfasts of seaweeds such as kelps or the adhesive disk of sea anemones.

anemometer   *See* wind.

Angiospermophyta (angiosperms; flowering plants)   A phylum of vascular seed plants that bear flowers. Most also produce fruits (seeds enclosed in carpels). The dominant generation is the sporophyte, which may be herbaceous or woody. In most angiosperm species the XYLEM contains vessels and the PHLOEM has companion cells associated with the sieve tube elements. The flower bears the reproductive organs, and often has associated brightly colored sepals and/or petals. The gametophyte generation is highly reduced, consisting of the female *embryo sac* inside

the *ovule* and the male cells of the pollen grain, contained in the anthers of the male stamens. The pollen may be transferred from one flower to another by wind, water, insects, or other animals (*see* pollination). The *ovary*, which contains the ovules, consists of one or more *carpels*, thought to have evolved from folded leaflike structures. A specialized extension of each carpel, the *stigma*, bears a receptive surface on which the pollen germinates, growing a pollen tube down to the ovule. The nonmotile male gametes are released into the ovule, and double fertilization takes place: one male gamete fertilizes the ovum, while the other fuses with two of the haploid female nuclei, the polar bodies, giving rise to a triploid tissue (having three sets of chromosomes per nucleus), the *endosperm*. The ovules develop into seeds, which contain stored food in the form of endosperm or swollen first leaves (*cotyledons*). The ovary wall and/or other flower parts may be modified to form specialized fruits adapted for animal, wind, water, or explosive dispersal. The production of the embryo inside a tough resistant seed coat and the elimination of the need for water to enable the male sex cells to reach the females enabled the angiosperms to become highly successful terrestrial plants. Angiosperms are divided into two major classes: *dicotyledons* (characterized by having two cotyledons in the seed and net-veined leaves) and *monocotyledons* (characterized by having one cotyledon in the seed and leaves with parallel veins).

angiosperms   *See* Angiospermophyta.

anhydrobiosis (cryptobiosis)   The ability to survive the drying up of the habitat by means of dormancy or estivation.

Animalia   The kingdom that contains animals. Animals are heterotrophic, multicellular EUKARYOTES. They are often motile and respond to stimuli rapidly. They lack cellulose cell walls and chlorophyll and growth is usually limited. *See* heterotroph.

anion   A negatively charged ion, e.g. the hydroxyl ion (OH^-). *Compare* cation.

9

Annelida A phylum of segmented worms that includes the earthworms, bristle worms, lugworms, and leeches. They are found in aquatic, marine, and terrestrial environments. Annelids are distinguished by series of external ringlike segments running the length of the long soft cylindrical body. The body is covered in a cuticle of chitin, and most annelids have segmentally arranged bristles called chaetae, which are used in locomotion. The body wall contains layers of longitudinal and circular muscle and the body cavity (coelom) isolates the gut from the body wall. The gut runs from the mouth to anus, there are well-developed blood and nervous systems, and nephridia for excretion. Many annelids are hermaphrodite.

The class *Oligochaeta* comprise the terrestrial earthworms, and many freshwater species. Ranging in length from a few millimeters to over 3 m they have relatively few chaetae. All oligochaetes are hermaphrodite. Most are scavengers, feeding on decaying organic matter. Earthworms (e.g. *Lumbricus*) live in burrows and eat soil, extracting the digestible matter and expelling mineral particles. Their burrows and feeding habits are extremely important in maintaining aeration and drainage in soils, turning over soil layers and mixing nutrient-rich and oxygenated layers with the rest of the soil. Some aquatic species also feed on living microscopic organisms.

annual 1. Occurring every year or once a year.
2. Covering a 12-month period, as in rainfall statistics, for example.
3. A plant species that completes its life cycle within 12 months, especially a plant that germinates, flowers, sets seed, and dies within a single year or growth season. *Compare* biennial; ephemeral; perennial.

annual rings Rings of wood and bark produced by a tree during the growing season each year. In temperate climates, the production of new wood begins in spring and ends in the fall. Wood is made up of XYLEM, a water-conducting tissue whose cells lose their protoplasm at maturity, becoming hollow. In many species of north temperate trees the xylem vessels produced at the start of the growing season are much larger than those developed toward the end of the season, their larger lumens allowing more rapid conduction of water. This difference is partly the result of better water availability and lower transpiration rates, and partly because of hormone changes. Thus, early wood has large diameter cells with thin cell walls and later wood the converse. This gives rise to alternating light and dark rings of wood respectively, one light and one dark band usually representing one year's growth. Annual rings give valuable information about past climates and may also be used to date wood. *See* dendrochronology; dendroclimatology.

ANOVA *See* analysis of variance.

antagonistic pleiotropy *See* senescence.

antagonistic resource *See* resource.

Antarctic The entire area of the Earth's surface south of latitude 66°30′ S. *Compare* Arctic.

Antarctic Circle The line of latitude around the Earth at 66°30′ S. It marks the northern limit of the area within which for one day or more each year the sun does not set (winter solstice, December 21 or 22) or rise (summer solstice, June 21 or 22). The period of continuous day or night increases southward from the Antarctic Circle, reaching six months at the South Pole. *Compare* Arctic Circle.

Anthozoa A class of the phylum Cnidaria containing the *corals* and sea anemones, in which the polyp is the only body form. Sea anemones are solitary animals that trap food using a ring of feathery tentacles that eject stinging cells to paralyze their prey. Corals are colonial, polyps being embedded in a gelatinous, horny, or calcareous matrix. Accumulations of colonial corals with calcareous skeletons form reefs in warm shallow waters.

anthropogenic Resulting from the actions of humans. The term is usually applied to environmental changes such as habitat change or pollution.

anthropogenic extinction   *See* extinction.

anthropology   The study of humans, including their diversity of culture and physical characters, evolutionary history and phylogeny, and geographical distribution.

antibiosis   *See* amensalism.

antibiotic   A specific chemical produced by a living organism, usually a microorganism, that inhibits the growth of or kills another microorganism.

anticyclone   A circulating air mass in which atmospheric pressure increases toward the center, around which winds circulate in a clockwise direction in a northern hemisphere and a counterclockwise direction in the southern hemisphere. Anticyclones give rise to calm fine weather, perhaps with fog in winter. *Compare* depression.

antifouling paints   Paints applied to the hulls of boats to prevent build up of algae and marine invertebrates. Some of these, such as TRIBTUYL TIN OXIDE (TBTO), are causes of reproductive disruption in marine invertebrates.

antioxidant   A compound that prevents or delays the oxidation of other compounds. Industrial antioxidants include vegetable oils and phenol derivatives used to reduce the speed of drying of paints or delay the oxidation of plastics, rubbers, drugs, and other synthetic substances as well as foodstuffs.

ants   Small insects of the order Hymenoptera, family Formicidae, distributed worldwide, but especially in warm climates. Most ants are social insects living in colonies of nonbreeding workers dominated by a reproductive queen ant. Male reproductives are produced usually once a year, and die after the mating flight. The fertilized female removes her wings and founds a new nest. Larger soldier ants guard the colony and the foraging workers. Some ants build moundlike nests of soil, sand, plant material, and even dung. Others live in complex underground nests.

Ants form more than half of the total insect biomass in the world, and almost a third of the total animal biomass in tropical rainforest. In certain habitats, they are the main predators, scavengers, and turners of soil, playing an important role in decomposition and in maintenance of soil aeration, drainage, and fertility.

aphids   Small sap-sucking, soft-bodied insects of the order Hemiptera, family Aphididae, of average length about 2.54 mm. They include greenfly and blackfly. Aphids can reproduce rapidly by PARTHENOGENESIS. In most species wingless females produce live female young by parthenogenesis throughout the summer. When overcrowding occurs, winged forms arise and colonize new plants. Aphids are serious economic pests of crops and greenhouse plants, stunting plant growth or inhibiting flower and fruit production, and transmitting pathogenic bacteria and fungi or viruses. They can be controlled by insecticides, or by biological control, usually involving ladybirds, the larvae and adults of which prey on aphids.

aphotic zone   The part of a body of water in which light intensity is too low for photosynthesis to take place. *Compare* photic zone.

apical meristem   *See* meristem.

apomixis   1. Any kind of ASEXUAL REPRODUCTION, including VEGETATIVE PROPAGATION and PARTHENOGENESIS. 2. The production of seeds without prior fertilization. This may occur by means of the formation of a diploid embryo or embryo sac by a somatic (non-reproductive) cell, or from an ovum that has not undergone meiosis. A plant that undergoes apomixis is termed an *apomict*. Apomixis is common among polyploid plants (*see* polyploidy) (e.g. triploids in which homologous chromosomes cannot pair properly at meiosis) that are unable to reproduce sexually; these are *obligate apomicts*, e.g. some cultivated varieties of banana (*Musa*) and garlic (*Allium*). It also occurs in species near the edge of their geographical range, where environmental conditions such as daylength are not favorable for sexual reproduction or where insect pollinators are absent; these are *facultative apomicts*, e.g. some *Potentilla* species. Apomixis also permits the formation of HYBRID SWARMS. *See* asexual reproduction; hybrid.

aposematism The means by which a poisonous or otherwise dangerous organism advertises its dangerous nature to potential predators. The commonest form is warning coloration: bright contrasting colors such as black and yellow, orange, or red makes it easy for a potential predator to recognize the organism as unfavorable prey, deterring it from attacking. Some species instinctively avoid animals with such coloring.

apostatic prey selection *See* predator–prey relationships.

apparent competition *See* competition.

aquaculture The culture of fish, shellfish, or underwater plants in natural or controlled freshwater or marine environments. Aquaculture may take place in enclosed shallow coastal areas, in cages at or below the surface of the open sea, in freshwater ponds, or in large tanks. It is a commercially important method of producing seafood such as mussels, oysters, clams, lobsters, crawfish, and shrimps, and fish such as salmon, tilapia, and catfish. Seaweeds may be grown in similar situations. The method of growing plants in nutrient solution instead of soil is called HYDROPONICS.

aquatic 1. Living or taking place in water.
2. A plant or animal that lives in water.

aquatic biome see biome.

aquiclude A mass of impermeable rock that forms a barrier to the movement of groundwater. *Compare* aquifer.

aquifer A body of permeable rock or drift saturated with GROUNDWATER and through which groundwater may move. An aquifer that is overlain by an impermeable rock layer is said to be *confined*. Where the path from the aquifer to the atmosphere is through permeable material, the aquifer is aids to be *unconfined*. Aquifers are important sources of water for drinking, irrigation, and industry, but unless the rate of extraction is equal to or less than the rate of replenishment from precipitation the supply will eventually become exhausted.

Archaea (archaebacteria) A kingdom or DOMAIN of PROKARYOTES that includes the halophilic and methanogenic bacteria and the thermoacidophilic bacteria – the EXTREMOPHILES. They are found in extreme environments, such as hot springs, boiling muds, volcanic craters, saline sediments, and deep-sea (hydrothermal) vents. The Archaea are distinguished from the BACTERIA by the nucleotide sequences of their ribosomal RNA; their ribosomes, which have a distinctive shape and resemble those of EUKARYOTES rather than prokaryotes; cell walls lacking a peptidoglycan layer; and the composition of their lipids and RNA polymerase enzyme. *See also* hydrothermal vent.

archaebacteria *See* Archaea.

Archean (Archeozoic) *See* Precambrian.

archipelago A group of islands or an expanse of water with many islands scattered across it.

Arctic The region north of the northernmost limit of tree growth (the treeline) in the northern hemisphere, which includes Greenland (Kalaallit Nunaat), Svalbard and other polar islands, the northernmost parts of Canada, Alaska and Siberia, Scandinavia, and Iceland, and the Arctic Ocean. Most of the region consists of TUNDRA vegetation, permanent ice or snow cover, or ocean, most of which freezes in winter. On land, the ground below a certain depth remains permanently frozen. It experiences extremely long days in summer (24 hours at the pole), and long nights in winter (24 hours at the pole). The term is sometimes restricted to the area north of the ARCTIC CIRCLE (66°30′N), in which there is at least one 24-hour period each year when the sun does not set and one when it does not rise. *See also* Arctic-alpine species. *Compare* Antarctic.

Arctic-alpine species A species found both in the Arctic and on high mountains in temperate latitudes such as the Rockies, Alps, and Himalayas, e.g. least willow (*Salix herbacea)*. This relict distribution indicates that these species were once part of the TUNDRA vegetation that was widespread during the PLEISTOCENE: as the climate warmed they survived only in cooler latitudes and altitudes.

Arctic Circle The line of latitude around the Earth at 66°30′ N. It marks the southern limit of the area within which for one day or more each year the sun does not set (winter solstice, December 21 or 22) or rise (summer solstice, June 21 or 22). The period of continuous day or night increases northward from the Arctic Circle, reaching six months at the North Pole. *Compare* Antarctic Circle.

arctic vegetation *See* tundra.

area effect The effect of HABITAT area on species diversity and on the number of individuals of each species present. In general, the number of species decreases with a decrease in the area of the habitat patch. *See* island biogeography theory.

arid Describing an area where rainfall is insufficient to support agriculture. In most arid areas, annual PRECIPITATION is less than 25 centimeters. Areas with 25–51 cm of annual precipitation are termed *semi-arid*.

arithmetic mean *See* mean.

arrested succession *See* succession.

Arrhenius, Svante August (1859–1927) Swedish physical chemist. Arrhenius is noted for his work in chemistry on the theory of electrolytes and the rates of chemical reactions. He was also the first to speculate on the effect of carbon dioxide on atmospheric temperature (what is now known as the GREENHOUSE EFFECT). In 1895 he calculated that a doubling in the level of carbon dioxide in the atmosphere would lead to a 10°C rise in average temperature.

arsenic (As) A chemical element found both free and combined in many minerals, usually associated with silver and antimony. Arsenious oxide, A_2O_3, formed by roasting arsenic-containing metal ores, is highly poisonous and is used in pesticides. Arsenic pentoxide, A_2O_5, is used in herbicides and insecticides. In parts of India and Bangladesh, arsenic compounds in well water cause major health problems, especially cancers. These chemicals derive from arsenic-containing rocks through which the deep well water passes. Similar arsenic contamination of groundwater also occurs in other parts of the Far East and the United States.

artesian well A spring into which water flows without the need for pumping. Artesian wells obtain their water from AQUIFERS in which the GROUNDWATER is trapped buy impermeable rock layers above and below the water-holding strata. If the aquifer is being topped up by water entering the aquifer at a level higher than the artesian well, the hydrostatic pressure of the water in the aquifer forces water to the surface.

Arthropoda The largest phylum of the Animal kingdom, with segmented bodies, jointed appendages, and a hard external skeleton (*exoskeleton*) made of chitin. It includes the crustaceans, insects, spiders, scorpions, mites, ticks, horseshoe crabs, centipedes, and millipedes. In the Five Kingdoms system of classification, the Arthropods are subdivided into three distinct phyla – the *Chelicerata* (spiders, scorpions, mites, ticks, harvestmen or daddylonglegs, and horseshoe crabs), in which the anterior (front) pair of appendages (*chelicerae*) are clawed, and which have no mandibles or antennae; the *Mandibulata*, (insects, millipedes, and centipedes), whose bodies are divided into distinct head, thorax, and abdomen, and which have mandibles; and the *Crustacea*, which have two pairs of antennae on the head.

Arthropods are found in almost every habitat on the planet, and are of great economic importance. Shrimps, lobsters, and crabs are important sources of food worldwide. Planktonic crustaceans and insect larvae are major links in the food chains of both freshwater and marine habitats, insects are vital pollinators of crops, and decomposers aid the formation of humus, enhancing the fertility of the soil. However, many arthropods are serious pests of livestock, crops, and stored food: insects and mites attack crops and timbers, marine arthropods parasitize farmed fish, sapsucking insects drain the host plants of resources and may transmit pathogenic bacteria, fungi, or viruses, and ticks and insects such as mosquitoes and flies spread diseases through the human population and its livestock.

artifact An object created by humans or an attribute characteristic of or resulting from a human activity or institution, e.g. a piece of prehistoric pottery.

artifical rain Rain produced by nonnatural means. A number of experiments have been tried using chemicals distributed from aircraft or rockets, either to encourage cloud formation or to induce rain formation from existing clouds.

artificial selection Manipulation by humans of the breeding between different individuals of a population in order to produce a desired evolutionary outcome.

asbestos A fibrous mineral used to make heat-resistant products such as insulation materials, brake linings, electrical insulation, and fireproof fabrics. Asbestos dust is highly toxic when inhaled, giving rise to mesothelioma, lung cancer, and *asbestosis*. Blue asbestos is the most toxic form, and its use has been banned in many countries.

asexual reproduction The formation of new individuals from a single parent without the production of gametes or special reproductive structures. It occurs in many plants, usually by vegetative propagation or spore formation; in unicellular organisms usually by fission; and in multicellular invertebrates by fission, budding, fragmentation, etc.

aspect Morphological appearance, such as shape, size, color, and pattern. Aspect is particularly important in cryptic species seeking to avoid detection by PREDATORS. If the HABITAT contains a variety of backgrounds against which an animal might be viewed, evolution may lead to an increase in *aspect diversity* in the species living there, each being camouflaged against a particular feature of the background. For example, certain moths have evolved to match the pattern of bark of the tree species they frequent, the shape of broken twigs, dead or living leaves complete with patches of decay, and so on. Thus an increase in aspect diversity may reflect the diversity of the habitat, and is related to the characteristics of the available resting places and the searching techniques of the predator(s).

assemblage A collection of plant and animal species typical of a particular environment. Species assemblages can serve as indicators of environmental conditions.

assimilation The process of incorporation of simple molecules of food that has been digested and absorbed into living cells of an animal and conversion into the complex molecules making up the organism.

association A group of species living in the same place. More specifically, a large CLIMAX COMMUNITY of plants usually named after the dominant plant species (usually of the uppermost layer of vegetation), e.g. heath association. Associations usually have more than one dominant species. The term *alliance* is used to describe a community with only one dominant species.

assortative mating Reproduction of animals in which the males and females appear not to pair at random but tend to select partners of a similar phenotype.

atavism The reappearance of a characteristic, known as a *throwback*, in an individual after several generations of absence, usually caused by a chance combination of particular genes.

Atlantic period The period from about 6000 to 3000 BC that was characterized by a warming of the climate worldwide and an increase in rainfall in north temperate regions. During this period deciduous broad-leaved forests spread northward. The early Atlantic period was the warmest phase so far of the present interglacial period.

atmometer An instrument that measures the evaporating capacity of the air by measuring water loss from a porous nonliving wet surface.

atmosphere 1. The gaseous envelope surrounding the Earth, a mixture of invisible, odorless and tasteless gases containing almost constant amounts of nitrogen (N_2) and oxygen (O_2) with lesser amounts of helium (He), methane (CH_4) (derived mainly from ruminant animals), hydrogen (H_2), nitrous oxide (N_2O), and the inert gases argon (Ar), neon (Ne), krypton (Kr), and xenon (Xe). Other components present in variable concentrations include water vapor (H_2O), carbon dioxide (CO_2), ozone (O_3), sulfur dioxide (SO_2), nitrogen dioxide (NO_2), various anthropogenic pollutants such as chlorofluorocarbons (CFCs) and chlorine (Cl_2), and microscopic solid and liquid particles in suspension. The atmosphere extends thousands of kilometers above the Earth, and is retained by the Earth's gravity. Its composition is more or less constant up to about 100 km above the ground, although its density decreases with

distance from the ground – hence the shortage of oxygen for mountaineers at high altitudes. At 100 km there is a high level of incident solar radiation, and an increasing proportion of charged particles (ions), giving rise to the auroras.

The atmosphere has several distinct layers. The lowermost, extending up to 15 km, is the *troposphere*. In this layer convection currents mix air warmed at the Earth's surface with high air, maintaining a vertical temperature gradient of about 6°C (10.8°F) per kilometer. The troposphere is the zone in which water vapor is found and weather occurs. At the top of the troposphere is the *tropopause*, where the temperature falls to about –60°C (–76°F). Above this the *stratosphere* extends for some 50 km, a dry layer lacking strong convection currents, with a marked increase in temperature with increasing altitude. In the upper part of this zone is the OZONE LAYER, formed by action of ultraviolet light from the Sun on oxygen molecules (O_2), which break down, allowing oxygen atoms to recombine as ozone (O_3). Ozone forms a barrier restricting the passage of harmful ultraviolet radiation to the Earth's surface. Above the *stratopause* (mesopeak) at 50 km is the *mesosphere*, where temperatures once again decline with altitude. Its outer limit is the *mesopause*, at a temperature of

COMPOSITION OF AIR (% by volume)		
nitrogen	(N_2)	78.08
oxygen	(O_2)	20.95
carbon dioxide	(CO_2)	0.03
argon	(Ar)	0.93
neon	(Ne)	1.82×10^{-3}
helium	(He)	5.24×10^{-4}
methane	(CH_4)	1.5×10^{-4}
krypton	(Kr)	1.14×10^{-4}
xenon	(Xe)	8.7×10^{-5}
ozone	(O_3)	1×10^{-5}
nitrous oxide	(N_2O)	3×10^{-5}
water	(H_2O)	variable, up to 1.00
hydrogen	(H_2)	5×10^{-5}

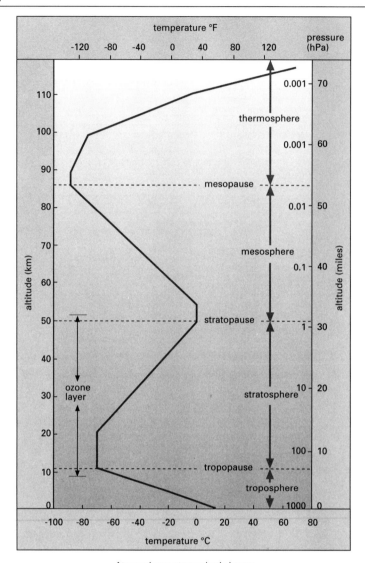

Atmosphere: atmospheric layers

about −85°C. Beyond this, in the *thermosphere*, temperature rises with altitude. Higher still the density of the gases decreases still further as they merge with the solar wind, a stream of charged particles emanating from the Sun.

2. (standard atmosphere) A unit of pressure approximating to the mean atmospheric pressure at sea level or to the pressure exerted by a vertical column of mercury 760 mm high. It is defined as 101 325 pascals (about 14.7 lbs per square inch).

atoll A coral reef surrounding a lagoon. Many atolls are derived from the crater rims of sunken volcanoes, coral growth on the rim keeping pace as a volcano sinks below the surface of the ocean to enclose the lagoon.

atomic energy   *See* nuclear energy.

ATP   *See* respiration.

auger   A device used to sample soils. The commonest type is the *corkscrew auger*, which resembles a large corkscrew that is twisted down into the soil, then withdrawn. Ideally, an auger is marked at intervals, so the depth of the source of each part of the sample is known. A corkscrew auger can also be used horizontally to extract a sample from a tree trunk for dating purposes (*see* dendrochronology).

augmentation   A form of PEST CONTROL in which extra individuals of an indigenous natural predator are released to help suppress a rapidly-growing pest population.

Australasia   One of the six main zoogeographical areas, composed of Australia and the islands of its continental shelf, Tasmania, New Guinea, and New Zealand. Marsupial (pouched) and monotreme (egg-laying) mammals are particularly characteristic, but many other unique vertebrates and invertebrates are also found.

autecology (species ecology)   The study of the interaction of a single species or individual organism with its environment. It involves a detailed study of the life cycle, coupled with measurements of the various living and nonliving environmental factors that affect the species. *Compare* synecology.

autochthonous   Describing material produced within a community or ecosystem, such as organic matter production or mineral cycling.

autocidal control   A form of PEST CONTROL in which the pest contributes to its own demise. An example is the eradication of the screw-worm fly (*Cochliomyia hominivorax*) from the southern USA by releasing sterile (irradiated) males into populations to compete with fertile males and decrease the birth rate.

autogamy   *See* sexual reproduction.

autogenic succession   *See* succession.

autopolyploid   *See* polyploidy.

autotroph   An organism that requires only simple inorganic compounds for growth. For example, most plants and algae are autotrophs, using carbon dioxide and water to synthesize organic compounds with the aid of light energy (*photoautotrophs*). Some photosynthetic bacteria use hydrogen sulfide or other inorganic compounds instead of water as the hydrogen source/electron donor. Other bacteria (*chemoautotrophs*) derive the energy for synthesis from inorganic compounds. Autotrophs are *primary producers*, forming the base of FOOD CHAINS. *Compare* heterotroph. *See* producer.

autotrophic succession   *See* succession.

average reproductive rate   *See* reproductive rate.

Aves   The class of vertebrates (phylum Chordata) that contains the birds, most of whose characteristics are adaptations for flight. Birds are endothermic (warm-blooded) vertebrates with a body covered in feathers, a four-chambered heart, forelimbs modified to form wings, the sternum modified to form a keel for the attachment of the wing muscles, the pelvic girdle and hind limbs modified so as to support the bird when standing on two legs, a horny bill with no teeth, and a yolked egg with a calcareous shell.

  Birds have a well-developed social life, including territorial and courtship displays, nesting, parental care, and song. Many undertake long migrations. There are 28 orders. The perching birds (passerines) account for 60% of all birds and the remaining flightless birds (ratites) tend to be swift runners.

B

Bacillariophyta *See* diatoms.

backcross A cross between an individual and one of its parents or an individual of identical genotype to that of one of its parents. *See* heterozygous; homozygous; plant breeding.

background extinction *See* extinction.

background radiation RADIATION coming from sources other than that under investigation. The background radiation may increase at altitude, owing to the greater intensity of cosmic rays, and in areas where the rock or soil contains radium, whose decay causes release of the radioactive gas radon. *See* radioactive decay.

Bacteria In the Five Kingdoms system of classification, a KINGDOM containing all the PROKARYOTES, formerly called the Prokaryotae or Monera. Most taxonomists now divide the prokaryotes into two distinct kingdoms or *domains* (a taxonomic level above kingdom), the Bacteria and the AR-CHAEA.

Bacteria are found over almost all the planet, from the interstices of rocks deep underground to icebergs, mountains, deserts, deep oceans, and the insides of other organisms. Most are single cells, ranging from rounded cocci to rods and spirochetes. A few, such as the cyanobacterium *Anabaena*, form filaments of cells, clusters, or dense colonies (e.g. the cyanobacteria that form stromatolites). Bacteria play an important role in the recycling of nutrients in soil and water, in DECOMPOSITION, and in NITROGEN FIXATION (e.g. the symbiotic bacterium *Rhizobium* in the root nodules of legumes).

Bacterial Code of Nomenclature *See* International Codes of Nomenclature.

bacteriophage (phage) Any VIRUS that infects bacteria. Phages have protein coats (*capsids*) consisting of a head containing the nucleic acid (double- or single-stranded DNA or RNA), and a helical tail through which the virus injects nucleic acid into its host. RNA-containing phages are always lethal to the host cell, but some DNA phages may become integrated into the host DNA, being replicated when it replicates. At a later time the host may be triggered to reproduce new viruses, which escape by causing the breakdown of the host cell wall (lysis), killing the host. Phages can sometimes transfer bacterial genes between bacteria, when segments of nucleic acid become incorporated into the phage genome, a process called *transduction*. Phages are used in GENETIC ENGINEERING to transfer nonviral DNA (deliberately inserted into the phage nucleic acid) to bacteria.

bacteroid A *Rhizobium* bacterium in the root nodule of a legume after it has increased in size about 40-fold and changed shape to a Y- or X-shaped cell, becoming capable of NITROGEN FIXATION. In this form, the bacterium is incapable of independent growth, and is dependent on its host for part of its nutrition. *See* symbiosis.

balanced polymorphism *See* polymorphism.

bar chart *See* graph.

barren A fairly level area of land with little vegetation and few trees, usually on sandy or serpentine soils. Plants are often smaller and stunted when compared to individuals of the same species from more fertile habitats, and there are often groups of specialized endemic species. The term is used especially in the eastern United States, for example, the pine barrens of New Jersey.

basal metabolic rate (BMR)   The minimum amount of energy an animal needs to sustain itself for a given time. It is sometimes expressed as the minimal quantity of heat produced (by metabolism) by an individual at complete physical and mental rest (but not sleeping) 12–18 hours after eating, expressed in milliwatts per square meter of body surface.

base   1. A substance that reacts with an ACID to form a salt and water. In solution, it releases ions that can combine with hydrogen ions.
2. A nitrogenous (nitrogen-containing) molecule that forms part of a nucleic acid, and is the basic unit of the genetic code. Such bases are either PURINES or PYRIMIDINES.

baseflow   See dry weather flow.

Batesian mimicry   See mimicry.

bathymetry   The measurement of water depth, especially in the oceans. This is usually done by echo sounding, sometimes supplemented by the use of underwater cameras. See echo sounder.

baumgrenze   See tree line.

bearing   The horizontal angle between a survey line and a specific reference direction.

Beaufort scale   A numerical scale ranging from 0 (calm) to 12 (hurricane) that was introduced by Admiral Sir Francis Beaufort (1774–1857) in 1805 to describe variations in wind force at 10 m (30 ft) above the ground and originally based on the effect of wind on a fully rigged man-of-war. See table overleaf.

becquerel (Bq)   A unit of radioactivity: one becquerel is the activity of a quantity of a radioactive substance in which there is one spontaneous atomic disintegration per second. The becquerel is an SI unit, and supersedes the curie (Ci): 1 becquerel = 2.7 x 10^{-11} Ci. See radioactive decay.

bedding plane   A surface that indicates where successive phases of deposition of a rock occurred. Many SEDIMENTARY ROCKS show successive parallel bedding planes that represent times when deposition stopped then started again, or when the size of particles deposited changed significantly. Visually bedding planes may be detected as cracks at the surface of a rock exposure, or changes in color or rock type. The angle of the bedding planes may indicate whether the rocks have been tilted or folded since their formation. Rocks cleave most readily along their bedding planes.

bees   Insects of the order Hymenoptera. Unlike the related wasps, they feed their larvae on a mixture of pollen and honey, and some of their hairs are branched or feathered to trap pollen. Bees feed on pollen and nectar, and store some of the nectar as honey. Some species, such as honeybees (Apis mellifera), have basketlike structures on their rear legs in which to carry pollen from the flower to the nest. The main economic importance of bees is as pollinators: without them many major crops would fail.

beetles   Insects of the order Coleoptera. Their forewings are modified to form hard leathery wing-cases (elytra) that protect the abdomen and membranous hind wings while resting. The head has biting mouthparts. Beetles undergo complete metamorphosis, the eggs hatching into legless grubs or caterpillar-like larvae, which pupate before emerging as adult insects. Beetles are found in a wide range of terrestrial and freshwater habitats. Many are predators, but some feed on plant material (including wood) and others on detritus or carrion (e.g. carrion beetles). Some are serious pests of crops (e.g. Colorado potato beetle), trees (e.g. elm bark beetle), stored

THE BEAUFORT SCALE ON LAND

Beaufort number	Descriptive term	Wind speed		Visual effect
		knots	mph	
0	calm	< 1	< 1	smoke from factories rises vertically
1	light air	1 – 4	1 – 4	wind direction shown by smoke, but not by a weather vane
2	light breeze	4 – 6	4 – 7	wind felt on face
3	gentle breeze	7 – 10	8 – 12	broken twigs from trees are in constant motion
4	moderate breeze	11 – 16	13 – 18	dust and loose paper raised from sidewalk
5	fresh breeze	17 – 21	19 – 24	swaying of small trees
6	strong breeze	22 – 27	25 – 31	large branches of trees in motion
7	near gale	27 – 33	32 – 38	whole trees vibrate
8	gale	34 – 40	39 – 46	twigs detached from trees
9	strong gale	41 – 47	47 –54	chimney pots and roof tiles fall from houses
10	storm	48 – 55	55 – 63	trees uprooted; much structural damage
11	violent storm	56 – 63	64 – 72	considerable damage. Usually experienced along the coast; very infrequent inland
12	hurricane	> 64	> 73	mostly confined to the tropics. Disastrous; loss of life, towns flattened

(As the Beaufort scale is subjective, wind speeds assigned to each Beaufort number vary; as a result some values appear in more than one wind force class.)

goods (e.g. grain weevils), and household goods (e.g. woodworm). Others are beneficial – ladybirds (*Coccinella*) feed on aphids and are used in the BIOLOGICAL CONTROL of pests in greenhouses.

behavioral ecology   *See* ecology.

bell curve   *See* normal distribution.

benthic zone   The sediments at the bottom of streams, ponds, lakes, and oceans. The benthic zone is an extremely variable environment, depending on the latitude, the depth of the water (and hence temperature, light intensity, and pressure), its salinity, and the nature of the sediments (whether they are well aerated or anoxic). In shallow waters light supports photosynthesizing organisms, including diatoms and photosynthetic bacteria, but in deeper water the main source of nutrients is the rain of detritus from above – the remains and excreta of plankton and larger animals and plants, as well as the products of the decomposition of larger carcasses that sink down to the sediments. This fallout varies with the seasons and currents, and with certain weather patterns.

benthos The organisms that live on or in the floor of a stream, pond, lake, or ocean. The benthos is made up of the many microorganisms and invertebrates living in and on the surface layers of the sediments, and includes: bacteria, diatoms, foraminiferans, ciliates, amebas, flagellates, copepods, and other small crustaceans; attached forms such as crinoids, hydroids, sea anemones, sponges, sea squirts (ascidians), and bivalves; burrowing animals such as polychaete worms and nematodes; creeping animals such as starfish, brittlestars, sea urchins, snails, and flatworms (platyhelminths); bottom-dwelling fish such as flounders; and plants and algae with roots or holdfasts fixed in the sediment, such as pondweeds, water lilies, and many seaweeds (e.g. kelps). Organisms larger than 1 mm are called *macrobenthos*, those between 0.1 and 1 mm *meiobenthos*, and those smaller than 0.1 mm *microbenthos*. Some organisms filter detritus from the water, or from water drawn through burrows in the sediments; others scavenge, prey on the detritus feeders, or graze on bottom-rooted plants or algae. *See* detritus. *See also* benthic zone; hydrothermal vent.

Bergmann's rule The generalization that warm-blooded animals of the same species living in cold regions tend to be larger than those living in warmer climates. It is suggested that the smaller ratio of surface area to volume in the larger animals reduces the rate of heat loss, thus saving energy. *See also* Allen's rule.

beta decay (*β*-decay) The disintegration of a radioactive element (with the emission of an electron or positron) that results in the mass number of the nucleus remaining unchanged, but the atomic number (proton number) increasing or decreasing by one. Beta radiation is more penetrating than alpha radiation. *See* radioactive decay.

beta diversity *See* species diversity.

Bhopal A town in central India, where in 1984 a catastrophic release of some 44 tonnes of methyl isocyanate (MIC) occurred at the Union Carbide factory. MIC was an ingredient of the insecticide *carbaryl*, which was produced at the factory. Water seeping into a storage tank of MIC triggered a runaway chemical reaction that blew open the safety valve on the tank, releasing MIC. The wind carried a cloud of toxic vapor over an area housing some 100 000 people. At least 2500 people died, and 50 000 others were disabled for a time by respiratory and eye problems.

bias A nonrandom distortion in measurements due to such factors as poorly calibrated instruments, changes in a specimen during an experiment, or the subjective judgment of the experimenter. Personal bias can be eliminated by 'blind trials', in which the identities of individual specimens are not known to the person making the measurements or observations. Other types of bias can be detected by means of trial runs using different methods to measure single variables and comparing the results.

bicentric distribution A geographical DISTRIBUTION of a species in which there are two separate centers of distribution. It is an example of a *disjunct distribution*. It may be explained by the contraction of the range of a formerly more widespread species because of changing environmental conditions, often because of climate change. In cases in which the centers are not too far apart it may be caused by the ability of the species to disperse its offspring across unfavorable areas to reach areas where conditions are favorable for survival.

biennial A plant that takes two years to complete its life cycle. In the first year it grows vegetatively, storing food in *perennating organs* (storage organs that enable the plant to survive an unfavorable season such as winter or a dry season). In the second year it uses this stored food to produce more leaves, and also flowers and seeds. *Compare* annual; ephemeral; perennial.

bimodal distribution   A frequency distribution that has two major peaks. These do not have to be of equal height.

binary fission   *See* asexual reproduction.

binomial distribution   *See* distribution.

binomial system of nomenclature   A system of classification of living organisms devised by the Swedish botanist Carl Linnaeus (1707–78), in which each species is defined by two names. The first, written with a capital letter, is the genus name, and the second is the species name. Both names are in Latin or are Latinized, and are normally printed in italic type. For example, humans belong to the genus *Homo* and species *sapiens*, so the species is *Homo sapiens*. Such a two-part name is called a *binomial*. *See* International Codes of Nomenclature.

binomics   (bioecology; environmental biology)   The study of the relationships between organisms and their environment.

bioaccumulation   The accumulation of POLLUTANTS and other toxins in the tissues of living organisms. For example, the insecticide DDT and its derivatives accumulate in the fatty tissues of mammals and birds. These toxins are passed up the food chain, accumulating in greatest concentrations in top predators such as mammals and birds of prey. Certain species or varieties of plants accumulate toxic ions such as cadmium, lead, nickel, and zinc and are able to tolerate them, and a few species, called *hyperaccumulators*, can accumulate more than 100 times more toxins than most plants (hyperaccumulators are defined as organisms that can accumulate over 10 000 mg of the toxic element per kilogram). *See also* bioremediation; bioconcentration; biomining.

bioassay   A experimental technique for determining the strength of a biologically active chemical by its effect on a living organism or group of organisms. Examples of this include the use of caged fish to determine the rate at which they accumulate poisons or even die in polluted waters – an indication of the level of pollution (this process is called *biomonitoring*); the use of simple animals as indicators of the hormone activity in extracts from mammalian endocrine glands; and the use of the rate of increase in numbers of cultured bacteria to measure the activity of vitamins.

bioaugmentation   The reinforcing of natural biological processes using addition organisms. For example, decomposition of organic material such as sewage is augmented by the addition of various species of bacteria. It is essentially the concentration of organisms that grow naturally in the environment, not the addition of genetically modified organisms. Bioaugmentation is important in converting human or industrial waste to nonpolluting products.

biochemical oxygen demand   *See* biological oxygen demand.

biocide   A substance that kills living organisms. *See* fungicide; herbicide; insecticide; pesticide.

bioclimatology   The study of climate in relation to the environments of living organisms.

bioconcentration   The increase in bioaccumulated chemicals in the tissues of organisms as they pass up the FOOD CHAIN. It is sometimes used specifically to identify the net accumulation of a chemical directly from the aquatic environment by gills or other epithelial tissue and its subsequent retention in the tissues, as opposed to uptake through the ingestion of a chemical in food or water. *See also* bioaccumulation.

biodegradable   Describing a substance that can be broken down by microorganisms.

biodiesel   A vegetable oil, such as rapeseed oil, that can be used as a fuel in modified diesel engines. The methyl ester of rapeseed oil (rapeseed methyl ester, or RME) can be used in unmodified engines.

biodiversity The variety of organisms present in the living world. More specifically, *species biodiversity* is the number of species present in a particular area or ecosystem. In general, biodiversity tends to be highest in complex and highly productive ecosystems, such as topical rainforests. Biodiversity is often used as an indicator of the health of such ecosystems.

bioecology See binomics.

bioenergetics The study of energy transformations in living organisms. Energy transfer from the environment to living organisms and from organism to organism via the FOOD CHAIN is inefficient: about 2% of the energy in incident radiation is converted to chemical energy in photosynthesis, and when heterotrophic organisms consume other organisms, only about 10–20% of the energy is passed on from level to level in the food chain. See energy transformation; trophic level; respiration. See also bioenergy.

bioenergetic web See food web.

bioenergy The energy contained in living organisms. The term is used to mean energy, fuels, or energy-related products derived from renewable biological resources such as plants (but not FOSSIL FUELS), microorganisms, and biological wastes. Typical sources are fast-growing crops and trees grown as a fuel sources, aquatic plants, agricultural feed crops, waste and residues from timber mills and related industries, animal wastes, and municipal and industrial wastes. These may be burned in coal- or wood-fired boilers to generate electricity, or in high-efficiency gasification plants that drive turbines, and fuel cells.

biofilms Thin layers of bacteria and other microorganisms, including protoozoans, algae, and fungi, that adhere to surfaces in aqueous environments by means of gluelike mucilages. They are found naturally coating rocks, stones, and other objects in streams and rivers and clogging drains. They are also the main component of plaque on teeth. Biofilms cost industry large sums of money, because they not only clog up working parts of machinery, but also produce corrosive chemicals that damage metal, plastic, and other surfaces.

biofuel A fuel derived from renewable organic resources such as fast-growing crops and trees; animal, industrial, and municipal wastes; and the reactions of microorganisms (see bioenergy). They include BIODIESEL, BIOGAS, and GASOHOL.

biogas Gas produced by the degradation of organic material by microorganisms in the absence of oxygen. It is a mixture of methane and carbon dioxide. Biogas results from the treatment of organic waste, and can be a problem in garbage dumps and septic tanks if not piped away. Biogas is also produced in the digestive systems of animals – the biogas emitted by livestock worldwide has a significant effect on GLOBAL WARMING, since both carbon dioxide and methane are greenhouse gases. Biogas from garbage dumps can be piped to power electricity generation for local homes and businesses, and small-scale digesters are available for home use.

biogeochemical cycle The cycling of chemical elements between organisms and their physical environment, for example the cycling of nitrogen in the NITROGEN CYCLE. Various feedback controls operate to control biogeochemical cycles, and these may be disrupted by human activity, leading to pollution. For example, the present rapid accumulation of carbon dioxide in the atmosphere is in part due to the burning of fossil fuels, which unbalances the CARBON CYCLE.

biogeographical region Any geographical region that has a distinctive flora and fauna.

biogeography The study of the distribution of plants (including world vegetation types) and animals past and present, and of the changing relationship between humans and their natural environment.

biohazard A biological agent (e.g. a pathogenic microorganism) or condition (e.g. inadequate laboratory procedures) that poses a risk to life or health. There is an internationally recognized biohazard warning symbol.

biological clock The internal mechanism of an organism that regulates CIRCADIAN RHYTHMS and various other periodic cycles.

biological conservation Active management aimed at preserving habitat and species diversity and the genetic diversity within species. It also involves management of the processes upon which species diversity depends, such as biogeochemical cycling, and the control of biotic and abiotic factors that affect diversity and the survival of species and varieties. Biological conservation may include allowing the sustainable use of resources. *See also* conservation; sustainable development.

biological control The control of pests and diseases by introducing or substantially increasing the numbers of their natural predators or pathogens, or by interfering with their natural processes. For example, ladybirds (*Coccinella*) or parasitic wasps may be introduced to greenhouses to control aphid populations, and spores of fungal diseases are sprayed on invasive waterweeds such as the water hycainth (*Eichhornia*). Introduced predators or disease organisms may be genetically engineered to make them more potent predators/pathogens. Sterile males may be released in large numbers to reduce the reproductive success of some wild insect pests, or sex-attractant chemicals (pheromones) may be used to trap males or females. Biological control may also be indirect, as in the addition of organic nutrient supplements to soils to increase the numbers of saprotrophic microorganisms

to combat damping-off fungi (*Pythium, Rhizoctonia*). *See also* autocidal control.

biological efficiency *See* ecological efficiency.

biological oxygen demand (biochemical oxygen demand; BOD) The standard measurement for determining the level of organic pollution in a sample of water. It is the amount of oxygen used by microorganisms feeding on the organic material over a given period of time, usually 5 days, typically expressed as milligrams of oxygen per liter of water. Sewage effluent must be diluted to comply with the statutory BOD before it can be disposed of into clean rivers.

biological species concept The definition of a SPECIES as a population or group of populations whose members are capable of interbreeding to produce fertile offspring. In practice, some so-called species that are normally separated by geographical barriers may interbreed with each other if brought together. The definition also breaks down in areas where HYBRID SWARMS occur, leading to AGGREGATE SPECIES, especially where subsequent POLYPLOIDY renders the offspring fertile. *See* geographical isolation; reproductive isolation; speciation.

bioluminescence The emission of light by living organisms. It may be produced by internal chemical reactions or by the re-emission of absorbed energy as light radiation. In certain deep-sea fish light is produced by symbiotic bacteria living in special tissues. A wide range of organisms exhibit bioluminscence, from bacteria, fungi, algae, and plants to insects, squid, and fish. In deep-sea animals it can serve to illuminate prey or to conceal an animal's silhouette from predators below. The flashing luminescence of fireflies is used in species recognition when courting.

biomagnification The increase in concentration of accumulated chemicals in the tissues of organisms at higher levels of the FOOD CHAIN due to the processes of BIOAC-

CUMULATION and BIOCONCENTRATION. For example, the concentration of organochlorines in sea water is in picograms per liter to a few nanograms per liter, in marine invertebrates this rises to tens of nanograms per liter, in mussels several milligrams per liter, and in the fatty tissues of marine mammals and predatory birds, hundreds of milligrams per liter. *See* trophic level.

biomanipulation   The management of an ECOSYSTEM by manipulating the top end of the FOOD CHAIN (top-down control) rather than the nutrient input. For example, EUTROPHICATION can be ameliorated by removing plankton-eating fish, thus allowing herbivorous zooplankton that graze on algae to increase in numbers. The decrease in numbers of algae in turn allows greater light penetration, promoting the growth of aquatic plants and the reoxygenation of the water. *Compare* bottom-up control.

biomarkers   Physiological, histological, or biochemical changes in organisms that can be used as indicators of exposure to toxic chemicals. For example, the concentration of vitamin A in the livers of otters (*Lutra*) shows strong negative correlation with the concentration of PCBs (polychlorinated biphenyls) in the waters they live in. *See* hazard indicator; indicator species.

biomass   The total mass of all the living organisms in a particular ECOSYSTEM or area or at a particular TROPHIC LEVEL in a FOOD CHAIN. It is usually expressed as dry weight per unit area. The ratio of biomass to annual production is called the *biomass accumulation ratio*:

$$\text{biomass } (g\ m^{-2})/\text{net productivity } (g\ m^{-2}\ yr^{-1})$$

See net primary production.

biome   A major regional community characterized by distinctive vegetation and animal forms. Biomes are related to climate, latitude, topography, and soils, and merge into one another at their boundaries. They are the largest geographical units, and are often named after the dominant type of vegetation in the case of *terrestrial biomes*, e.g. *grassland. Aquatic*

biomes include coral reefs, lakes, and the marine rocky shore.

biometry   The statistical analysis of biological phenomena.

biomining   The use of genetically engineered microorganisms to recover toxic or economically important metals. *See also* bioremediation.

biomonitoring   *See* bioassay.

bioremediation   The use of living organisms to break down pollutants or wastes, such as industrial effluents, mining spoil, or oil spills, and to restore contaminated ecosystems. Plants may be used (*phytoremediation*) to extract heavy metals from contaminated soils and water. Some crop species can be genetically modified to accumulate toxic ions, e.g. *Arabidopsis* has been altered to express the enzyme mercuric ion reductase, which converts Hg^{2+} to Hg, which is volatilized and released into the atmosphere. Uptake may also be assisted by the use of chemical chelating agents to immobilize the toxins. Contaminated water is treated by RHIZOFILTRATION using plants with high transpiration rates and extensive root systems, such as willows (*Salix*) or reeds (*Phragmites*), or by the use of aquatic plants that are removed and destroyed once they have extracted the toxins. Organic wastes are usually tackled by bacteria and protozoans, and occasionally fungi (certain fungi are capable of breaking down POLYCHLORINATED BIPHENYLS).

biosensor   An analytical device that uses biological interactions to provide qualitative or quantitative data about chemical or biological molecules or processes. Biosensors are composed of biological material such as enzymes, antibodies, nucleic acids, cells, or microorganisms, usually intimately associated with a transducer that outputs electrochemical, optical, or other signals proportional to the concentration of the substance being analyzed or to changes in it. Such sensors have wide applications in clinical diagnosis and medical and veterinary research and in drug analysis and detection. They are also used in the control of industrial effluents and in the monitoring and control of pollution.

biosphere *See* ecosphere.

biosphere reserve A type of nature reserve of global importance, designated by UNESCO's Man and the Biosphere Programme. Typically, a biosphere reserve has a zoned region of management consisting of a highly protected core area. This is surrounded by buffer zones of varying degrees of restriction on human activity in which land use such as forestry, agriculture, or grazing may be permitted, provided they have no significant effects on the core area. Around the periphery are zones of cooperation, zones of influence, and transition areas where human activity may affect the reserve. By preserving a relatively undisturbed core area, the effects of land use and human activity on the remainder of the ecosystem can be assessed.

biosynthesis The chemical reactions by which a living cell builds up its necessary molecules from other molecules present. *See* anabolism.

biota The flora and fauna of a particular region or geological period.

biotechnology The use of biological processes for industrial, agricultural, and medical purposes. Examples are: the production of antibiotics by bacteria such as *Penicillium*; the use of FERMENTATION by yeasts and bacteria in the production of beer, wine, and yogurt; and the genetic modification of farm crops and livestock, including the production of human hormones and blood proteins by genetically engineered animals, and the production of monoclonal antibodies by microorganisms. *See* genetic engineering.

biotelemetry The remote electrical measurement and recording of physiological variables and actions of organisms and substances, allowing continuous monitoring of animal movements or physiological variables such as body temperature, blood pressure, heart rate, food and water intake, and nerve activity. The data are usually transmitted by radio, infrared, or ultrasonic signals from a remote site, which may range from a few centimeters to thousands of kilometers away. It includes the tracking of wild mammals, birds, fish, and other animals. Distant signals, for example from migrating birds, may be transmitted via satellite.

biotic Relating to life, especially describing living components of the environment that through their presence or activities affect the life of organisms in that environment or alter other aspects of the environment. For example, the presence of tall trees in a forest affects abiotic (nonbiotic) factors such as temperature, light intensity, humidity, and wind speed. The presence of several similar species leads to competition between them for resources such as nutrients and water.

biotic climax *See* climax community.

biotic potential The maximum capacity of an organism or population to increase under optimum, nonlimiting environmental conditions. It is determined by the age of sexual maturity, the frequency of reproduction, and the number of offspring born at a time. Optimum environmental conditions are those in which food, space, water, and other abiotic factors are not limiting and there are no predators, parasites, or diseases. Deviations from these conditions constitute *environmental resistance*.

biotope A geographic region characterized by certain environmental conditions and populated by a characteristic flora and fauna. The term is applied particularly to microhabitats, such as cow droppings in a pasture.

biotype 1. A naturally occurring group of individuals all with the same genetic composition, i.e. a clone of a pure line. *Compare* ecotype.

2. A physiological race or form within a species that is morphologically identical with it, but differs in genetic, physiological, biochemical, or pathogenic characteristics.

birds *See* Aves.

birth rate The frequency of live births in a population, usually expressed as the number of live births per 1000 individuals per year, or as the average number of offspring produced per individual per unit of time. This may be expressed as a function of age (b_x). *Compare* natality rate.

bite density The number of bites a herbivore takes while foraging. This is affected by the density of the plants in that area, and is directly related to the rate of food intake and the method of foraging. Where plant density is high, bite size becomes more important than bite density in determining food intake.

black earth *See* chernozem.

blanket bog A peat-forming plant community that occurs, even on slopes, in areas of high rainfall, which causes a constant downward movement of water, so that minerals from the soil or rock below are prevented from reaching the vegetation at the surface. This appears to prevent the establishment of trees and shrubs. It is thus a climatic climax vegetation type (*see* climax community). Blanket bogs are dominated by bog mosses (*Sphagnum* spp.), often associated with cotton grass (*Eriophorum*) or species of *Scirpus*. Low-growing ericaceous plants (Ericaceae), rushes (Juncaceae), and sedges (Cyperaceae) may also be common. In parts of north temperature regions blanket bogs have been growing since the start of the ATLANTIC PERIOD, and some have accumulated great thicknesses of highly acid PEAT, often with a pH approaching 4, even over limestone. *See* bog.

blastula *See* Animalia.

blight One of a range of plant diseases that causes sudden, serious leaf damage. The cause may be a pathogenic fungus,

oomycete, or bacterium. For example early blight of potatoes is due to the oomycete *Phytophthora infestans*, late blight to the fungus *Alternaria solani*, and fire blight of pears to the bacterium *Erwinia amylovora*.

bloom (water bloom; algal bloom) A visible sharp increase in the numbers of a species of plankton, usually an alga (e.g. diatoms) or dinoflagellate (Dinomastigota). Blooms of diatoms often occur in spring, when nutrients are plentiful. As the growth of increasing numbers of algae in the water uses up the nutrients (especially the silica needed for diatom tests), reproductive rates slow and numbers decline. Other blooms may be due to EUTROPHICATION.

blow-out An area of DUNES where the vegetation has been eroded away, usually by human feet or vehicles, leading to further erosion by the wind and sometimes the destruction of the dune or dune system. In coastal areas this may be exacerbated by high tides or storm tides, which allow the sea to rush the blow-out and flood the hinterland, further eroding the dunes as it does so. Such destabilized dunes may also begin to migrate, burying the vegetation (and sometimes also human settlements) in their path.

blue-green algae *See* Cyanobacteria.

blue-green bacteria *See* Cyanobacteria.

BMR *See* basal metabolic rate.

BOD *See* biological oxygen demand.

body residue The total amount of a particular chemical in an individual organism.

bog A climax plant community with no trees, in which organic matter accumulates as PEAT. Bogs are found in regions of poorly drained and permanently wet land in areas of high year-round rainfall and humidity. Decomposition rates are slow, so peat builds up. The vegetation is often dominated by bog mosses (*Sphagnum* spp.)

and sedges; other plants may include such species as bog myrtle (*Myrica gale*), bog rosemary (*Andromeda polifolia*), Labrador tea (*Rhododendron tomentosum*), and various insectivorous plants. Mats of bog vegetation may develop around the shores of lakes and large pools, gradually extending over the water surface, often to a depth of several feet. These can be dangerous, because what appears to be solid vegetation in fact covers deep water. There are several types of bogs: BLANKET BOG develops in areas of constant high rainfall. *Raised bogs* form from fens, where rain leaches nutrients from the top of the fen peat, making it acidic, especially in the center, so peat accumulates faster than at the edges. *Valley bogs* occur in mountain valleys where run-off and snow meltwater accumulate. *Kettle bogs* are found where bog mats creep over the surface of lakes formed by the melting of buried blocks of ice stranded at the end of the last glacial period.

boom and bust cycle The cycle of adaptation that drives the evolution of parasites when attempts are made to cultivate resistant varieties of plants. Initially, a disease-resistant crop variety is widely adopted, but then a new variant of the pathogen arises and the resistant variety suffers a crash in numbers. A new resistant strain of crop is then developed, and eventually the pathogen evolves a new resistant race, and so the cycle continues.

boreal forest (cold forest; taiga; northern coniferous forest) A major BIOME south of the TUNDRA dominated by tall evergreen conifers, mainly spruce, fir, and pine. Boreal forest is found in Canada and in northern Eurasia, stretching from Scandinavia across to Siberia. The summers are cool and the winters are very severe with temperatures as cold as −60°C. The conifers retain their leaves throughout the year and are adapted to cold conditions but in the northernmost areas bordering the tundra deciduous larches and birches replace them. In these areas, the temperatures are too low and the landscape too exposed for conifers to survive. The needlelike leaves of the boreal forest create a nutrient-poor PODZOL. The low levels of light result in few plant species on the forest floor, but the deep leaf litter supports a more diverse community of invertebrates and microorganisms. Reindeer and wolves migrate southward from the tundra into the boreal forest in winter. *Mountain coniferous forest* and *pine forest* for the south-east United States, the Andes, the Alps, and the Himalayas are also boreal forests.

boron (B) An element found in low concentration as boric acid in the soil solution and in plant tissues. It is an essential element for plant growth; alfalfa yellows and heart-rot of beets have been attributed to boron deficiency. Its role in plants is poorly understood, but it is thought to be involved in carbohydrate transport.

Botanical Code of Nomenclature *See* International Codes of Nomenclature.

botanical insecticide (botanical) An INSECTICIDE derived from plant material. Botanicals are accepted for use in organic crop production, but they can leave residues and may be disruptive to natural predators of pests and also be toxic to humans. Pyrethrum, a broad-spectrum insecticide derived from the dried flower heads of *Chrysanthemum cinerarifolium*, disrupts an insect's nervous system on contact. Many botanicals degrade rapidly in sunlight and may cause allergic reactions and dermatitis in humans. Some are more toxic, e.g. rotenone, derived from the roots of various tropical legumes, including *Derris*, a broad-spectrum insecticide. This is safe for honeybees, but does kill some other beneficial insects, and can be fatal if inhaled by mammals.

bottleneck (population) A temporary dramatic reduction in population size, resulting in a decrease in genetic diversity and gene frequency in succeeding generations, due partly to loss of individuals from the population and partly to the increase in number of deleterious homozygous recessives that are produced by subsequent inbreeding. An example is the cheetah, which today has very little genetic variation. A bottleneck causes the population to diverge from other populations of the same species, and may be the first step toward speciation or extinction. The decrease in genetic diversity may restrict the population's ability to adapt to future environmental change. *See* inbreeding; inbreeding depression.

bottom-up control   The regulation of a FOOD WEB or ECOSYSTEM by the availability of resources, whereby organisms on the same TROPHIC LEVEL experience limited availability of food, so are controlled mainly by competition for resources with other organisms in the same trophic level, rather than by predation. Thus the availability of organisms in the lower trophic levels regulates the numbers of organisms in the levels above. This is a useful approach when considering the productivity of each trophic level. *See also* biomanipulation.

brackish   Describing water that is more saline than fresh water, but less saline than sea water.

Braun-Blanquet scale   A system for describing an area of vegetation in terms of ABUNDANCE, COVER, and association devised by J. Braun-Blanquet in 1927. Scales of 1–5 are assigned to numbers of individuals and the proportion of the area covered by them, the degree of clumping of individuals, and the *degree of presence* of an individual plant (what proportion of the areas sampled it occurs in). It is now used mainly in a modified form as the *Domin scale*, which contains more divisions and is more accurate.

breeding system (mating system)   The pattern of matings between individuals in a population, including the method of mate selection or pollination, number of simultaneous mates and permanence of pair bond, degree of inbreeding, genetic determination of sex and incompatibility, and any other factors that place limits on the genetic diversity of the offspring.

broad-leaved evergreen forest   *See* temperate deciduous forest.

broken stick model   *See* random niche model.

bromomethane   *See* methyl bromide.

brood parasitism   The act of leaving eggs or young to be reared by an individual that is not the parent – usually an individual of another species. For example, many cuckoos lay their eggs in the nests of other birds.

brown earth   A type of SOIL found under deciduous forests in temperate regions. It is typically rich in humus and slightly acid, and is a fertile soil for agriculture.

brown tree snake   A species of snake (*Boiga irregularis*), native to Asia, which was accidentally introduced to the island of Guam in a military cargo and caused the EXTINCTION there of 12 species of birds and 6 species of lizards. *See* exotic.

browser   A vertebrate herbivore that feeds on trees and shrubs, e.g. giraffe (*Giraffa camelopardalis*) and tapir (*Tapirus*).

Bryophyta (mosses)   A phylum of simple plants, the mosses, mostly found in moist habitats. They have distinct gametophyte and sporophyte generations, the gametophyte being the dominant stage.

bryophytes (Bryata)   A general term for mosses (Bryophyta), liverworts (Hepatophyta), and hornworts (Anthocerophyta), plant phyla that were formerly grouped together in the phylum or division Bryophyta. In the Five Kingdoms classification, the term Bryophyta is reserved for the mosses, the other groups being placed in separate phyla.

bryozoans (moss animals; ectoprocts)   Members of the phylum Bryozoa – a group of aquatic invertebrates that form colonies composed of units called *zooids*. They are widely distributed in both fresh and salt water, mainly in shallow water, anchored to the seabed or forming crusts on rocks, shells, ships, seaweeds, and the shaded pilings of breakwaters. The zooids are little hydralike animals with a cilia-covered tentacle-bearing organ, the *lophophore*, filled with fluid, which can be extended to propel tiny food particles into the mouth.

budding 1. The production of buds on plants.
2. *See* asexual reproduction.

buffer A solution that resists any change in acidity or alkalinity (i.e. a change in H^+ concentration). Buffers are important in living organisms because they guard against sudden changes in pH. They involve a chemical equilibrium between a weak acid and its salt or a weak base and its salt. In biochemistry, the main buffer systems are the phosphate ($H_2PO_4^-/HPO_4^{2-}$) and the carbonate (H_2CO_3/HCO_3^-) systems. They are also useful for controlling pH in *in vitro* experiments and cultures. *See* pH.

bulb *See* perennating organ.

bulk density The ratio of the mass of a sample to its external volume (which includes the volume of air and water that may be contained within it).

buttress root A kind of PROP ROOT (*see* aerial root; stilt root) that is asymmetrically thickened, forming a planklike aboveground outgrowth at the base of the tree, providing extra support. Buttress roots are common in many tropical trees, including many species of fig (*Ficus*). *See also* mangrove.

cactus Any member of the large angiosperm family Cactaceae, which are mainly succulent xerophytic dicotyledons found in warm dry parts of the Americas. In many species the leaves are reduced to spines, which reduces transpiration, deters herbivores, and helps reflect the heat of the sun. Most species have succulent photosynthetic stems that store water. The root systems are shallow and spreading, exploiting occasional showers. *See* xerophyte.

cadmium (Cd) A bluish-white relatively soft metallic element found in trace amounts in the Earth's crust, and also in zinc ores and certain other ores. It is highly toxic if inhaled or ingested. Discharges of cadmium into air or water are strictly limited by law in most countries.

Caenozoic *See* Cenozoic.

Cainozoic *See* Cenozoic.

calcareous 1. Consisting of or containing calcium carbonate.
2. Growing on limestone or chalk or in soil rich in calcium carbonate.

calcareous soil (alkaline soil; basic soil) A SOIL with a pH greater than 6.0. Such soils are found where chalk (calcium carbonate) accumulates in the surface layers, usually derived from calcareous bedrock such as limestone, or from a calcareous sand deposit near coasts. Calcareous soils are commonest in regions of light rainfall, and are rich in calcium ions and often in other elements.

calcicole Describing a plant that thrives on neutral to alkaline soils high in calcium carbonate. *Compare* calcifuge.

calcifuge Describing a plant that grows on sandy or peaty soils, and cannot grow on basic substrates, such as most chalk and limestone soils. *Compare* calcicole.

calcium (Ca) A silver-white metallic element that does not occur free in nature, usually forming compounds such as carbonates (limestone and chalk) and sulfates (gypsum). The human body is 2% calcium, because calcium is a major constituent (70% by weight) of bones and teeth. The element is also involved in many metabolic and physiological processes. It also occurs in the shells of many animals, such as corals, mollusks, echinoderms, foraminiferans, and some planktonic algae. The main source of dietary calcium is milk and other dairy products, leafy green vegetables, and oily fish.

calcium carbonate ($CaCO_3$) A white or colorless compound commonly found in nature as chalk, limestone, and calcite. It makes up about 4% of the Earth's crust. The incorporation of calcium carbonate in the shells of marine plankton and corals, mainly as the mineral aragonite, removes carbon dioxide from the atmosphere.

calorie A unit of energy equal to the amount of heat required to raise the temperature of one gram of water by 1°C at one atmosphere pressure. It has now been replaced by the SI unit the joule (1 calorie = 4.1855 J), but is still often used to designate the energy-producing potential of foods, where the usual unit is the kilocalorie or Calorie (with capital letter 'C').

calorimetry The measurement of the amount of heat absorbed or evolved during a chemical or physical reaction, change of state, or dissolving of a solid or gas. For example, the calorific value of foods (the amount of heat released if the food is completely oxidized) is measured using a BOMB CALORIMETER.

Calvin–Benson cycle   *See* photosynthesis.

Cambrian   The earliest period of the Paleozoic era, about 590–510 million years ago. It is characterized by the appearance of algae and a proliferation of marine invertebrate animal forms, including ancestors of most modern animals – the so-called *Cambrian explosion. See also* geological time scale.

camouflage   The color, pattern, or shape of an animal that has evolved so that the animal can blend with its background. For example, many moths take on the color and pattern of the bark on which they rest (*see* industrial melanism). Other forms of camouflage include breaking up the recognizable body outline by large, bold blocks of color or stripes, concealing shadows by having darker undersides than back (countershading), or adorning the body with pieces of debris from the surrounding environment (e.g. decorator crabs). Such concealment is also called *crypsis*.

CAM plant   *See* Crassulacean acid metabolism.

campos   *See* savanna.

cannibalism   The eating of an animal by another member of the same species. This may serve various purposes. For example, if their temporary pool starts to dry up and food runs short, the tadpoles of desert spadefoot toads will start to eat each other; this ensures that at least some survive to breed. Male lions taking over a pride of females often kill and eat the existing young; this rapidly brings the mothers into reproductive condition, and the new pride males can sire their own young. In crowded conditions, animals such as aquarium guppies will eat their young, a means of controlling population size.

canopy   In a woodland, forest, or shrub community, the uppermost layer of vegetation, formed from the branches of the tallest plants. In complex forests, several zones of the canopy are recognized – the lower zone, the upper zone consisting of the intermeshed crowns of trees, and the emergent zone, where the crowns of a few taller species emerge into the light above the general canopy. *Compare* ground layer.

Cape Province   An area of southern Africa with one of the richest floras in the world, containing some 8550 species of flowering plants in more than 1500 genera, with a high proportion (30%) of endemic genera. Many ornamental plants and greenhouse plants originated in this region.

capillarity   *See* surface tension.

captive breeding   The breeding of animals in captivity, usually in order to preserve an endangered species or to improve the breeding stock by mating animals of different genetic backgrounds. If sufficient numbers can be bred in captivity, animals may be released into suitable habitats in the wild, helping re-establish populations in areas from which the species has disappeared, e.g. the black-footed ferret (*Mustela nigripes*) of the United States, and the Hawaiian goose (*Branta sandvicensis*) in Hawaii. In other cases, species may be protected by captive breeding while threats such as introduced predators are removed from their native habitats. Certain endangered species of partulid tree snails, rendered extinct in their native habitat by the rosy glandina snail, which was introduced to control yet another introduced predator, the giant African snail, survive only in captivity. Where there are insufficient numbers to establish a viable breeding population, eggs or young may be fostered – placed with parents of a closely related species.

capture–recapture   A method of estimating population size or density of mobile animals. This involves catching a random sample of a population, marking the individuals in some distinctive way, then releasing them back to mix with the rest of the population. At a later date another random sample is captured. Population size is estimated from the proportion of this second sample that bear the mark. Roughly speaking, the proportion of marked animals in the second sample is inversely proportional to the size of the population.

carbamates   Esters or salts of carbamic acid (NH_2COOH), certain of which are used as insecticides or parasticides. These act by inhibiting cholinesterase, which breaks down the neurotransmitter acetylcholine, vital to the transmission of many nerve impulses. Carbamates are contact insecticides that kill larvae, nymphs, and adults. Although high levels of exposure to these pesticides may cause carbamate poisoning in humans and livestock, they break down more quickly than organophosphate pesticides.

carbohydrates   A class of organic compounds that occur widely in nature and have the general formula ($C_x(H_2O)_y$). Carbohydrates are generally divided into two main classes: *sugars* (*saccharides*) and *polysaccharides*. They are both stores of energy (e.g. starch and glycogen) and structural elements in living systems (e.g. the cellulose of the plant cell wall).

carbon   (C)   A nonmetallic element found in all organic compounds. It is the basis of all living matter and an essential element in plant and animal nutrition. Carbon enters plants as carbon dioxide and is assimilated during photosynthesis into carbohydrates, proteins, and fats, forming the backbones of such molecules. The element carbon is particularly suited to such a role because it can form stable covalent bonds with other carbon atoms, and with hydrogen, oxygen, nitrogen, and sulfur atoms. It is also capable of forming double and triple bonds as well as single bonds and is thus a particularly versatile building block. Carbon, like hydrogen and nitrogen, is far more abundant in living materials than in the Earth's crust. *See also* carbon cycle.

carbonates   Compounds derived from carbonic acid or carbon dioxide, which contain the carbonate ion (CO_3^{2-}). They are weak bases: in solution, the carbonate ion can accept a hydrogen ion from water, forming a bicarbonate ion (HCO_3^-) and a hydroxyl ion. They react with acids, forming a salt and releasing gaseous carbon dioxide.

carbon cycle   The circulation of carbon from the atmosphere into living organisms and, after death, back into the atmosphere again. Plants, algae, and photosynthetic bacteria take in carbon dioxide from the atmosphere and during the process of PHOTOSYNTHESIS fix the carbon into living tissues. Carbon is returned to the atmosphere in carbon dioxide emitted by respiring organisms or resulting from decay. The atmospheric carbon dioxide concentration is also affected by the burning of FOSSIL FUELS, which release carbon fixed millions of years ago, so adding to current levels. Carbon dioxide is also given off during volcanic eruptions. The anaerobic breakdown of peat, especially the enhanced rates of breakdown in the tundra as a result of global warming, releases METHANE, another source of atmospheric carbon; even more significant is the release of methane by livestock. Smaller amounts are released by decomposition in landfill sites. The incorporation of carbon dioxide into calcium carbonate in the shells of marine plankton and invertebrates traps carbon that is not respired back into the atmosphere. *See illustration overleaf.*

carbon dating   (radiocarbon dating)   A method of estimating the age of organic material; it is based on the ratio of radioactive carbon atoms to stable carbon atoms in the material. The radioactive carbon-14 (^{14}C) is formed by the interaction of cosmic rays with atmospheric nitrogen. Together with the other two carbon isotopes, it is present in living organisms; once an organism dies it ceases to absorb ^{14}C, which steadily decreases, decaying at the

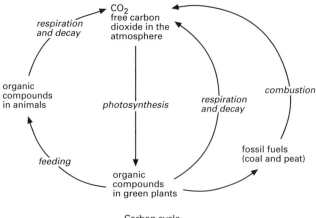

Carbon cycle

known rate of the half-life of 5730 ± 30 years. An estimate of the age of a material can be obtained from the residual ^{14}C present. It gives reasonably accurate results for about 40,000–50,000 years. However, since the method was originally developed in 1946–47 by Willard F. Libby (1908–80), uncertainties have arisen as to the actual ^{14}C content of the atmosphere during past ages. It is now accepted that the content varies in line with fluctuations in the cosmic ray activity that produces the ^{14}C. Corrections to dates within the recent past are now made through calibration with dates obtained through dendrochronology (tree-ring dating).

carbon dioxide (CO_2) A colorless odorless gas, heavier than air, that makes up at least 0.03% by volume in the atmosphere. It is taken in by plants during PHOTOSYNTHESIS and released during the RESPIRATION of living organisms and by the DECOMPOSITION of organic matter. It is also released during combustion and by volcanic activity.

Carbon dioxide is soluble in water, forming a weak solution of carbonic acid (H_2CO_3). The oceans contain large amounts of dissolved carbon dioxide, and act as buffers to counteract changes in atmospheric carbon dioxide, such as increases due to the burning of FOSSIL FUELS and forest trees. Carbon dioxide is a GREENHOUSE GAS, so contributes to GLOBAL WARMING. The contribution of different types of vegetation to atmospheric carbon dioxide levels can be demonstrated by measuring the *carbon dioxide flux* – the net movement of carbon dioxide between the land/soil surface and the atmosphere. This can vary with the season as well as with the vegetation type.

carbon dioxide equivalent (CDE) A measure used to compare the emissions of different greenhouse gases: the amount of carbon dioxide by weight that would cause the same amount of radiative forcing as a given weight of the other greenhouse gas being measured. Carbon dioxide equivalents are generally calculated by multiplying the mass of the gas of interest (in kilograms) by its estimated GLOBAL WARMING POTENTIAL.

carbon dioxide flux *See* carbon dioxide.

carbon equivalent (CE) A metric measure used to compare the emissions of different greenhouse gases. In the US, greenhouse gas emissions are usually expressed as million metric tons of carbon equivalents (MMTCE). Carbon equivalent (CE) units can be converted from CARBON DIOXIDE EQUIVALENT (CDE) units by multiplying the CDE by the ratio of the molecular weight of carbon to carbon dioxide (i.e. 12/44).

carbonic acid (H_2CO_3) A weak acid formed in small amounts when CARBON DIOXIDE dissolves in water:
$$CO_2 + H_2O \rightarrow H_2CO_3 \rightarrow H^+ + HCO_3^-$$
Carbonic acid occurs in rainwater. Although it is a relatively weak acid, in runoff water it is capable of dissolving CALCIUM CARBONATE in rocks such as limestone to form caves.

Carboniferous The second most recent period of the Paleozoic era, some 355–280 million years ago. It is named for the extensive coal deposits that formed from the remains of vast swamp forests which thrived in the warm, humid climate. In the United States the Carboniferous is often divided into the *Mississippian* or Lower Carboniferous and the *Pennsylvanian* or Upper Carboniferous. In the United States coal formation was restricted to the Pennsylvanian period. On land, amphibians and a few primitive early reptiles evolved, and in addition to treelike forms of clubmosses and giant horsetails, the first true trees appeared – primitive gymnosperms. Aquatic life included sharks and coelacanths. *See* geological time scale.

carbon monoxide (CO) A colorless flammable highly toxic gas produced by the incomplete combustion of carbon, especially during the burning of FOSSIL FUELS in vehicle engines. Motor vehicles may account for up to 98% of CO emissions in urban areas. Emissions can be reduced by fitting car exhausts with suitable CATALYTIC CONVERTERS. Vehicles running on liquefied petroleum gas also have lower CO emissions than gasoline-fueled ones. Carbon monoxide forms a complex with the red blood pigment hemoglobin, preventing the red blood cells from carrying oxygen, hence its toxicity.

carbon sequestration The net removal of carbon dioxide from the atmosphere into long-lived pools of carbon in terrestrial, marine, or freshwater ecosystems. Such pools (called *carbon sinks*) may be living biomass in soils and vegetation or inorganic forms of carbon in soils and rocks

and in the calcareous shells of marine plankton and invertebrates.

carbon sink *See* carbon sequestration.

carbon tetrachloride (CCl_4) (tetrachloromethane) A colorless toxic liquid used in some dry-cleaning processes and in certain fire extinguishers. Under the MONTREAL PROTOCOL, its use is being phased out, because its vapor is a GREENHOUSE GAS. *See* global warming.

carbon to nitrogen ratio (C:N ratio) The ratio of carbon to nitrogen in organic matter. When organic material is decomposed in the soil, the C:N ratio has an effect on the rate of DECOMPOSITION and hence on the amount of HUMUS formed and the rate at which soil nitrogen is released or immobilized. The rate of decomposition increases as the C:N ratio becomes smaller, as microorganisms in general require a ratio of about 30–35:1 for efficient digestion of compost or other organic material, materials with higher ratios having insufficient nitrogen for rapid decomposition. Decomposition of material with high C:N ratios results in release of nitrogen from the decomposing organic material into the soil. Most fresh plant material contains about 40% carbon, but species differ in their nitrogen content. During decomposition, up to 35% of the carbon present will be converted into humus if there is sufficient nitrogen present, the remainder being respired as carbon dioxide. The C:N ratio of humus is about 10:1. *See also* carbon cycle; nitrogen cycle.

carcinogen Any substance or agent that causes living tissues to become cancerous. Chemical carcinogens include many organic compounds and certain inorganic compounds such as asbestos. Physical agents that can cause cancer include radioactive materials and x-rays. Many carcinogens achieve their effect by causing mutations: they are *mutagens*.

Carnivora An order of mammals (*see* Mammalia) that contains flesh-eating mammals such as dogs (*Canis*), foxes and wolves (*Vulpes*), cats (*Felis*), otters (*Lutra*), badgers (*Meles*), and bears (*Ursus*). Their teeth are specialized for biting and tearing flesh and they have well-developed claws. A few members of the order are omnivorous, e.g. bears. The giant panda (*Ailuropoda melanoleuca*) is a herbivore.

carnivore Any animal that eats mostly flesh, especially a mammal of the order CARNIVORA. Insect-eating plants are also described as carnivorous. *Compare* herbivore; omnivore.

carnivorous plant *See* insectivorous plant.

carr A subclimax stage in the SUCCESSION of a HYDROSERE or HALOSERE, in which peat has built up but the soil is too wet for large trees to grow. Carr is dominated by plants tolerant of waterlogging, such as alders (*Alnus*) and willows (*Salix*). *See* fen; sere.

carrying capacity (*K*) The maximum population size that can be supported indefinitely by the available resources of a given environment.

Carson, Rachel Louise (1907–64) US science writer, who worked as a genetic biologist (1936–52) and later as editor for the US Fish and Wildlife Service. Her books, notably *The Sea around Us* (1951) and *Silent Spring* (1962), greatly increased public awareness of the natural environment and warned of the dangers of pollution.

caste A specialized form of a social insect that performs a particular function in the colony. Castes are distinguished by structural and functional differences. For example, honeybees have three castes: the queen (a fertile female) reproduces; workers (sterile females) gather food; drones (males) mate with the queen. Ants and termites often have in addition special soldier castes for defense. The caste system is an example of POLYMORPHISM in animals.

CAT *See* control action threshold.

catabolism The breakdown by living organisms of complex molecules into simpler compounds, as happens in respiration. The main function of catabolic reactions is to provide energy, which is used in the synthesis of new substances, for work (e.g. muscle contraction), for transmission of nerve impulses, and for maintenance of functional efficiency. *Compare* anabolism.

catadromous *See* diadromous.

catalyst A substance that changes the rate of a chemical reaction without being altered or used up in the reaction. Enzymes are highly specific biochemical catalysts.

catalytic converter A device fitted to vehicle exhaust systems that contains one or more catalysts that reduce the emission of pollutant gases such as carbon monoxide, nitrogen oxides, and unreacted hydrocarbons. Basic converters first reduce the exhaust gases to eliminate oxides of nitrogen, then oxidize them with additional air to eliminate carbon monoxide and unburned hydrocarbons. Modern 'three-way converters' use different catalysts to speed up the reactions with different types of gases, making it possible to deal simultaneously with different emissions. In many countries it is compulsory to fit catalytic converters to all new cars. *See* pollution.

catastrophe A major change in the environment that causes widespread damage and the death of many organisms. Some catastrophes are one-off events, such as the impact of a large asteroid on global climate, which may have contributed to the mass EXTINCTION at the end of the CRETACEOUS period. Others recur, but at intervals sufficiently far apart in time that species are unable to adapt to them by natural selection, for example large volcanic eruptions such as those of Mount St Helens in the United States.

catchment area (drainage basin) The area of land from which water runs off or drains into any given river valley or reservoir.

catch per unit effort (CPUE) In the harvesting of a natural resource such as fish, the total catch (biomass or numbers) divided by the total effort required (e.g. the number and size of ships and the number of days spent fishing).

cation A positively charged ION, formed by removal of electrons from atoms or molecules, e.g. H^+, Ca^{2+}. In electrolysis, cations are attracted to the negatively charged electrode (the cathode). *Compare* anion.

cation exchange capacity (CEC) The capacity of a soil to hold CATIONS (positively charged ions), usually expressed in milliequivalents per 100 grams (meq/100 g) of soil. It is an indication of the soil's ability to hold plant nutrients. A plant obtains the nutrients by exchanging one cation for another at the point soil colloids meet root colloids in the soil solution. CEC changes with pH, increasing with rising pH. The efficiency of CEC depends on the relative concentration of the cations, and their valences and diffusion rates. High cation exchange capacities are typical of clay minerals, in which spaces in the crystal lattices allow cations of similar size to be incorporated into the lattice. Clay soils have greater CECs than sandy soils. Organic matter also has a high CEC, so soils with plenty of HUMUS have high CECs.

catotelm *See* acrotelm.

CBD *See* Convention on Biological Diversity.

CDE *See* carbon dioxide equivalent.

CE *See* carbon equivalent.

CEC *See* cation exchange capacity.

cell The basic unit of structure of all living organisms, excluding viruses. Prokaryotic cells (typical diameter 1 μm) are significantly smaller than eukaryotic cells (typical diameter 20 μm). Some organisms consist only of single cells, e.g. amebae and diatoms. The largest cells are egg cells (e.g. ostrich, 13 cm in diameter), and the smallest are mycoplasmas (about 0.1 μm in diameter). All cells contain, or once contained, genetic material in the form of DNA, which controls the cell's growth and some of its activities; in eukaryotes this is enclosed in the membrane-bounded *nucleus*. All are filled with *cytoplasm*, containing various *organelles* (membrane-bound structures enclosing a specific environment in which certain cell processes are carried out, e.g. *mitochondria* are the sites of respiration, and *chloroplasts* are sites of photosynthesis), and are surrounded by a *plasma membrane* (cell membrane), which controls entry and exit of substances. Plant cells and most prokaryotic cells are surrounded by rigid *cell walls*, structures deposited outside the plasma membrane. In multicellular organisms cells become specialized for different functions; this is called *differentiation*. Within the cell, further division of labor occurs between the organelles.

Celsius scale *See* temperature.

Cenozoic (Caenozoic, Cainozoic) The present geological era, beginning some 65 million years ago, and divided into two periods, the Tertiary and the Quaternary. It is characterized by the rise of modern organisms, especially mammals and flowering plants. *See also* geological time scale.

center of diversity (gene center) A region where a certain taxon shows greater genetic diversity than it does anywhere else. The term is often used in relation to crop plants, especially to the region in which the particular crop is thought to have originated. For example, numerous wild relatives of wheat and other related grain crops occur in the Middle East, and it is believed that wheat was first domesticated there. Secondary centers of diversity represent areas where the crop has been cultivated for a long time, but where there are no wild relatives. Centers of diversity are important sources of genes for the development of new cultivars. *See* genetic resources.

centigrade scale *See* temperature.

central-place foraging Foraging in which the foragers radiate out from a central base. For example, the sorties of leaf-cutter ants (*Atta* spp.), which have a central underground nest. *See* foraging strategy.

Cephalochordata *See* Chordata.

cereal A plant that yields a starchy grain used as food for animals and humans. Cereals (e.g. barley) are also used in the brewing industry. Most cereals are grasses, e.g. barley, corn (maize), sorghum, oats, rice, rye, and wheat. Wheat is the world's most widely grown cereal, a staple food in temperate regions and one of the oldest cereals known. Milling of wheat dates back at least 75 000 years. Rice is the second largest cereal crop, grown mainly in tropical and subtropical regions, especially in Asia. Cereals grain are rich in carbohydrates but relatively low in protein.

cesium (Cs) A rare silvery-white soft metallic element, with a very low melting point (28°C). The weakly radioactive isotope cesium-137 has a half-life of 33 years, and is a waste product of nuclear power stations processing uranium and plutonium, and is also used as a coolant in nuclear reactors. If released into the atmosphere, it is carried into the soil by rainfall, and taken up by plants and grazing livestock, being found in their milk.

CFCs (chlorofluorocarbons) A group of hydrocarbon-based compounds in which some of the hydrogen atoms are replaced by chlorine or fluorine atoms. They are chemically inert and were formerly widely used as refrigerant liquids (freons) in refrigerators (especially $CFCl_3$ and CF_2Cl_2), aerosol propellants, and blowing agents in the manufacture of plastics. However, there is strong evidence that when they escape to the upper ATMOSPHERE they cause depletion of the OZONE LAYER, hence their use had been phased out since the late 1980s. Unlike most gases, CFCs are not broken down as they rise through the tro-

posophere, but once they reach the stratosphere the stronger light intensities there cause them to dissociate, releasing chlorine atoms that take part in the catalytic destruction of ozone. CFCs are also GREENHOUSE GASES, with global warming potentials 3000–13000 times greater than that of carbon dioxide.

chalk A white soft fine-grained limestone consisting of calcium carbonate and the remains of microscopic fossil plankton such as coccoliths and foraminifera. Some chalk deposits contain up to 99% pure calcite. Sponge spicules, diatoms, and radiolarian tests and grains of quartz contribute small amounts of silica. Extensive chalk outcrops of the upper Cretaceous period occur in western Europe, England, and the United States from South Dakota to Alabama and south to Texas. Soils over chalk are alkaline, and support a characteristic CHALK GRASSLAND in areas of grazing by sheep or rabbits. *See also* calcareous soil.

chalk grassland A species-rich grassland formed on thin CALCAREOUS SOILS overlying CHALK, where it forms a subclimax maintained by grazing, usually by sheep or rabbits. The flora is rich in herbs, especially rosette plants and other plants that grow close to the ground. It is found mainly in southern and eastern England, northern France, and a few other parts of western Europe.

chamaephyte *See* Raunkiaer's life-form classification.

chaparral *See* maquis.

Chapman cycle A cycle of reactions producing OZONE in the stratosphere by high-energy ultraviolet radiation from the Sun. When ultraviolet rays strike molecules of oxygen (O_2), they split the molecules into separate oxygen atoms. The free O atoms can then combine with oxygen molecules to form ozone (O_3) molecules. Ozone is a relatively unstable molecule, and if it absorbs certain wavelengths of ultraviolet or visible light it regenerates oxygen molecule and oxygen atoms, releasing heat and raising the temperature of the atmosphere locally. Ozone production is powered by ultraviolet radiation:

$$O_2 + UV \rightarrow 2O$$
$$O + O_2 \rightarrow O_3$$
$$O_3 + UV \rightarrow O + O_2$$
or
$$O_3 + O \rightarrow O_2 + O_2$$
See ozone layer.

character (trait) 1. In genetics, any recognizable attribute of the PHENOTYPE of an organism. Characters may be the result of the action of one or many genes. Heritable differences in a character shown by different individuals in a population are due to different forms (ALLELES) of the gene(s). 2. In taxonomy, attributes such as form, anatomy, and physiology that are used by taxonomist to compare taxa and construct phylogenies. Examples include leaf shape and the arrangement of petals in a flower.

character displacement The divergence of the characteristics of two similar species in areas where their ranges overlap. This reduces COMPETITION as each species evolves to exploit a different ecological niche. For example, the nuthatches *Sitta neumayer* and *S. tephronata* show differences in beak size in areas where their ranges overlap, allowing them to avoid competition by feeding on seeds of different sizes. This also occurs within a community, when similar species in the same community are compared. This is termed *community character displacement. See also* adaptive radiation.

chart datum The lowest mean water level measured over a period of time for a given area. Chart datum is used as a reference point against which to measure the height of tides, the chart datum point being 0.

checkerboard distribution In ISLAND BIOGEOGRAPHY THEORY a species distribution in which two or more ecologically similar species have mutually exclusive but interdigitating distributions such that any one island supports only one of the species, e.g. the cuckoo-doves *Macropygia mackin-*

layi and *M. nigrirostris* in the Bismarck archipelago off the coast of New Guinea, and the distribution of nocturnal species of lorises (*Nycticebus*) and tarsiers (*Tarsius*) on small islands west of the WALLACE'S LINE in Asia. If they occurred together, such species would probably compete for the same resources. These distributions are relatively rare.

chelating agent A chemical that forms ring complexes with metal ions, especially heavy metal ions, for example, EDTA, which is use to treat heavy metal poisoning. Chelating agents may be used in BIOREMEDIATION to bind toxic heavy metals to soil particles, a process known as *sequestration.*

Chelicerata *See* Arthropoda.

chemical competition *See* competition.

chemical oxygen demand (COD) The amount of oxygen consumed in the oxidation of organic and oxidizable inorganic matter in a sample of water, typically expressed as milligrams of oxygen per liter of water. COD is usually determined by incubating known volumes of water with known quantities of chemical reagents at about 150°C until oxidation is complete, then determining the amount of the reagent changed by means of colorimetry or spectrophotometry. COD is used in industrial and municipal laboratories dealing with industrial waste and chemically polluted water. *Compare* biological oxygen demand.

chemiluminescence *See* phosphorescence.

chemoautotroph An organism that derives energy for the synthesis of organic molecules from the oxidation of inorganic compounds. Only certain specialized bacteria are chemoautotrophs. For example, *Nitrobacter* is a chemoautotroph that oxidizes nitrite to nitrate.

chemoheterotroph An organism that

derives energy for the synthesis of organic molecules from the oxidation of organic compounds. All animals are chemoheterotrophs.

chemosterilant 1. A chemical used to sterilize an object or surface, such as surgical instruments, laboratory benches, animal housing, and air ducts. Seeds and soils may also be chemically sterilized. Chemosterilants are substances that kill bacteria and other pathogenic organisms. High-level sterilants are capable of killing bacteria, fungi, viruses, and spores. Some chemosterilants are available for household use, e.g. sodium hypochlorite (household chlorine bleach).
2. A chemical used to sterilize (in the reproductive sense) pests and thus reduce the number of offspring in the subsequent generation. For example, chemically sterilized laboratory-reared male fruit flies may be released into the wild population to compete for mates with normal males.

chemotaxonomy The classification of organisms according to their chemical makeup. The compounds involved are usually primary and secondary products of metabolism.

chemotroph An organism that derives energy for the synthesis of organic requirements from chemical sources. Most chemotroph organisms are HETEROTROPHS (i.e. CHEMOHETEROTROPHS) and their energy source is always an organic compound; animals, fungi, and most bacteria are chemoheterotrophs. In autotrophs (i.e. CHEMOAUTOTROPHS) the energy is obtained by oxidation of an inorganic compound; for example, by oxidation of ammonia to nitrite or a nitrite to nitrate (by nitrifying bacteria), or oxidation of hydrogen sulfide to sulfur (by colorless sulfur bacteria). *See* autotroph; chemoautotroph; chemoheterotroph; heterotroph. *Compare* phototroph.

Chernobyl nuclear accident A serious accident that occurred during a badly planned experiment at a nuclear power plant at Chernobyl, near Kiev, in the Ukraine, on April 25, 1986. Following an explosion, fire, and partial meltdown, radioactive substances, including radioactive isotopes of cesium, cobalt, iodine, krypton, and xenon, were released into the atmosphere and carried over a wide area on the wind. Some 30 workers and firefighters at the Chernobyl plant died within a few hours, 200 became acutely sick, and others exposed to the radioactive cloud have died since (some estimates put the deaths so far at 120 000). Farther from the center of the pollution, in the UK, projected excess deaths are only 50 over the 40 years following the accident. The area for miles around Chernobyl was heavily contaminated and crops can no longer be grown there. 135 000 local residents in a 30-mile radius of the plant have been permanently relocated. The toxic cloud reached Scandinavia, the UK, and continental Europe as far south as Italy, and the chemicals were brought down into the soil by rain, especially in upland areas, rendering livestock (including reindeer) and their milk unsafe for human consumption. Pasture remains contaminated in places today. Humans were exposed to gamma radiation from the pollutant cloud, gases, and contaminated particulates from the air, and other contaminants in food. The reactor has been buried in a concrete sarcophagus to contain the radioactivity. *See also* nuclear energy.

chernozem (black earth) A freely draining calcareous soil found in regions with high summer rainfall, characterized by humus distributed fairly evenly throughout the profile. The upper horizon of the soil is very thick and black with humus; it merges into a brown lime-rich horizon below. Chernozems are rich in plant nutrients and have a good crumb structure, allowing drainage and aeration. They are associated with grasslands in temperate climates such as the prairies of North America, the pampas of South America, and the Russian steppes, where they support highly productive agricultural systems.

chill factor *See* wind chill.

chimera   An individual or part of an individual in which the tissues are a mixture of two genetically different tissues. It may arise naturally due to mutation in a cell of a developing embryo, producing a line of cells with the mutant gene and hence different characteristics compared to surrounding cells. It may also be induced experimentally. For example, two mouse embryos at the eight-cell stage from different parents can be fused and develop into a mouse of normal size. Analysis of the genotypes of the tissues and organs of such a mouse reveals that there is a random mixture of the two original genotypes. In plants, chimeras between different species may arise as a result of grafting: a bud may develop at the junction between the scion and stock with a mixture of tissues from both. Many variegated plants are also chimeras: a mutation has occurred in a sector of tissue derived from a particular layer of the shoot apex, resulting in subsequent chlorophyll deficiency. For example, in a white-edged form of *Pelargonium*, the outermost layer is colorless, indicating a lack of chlorophyll due to a mutation, but in contrast to animal chimeras the genetic mixing is confined to this region, and does not extend throughout the plant.

chi-squared test ($\chi2$ test)   A statistical method for testing how well a set of experimental values fits the set of values that would be expected from some particular hypothesis. If a particular observed value is o and the corresponding expected (hypothesis) value is e, then for each observed value o one calculates $(o - e)^2/e$. These results are added for all values in the set of observations to give the parameter χ^2. The goodness of fit can be obtained from available tables of χ^2. Chi is the Greek letter χ.

chlorinated hydrocarbons   Hydrocarbons in which one or more hydrogen atoms are replaced by chlorine atoms. Some chlorinated hydrocarbons are contact poisons, and are used as INSECTICIDES, e.g. aldrin, dieldrin, chlordane, DDT, and methoxychlor. Use of these insecticides was banned in most Western countries after the discovery that they persisted through the food chain in the fatty tissues of animals and were toxic, particularly to birds of prey, which declined sharply in numbers. They are still used in some tropical countries. *See also* CFCs.

chlorination   The treatment of drinking water and swimming pools with chlorine or chlorine compounds such as sodium hypochlorite or calcium hypochlorite to disinfect the water, promote coagulation, or control tastes and odors. For small volumes of water, these compounds may be added in tablet form; for larger volumes gaseous chlorine is bubbled through the water from pressurized steel containers.

chlorine (Cl)   An element found in trace amounts in plants and one of the essential nutrients in animal diets. Common table salt, an important item of the diet, is made up of crystals of sodium chloride. The *chloride ion* (Cl^-) is important in buffering body fluids and, because it can pass easily through cell membranes, it is also important in the absorption and excretion of various cations.

chlorofluorocarbons   *See* CFCs.

chlorophyll   One of a group of photosynthetic pigments that absorb blue-violet and red light and reflect green light, imparting the green color to green plants. Chlorophylls are involved in the light reactions of PHOTOSYNTHESIS, absorbing light energy and initiating electron transport. Photosynthetic bacteria have a slightly different group of chlorophyll pigments, *bacteriochlorophylls*, similar to chlorophyll *a*, but absorbing at slightly longer wavelengths, including far-red and infrared light.

chloroplast   A photosynthetic organelle containing CHLOROPHYLL and other photosynthetic pigments. Chloroplasts are found in all photosynthetic cells of plants and protoctists, but not in photosynthetic PROKARYOTES (i.e. bacteria). A chloroplast has a membrane system containing the pigments, on which the light reactions of PHOTOSYNTHESIS occur. The surrounding gel-like ground substance, or *stroma*, is where the dark reactions occur.

chlorosis The loss of chlorophyll from plants, resulting in yellow (*chlorotic*) leaves. It may be the result of the normal process of senescence, lack of light, lack of key minerals for chlorophyll synthesis (particularly iron and magnesium), or disease. *See* deficiency disease.

cholera An acute infection of the small intestine caused by the bacterium *Vibrio cholerae*. The bacterium is taken into the body with contaminated food or drink, and is especially common in areas where there is poor sanitation and no clean water for drinking and cooking. Bacterial toxins released in the intestine trigger severe watery diarrhea, vomiting, and rapid dehydration. The skin pales, blood pressure falls, and severe muscular cramps set in. This is followed by coma and death. The disease can be rapidly reversed by administering salt solutions, further aided by antibiotics. Vaccination provides only partial, short-lived protection.

Chordata A major phylum of bilaterally symmetrical metamerically segmented coelomate animals, characterized by the possession at some or all stages in the life history of a dorsal supporting rod, the *notochord*. The dorsal tubular nerve cord lies immediately above the notochord and a number of visceral clefts (gill slits) are present in the pharynx at some stage of the life history. The post-anal flexible tail is the main propulsive organ in aquatic chordates. The phylum includes the subphylum *Craniata* (Verebrata), in which the notochord is replaced by a vertebral column (the backbone). There are two other subphyla, the *Urochordata* and the *Cephalochordata* (sometimes known collectively as the Acrania or Protochordata).

CHP plant *See* combined heat and power plant.

chromatid One of a pair of replicated chromosomes found during the prophase and metaphase stages of MITOSIS and MEIOSIS.

chromium (Cr) A hard silvery metal that occurs naturally in the mineral chromite ($FeO.Cr_2O_4$). Chromium is a micronutrient for animals, needed in trace amounts; it is involved in glucose metabolism. Dietary sources of chromium include meat, dairy products, pulses, and grains. Chromium can reach waterways and sewage sludge via effluents from metal-processing industries and tanneries, and is toxic above levels of about 2 mgl^{-1}, especially when also in the presence of other metals such as nickel.

chromosome One of a group of thread-like structures of different lengths and shapes in nuclei of eukaryotic cells. They consist of DNA with RNA and protein (mostly *histones*) and carry the GENES. Each chromosome contains one DNA molecule, which is folded and coiled. The name chromosome is also sometimes given to the genetic material of bacteria, though these have a rather different structure and, with the exception of a few Archaea, lack histone-like proteins. During nuclear division the chromosomes are tightly coiled and are easily visible through the light microscope. After division, they uncoil and may be difficult to see. Along the length of the uncoiled chromosomes are bead-like structures called *nucleosomes*, highly organized aggregations of DNA and histones. The number of chromosomes per nucleus is characteristic of the species (e.g. humans have 46).

chromosome mutation *See* mutation.

ciliates *See* Ciliophora.

Ciliophora A phylum of protoctists containing some of the best known protozoans. All have cilia for locomotion, a contractile vacuole, and a mouth. Most have two types of nuclei, the meganucleus controlling normal cell metabolism, and the smaller micronucleus controlling sexual reproduction (conjugation). Binary fission also takes place. Some (e.g. *Paramecium*) are covered with cilia. Others (e.g. *Vorticella*) have cilia only round the mouth, and in some (e.g. *Stentor*) these cilia are specialized for feeding.

circadian rhythm A daily rhythm of various metabolic activities in animals and plants. Such rhythms persist even when the organism is not exposed to 24-hour cycles of light and dark, and are thought to be controlled by an endogenous biological clock. Circadian rhythms are found in the most primitive and the most advanced of organisms. Thus *Euglena* (a single-celled protoctist) shows a diurnal rhythm in the speed at which it moves to a light source, while humans are believed to have at least 40 daily rhythms.

CITES (Convention on International Trade in Endangered Species) The Convention on International Trade in Endangered Species of Wild Fauna and Flora. It is an agreement signed between some 160 countries, prohibiting commercial trade in endangered species and products derived from them. As more species have become endangered, the list of protected species has lengthened. CITES sets out degrees of protection for different species, allowing threatened but not ENDANGERED SPECIES to be exported and imported under license.

clade In CLADISTICS, a branch of a lineage that results from the splitting of an earlier lineage. The split results into two new TAXA, each of which is represented as a clade (branch) in a phylogenetic diagram. *See* phylogeny.

cladistics A method of CLASSIFICATION in which the relationships between organisms are based on selected shared characteristics. These are generally assumed to have been derived from a common ancestor, in the evolutionary process of *cladogenesis*, although the 'transformed cladists' believe that shared characteristics alone provide a logical basis for classification without postulating evolutionary relationships. The patterns of these shared characteristics are demonstrated in a branching diagram called a *cladogram*. The branching points

of the cladogram may be regarded either as an ancestral species (as in an evolutionary tree) or solely as representing shared characteristics.

Cladistics assumes that the closeness of a relationship depends on the recentness of common ancestry, indicated by the number and distribution of shared characteristics that can be traced back to a recent common ancestor. Cladistics also regards only true natural groups as those containing *all* the descendants of a common ancestor.

cladogenesis *See* cladistics.

cladogram *See* cladistics; dendrogram.

clarification The removal of suspended or colloidal material from a liquid. For example, sewage sludge is clarified by settling in tanks, followed by the FLOCCULATION of suspended and colloidal matter. An alternative method, used in the treatment of industrial wastes, is FLOTATION, in which suspended particles, such as metal hydroxides, are carried to the surface of the liquid by gas bubbles that adhere to the particles. The scum is then skimmed off. FILTRATION is seldom a suitable option, because the filters tend to clog up quickly, but it may be used to further clarify a liquid that has already been processed by the above procedures.

class A taxonomic rank that is subordinate to a phylum (or sometimes a division in plant taxonomy) and superior to an order.

classification The grouping and arrangement of organisms into a hierarchical order (*hierarchical classification*). Each level of organization is called a TAXON. In the most widely used classification systems for living and fossil organisms, the major taxa, in descending rank, are DOMAIN, KINGDOM, PHYLUM, CLASS, ORDER, FAMILY, GENUS, SPECIES, and RACE or VARIETY. The aim of classification is usually to aid in identification of organisms or to represent their phylogenetic relationships or, ideally, both. An important aspect of classifications is their predictive value. For example, if a characteristic is found in one member of a group of plants, then it is also likely to be found in the other members of that group even though the characteristic in question was not used in the initial construction of the classification. The validity of existing classification systems is now being tested by mRNA studies. *See* Five Kingdoms classification; phylogeny.

clay An extremely fine-textured soil made up of small mineral particles less than 0.002 mm in diameter, formed mainly from hydrated aluminum and magnesium silicates. Clay soils are derived from granitic rocks. *Clay minerals* have high cation exchange capacities: spaces in the crystal lattices allow cations (positive ions) of similar size to be incorporated into the lattice. However, there is no such capacity to hold or exchange anions (negative ions) such as nitrate (NO_3^-) and sulfate (SO_4^{2-}), and these tend to leach from clay soils. The main source of anions in clay soils is humus. Clay soils are poorly aerated, since the air spaces between the particles are very small. They may also become very sticky and difficult to work when wet and can easily become waterlogged. Nutrient availability to plants can be a problem because the nutrients may become chemically bound to the surfaces of the particles. *See also* soil structure; soil texture.

Clean Air Act Legislation initially introduced in the USA in 1970 that aimed to reduce air pollution, especially sulfur dioxide emissions, from industries and motor vehicles. It was initially administered by the then newly formed ENVIRONMENTAL PROTECTION AGENCY (EPA). The act identified air pollutants and set out primary and secondary standards for each. For each pollutant, there are specific limits, procedures for reducing levels, and timescales for compliance. The primary standard protects human health, and the secondary standard encompasses potential damage to property and the environment. Similar Clean Air Acts were introduced in the UK in 1956 and 1958, outlawing the burning of coal in defined city areas; these made a

significant impact on reducing London's infamous smogs. 1990 amendments to the USA's Clean Air Act permitted transferable-allowance schemes that allow companies to buy and sell *pollution credits* (the legal rights to produce specific amounts of pollution, in particular sulfur dioxide, nitrogen oxides, carbon monoxide, lead, ozone, and particulate matter, as well as hazardous air pollutants (HAPs) such as the vapors released by dry cleaning plants, chemical plants, and printing plants). By purchasing these credits from cleaner companies, companies with high emissions can remain in business while they convert to cleaner technologies. The Clean Air Acts that apply today control emissions by factories, motor vehicles, and aircraft, and also the production and use of chemicals that impact on air quality, the OZONE LAYER (e.g. CFCS), and ACID RAIN deposition.

clean fuels Fuels that cause relatively low levels of polluting emissions. They include such power sources as FUEL CELLS, natural gas, alcohol, and BIOFUELS.

cleaning station A location to which fish come so that one or more individuals of another species (usually a fish or shrimp) may seek out and remove dead surface material and external parasites, including those from inside their mouths. This is a symbiotic relationship, and is often accompanied by specific displays by which the cleaner indicates its willingness to clean or displays coloration that the 'customer' can recognize, and the 'customer' signals its desire to be cleaned. This enables even predatory fish to be cleaned by potential prey species. *See* symbiosis.

clean technology A means of providing a human benefit that, overall, uses fewer resources and causes less environmental damage than alternative means with which it is economically competitive. For example, the use of CLEAN FUELS to generate electricity or automotive power.

clean-up technology (end-of-pipe technology) Technology used to clean up pollution, in contrast to *clean technology*, which is technology designed to generate less pollution in the first place. Flue gas desulfurization is an example; it involves greater costs than using fuels that release less sulfur dioxide, and also reduces the efficiency of the process. *Compare* clean technology.

Clean Water Act   A legislative measure introduced in the United States in 1972 'to restore and maintain the chemical, physical, and biological integrity of the nation's waters' so that they can support 'the protection and propagation of fish, shellfish, and wildlife and recreation in and on the water.' The Act aims to regulate discharges of POLLUTANTS into waterways, finance treatment plants for municipal wastewater, and manage polluted runoff. It does not deal directly with groundwater issues. A recent development has been trials of emissions trading for nutrients such as phosphates, nitrates, and ammonia discharges from farms and industrial premises (*see* Clean Air Act).

clear-cutting   The felling of timber in which all the trees are removed from a site. This greatly increases runoff and the LEACHING of nutrients, especially nitrates.

clear-felling   The cutting down of all the trees in the forest.

Clements, Frederic Edward   (1874–1945) American ecologist, noted for his original work *Research Methods in Ecology* (1905), in which he described the main techniques of field work.

climate   The average pattern of WEATHER at a place, including solar radiation, temperature, humidity, precipitation, wind velocity, and atmospheric pressure. Climate depends on the variation in atmospheric conditions at a location over a period of years. As well as the conditions that might be expected at different times of year, it also encompasses the extremes reached and the frequency of less common weather events. Thus the precise climate of a place

depends on the timescale over which it is being considered.

Solar radiation drives the atmospheric circulation, which is modified by the CORIOLIS FORCE (the effect of the Earth's rotation) and topography. Solar radiation varies with latitude: at the equator the Sun is overhead, and day length varies little all year round. At the poles the Sun's rays strike the Earth at a considerable angle, so incoming solar radiation is less, and daylength varies from 24-hour daylight in midsummer to 24-hour night in midwinter. Because of the tilt of the Earth's axis, solar radiation at a given site varies through the year, and the farther from the equator, the greater the seasonal variation in radiation. Altitude also affects climate, since atmospheric pressure decreases, solar radiation increases, and wind speed is usually greater at higher altitudes. Warm air holds more moisture than cold air, and this affects humidity. Humidity is also affected by distance from the sea.

Land and water have different specific heats: they heat up and cool down at different rates, land warming and cooling being much faster than that of water. Thus the continental interiors have more extreme climates than coastal areas, with hotter summers and colder winters. Distance from the sea also affects precipitation, as does the distribution and orientation of mountain ranges. Major ocean currents also affect climate. For example, the presence of the GULF STREAM, a warm current that flows across the north Atlantic and warms the coast of western Europe, especially the British Isles, means that temperatures in London, for instance, are warmer than those in Montreal in winter, although London is farther north than Montreal.

The *Köppen climate classification system* is widely used and is based on the major vegetation zones and their climatic requirements. It recognizes five major climate types, which are further subdivided on the basis of precipitation quantity and distribution through the year. These are:
A *Tropical moist climates*: the average monthly temperature remains above 18°C (65°F) all year round, and there are no sea-

sons. Annual rainfall is greater than 1500 mm, and is greater than total evaporation, so there is high humidity. These climates are between the equator and 15-25° latitude. They include *tropical wet* climates, where rainfall occurs all year; *tropical monsoon* climates, when most of the rainfall is during the hottest months and there is a dry season of 3-5 months, and *tropical*

wet and dry (*savanna*) climates, with a long, dry winter.

B *Dry climates*: on average, evaporation exceeds precipitation throughout the year, so there are no permanent streams. These climates are found between 20° and 35° of latitude.

C *Mild subtropical mid-latitude* (*warm temperate*) *climates*: The coldest month

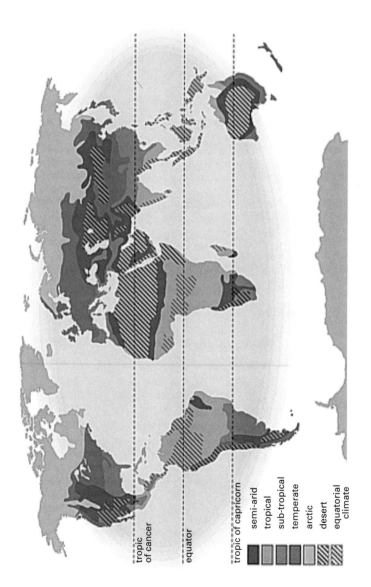

Climate: climatic zones

tropic of cancer

equator

tropic of capricorn

semi-arid
tropical
sub-tropical
temperate
arctic
desert
equatorial climate

has an average temperature below 18°C (65°F) but above –3°C (27°F), and at least one month has an average above 10°C (50°C). There are pronounced summer and winter seasons. These climates are found between 30° and 50° of latitude. They include the *humid subtropical climate* with hot warm summers with precipitation from frequent thunderstorms and mild winters with precipitation from cyclones; *Mediterranean climates* with rainfall mainly in winter from cyclones and very dry summers due to the sinking air of subtropical highs; and *marine (oceanic) climates* that are humid, with mild winters, during which heavy precipitation falls from depressions, and short dry summers, as along the western coasts of continents in the mid-latitudes.

D *Moist continental mid-latitude (cool temperate) climates*: the coldest month has an average temperature below –3°C (27°F) but more than –30°C, and the warmest month has an average above 10°C (50°C). These climates are found in the mid-latitudes north or south of 50°. The winters are severe, with heavy snow and strong winds blowing from continental or polar regions.

E *Polar climates*: the average temperature of the warmest month is below 10°C (50°C). There are no marked seasons. In the polar tundra there is *permafrost* – soil frozen permanently to depths of hundreds of meters; the polar ice caps are permanently covered in ice and snow.

climate change In general, the long-term fluctuations in the Earth's climate. More particularly the term is used to describe a significant change from one climatic condition to another on a global scale; for example the shifts from ice ages to warmer periods that have taken place during the Earth's geologic history. Such changes occur naturally, sometimes abruptly but at other times over considerable timescales, from changes, for example, in the Earth's orbit, the energy output from the Sun, volcanic eruptions, orogeny (mountain building), meteor strikes, or lithospheric motions. Over considerably shorter timescales more recent climate

changes include the periods of warmer-than-average temperatures and cooler than average periods. More recently, the study of climate change has sometimes been dominated by the concept of GLOBAL WARMING and other climate changes caused by human activities.

climate zone A division of the globe into a sequence of geographical regions defined by their climates, in particular the temperature and amount of precipitation and their distribution through the year. These zones are typically associated with particular types of natural vegetation.

climatic climax A climax community that is directly due to the influence of climate. *See* climax community.

climatic factor An influence that is due to climate, for example, temperature, humidity, wind speed, precipitation, atmospheric pressure, solar radiation. Aspects of climate affect living organisms. *See also* edaphic factor.

climatic optimum The period of highest prevailing temperature since the last ICE AGE. In most parts of the world this occurred about 4000–8000 years ago.

climax adaptation number A phytosociological ordering system in which species are assigned a value in the range 1 to 10 according to their *importance values* (the sum of each species' relative density, relative frequency, and relative abundance expressed as percentages). The importance value may range from 0 to 300 and enables species to be ranked according to their phytosocoiological importance.

climax community The final COMMUNITY in a SUCCESSION of natural plant communities in one area under a particular set of conditions. For example, the temperate rainforests of the Pacific northwest of the United States are probably climax communities. A climax is in a self-perpetuating steady state, at least for a time, although there are constant small changes, such as when a tree dies and pioneering plants in-

47

vade or a different species replaces it. A climax may be temporarily disrupted by natural disasters such as hurricanes or volcanic activity. If there are changes in the environment (for example, changes in temperature, weather patterns, or drainage) or in the local species pool (e.g. if exotic species invade), a different climax community will evolve. Succession to a climax may also be arrested and held at any stage by human intervention; for example, heathlands and many grasslands result from and are maintained by grazing, which prevents the growth of trees and shrubs. Such an equilibrium is called a *subclimax*, *plagioclimax*, or *biotic climax*.

Early proponents of the climax concept thought that a single climax community – a *monoclimax* – would dominate in a given climatic region. However, the nature of the climax community is also affected by local soil types, topography, and aspect, and the frequency of natural phenomena such as fire, and the composition of the local species pool. So a single climatic region may contain a mosaic of several different climax communities, often grading into one another – a *polyclimax*. *See* clisere. *See also* climax theory; climate zone.

climax theory The theory that vegetation SUCCESSION eventually reaches a state of equilibrium with its environment. There are two basic versions of the theory: *monoclimax*, proposed by F. E. Clements in 1904 and 1916, which predicts a specific end-point determined by climate, and *polyclimax*, proposed by A. G. Tansley in 1916 and 1920, which predicts a CLIMAX COMMUNITY determined by edaphic conditions or other conditions such as fire. The polyclimax version is most widely accepted today, since the climate rarely remains sufficiently stable for long enough to enable a single vegetational climax to be reached.

cline A gradual change in phenotype, especially morphology, shown by a species of other related groups of organisms across its geographical range, usually along a line of environmental or geographical transition. It reflects underlying changes in genotype.

clint *See* limestone pavement.

clisere A succession of CLIMAX COMMUNITIES in a given area over time, each giving way to the next as a result of climatic changes. *See* sere.

clonal dispersal In a MODULAR ORGANISM, the growth or movement away from each other of the component modules. This may or may not involve separation from the parent. For example, the polyps of a coral; the buds of a yeast colony; or the separation of successive whorls of leaves on a stem as the internodes elongate; or the growth of runners (e.g. strawberry) that produce shoots at intervals, which become independent plants. Such plants may cover large areas – a single clone of aspen has be known to cover 14 hectares.

clone 1. A group of organisms or cells that are genetically identical. In nature, clones are derived from a common ancestor by mitotic cell division, asexual reproduction, or parthenogenesis.
2. Identical copies of a segment of DNA that has been cloned in a cloning vector such as a plasmid, bacteriophage, or bacterium. Such DNA clones form the basis of DNA libraries (GENE BANKS).

cloning The production of genetically identical organisms (CLONES) or cells from a parent organism or cell. This may happen naturally by asexual reproduction, or it may be done by tissue culture, embryo culture, and other techniques. Cloning is an important part of GENETIC ENGINEERING.

cloning vector *See* vector.

closed canopy A CANOPY in which the crowns of individual trees overlap to form a continuous layer.

closed community *See* community structure.

closed population A population in which OUTBREEDING with individuals from other populations is prevented by a genetic barrier. This is an early step in SPECIATION.

cloud A suspension of tiny water droplets or ice particles in air, produced by the condensation of water vapor. Condensation happens around *condensation nuclei* – solid particles (e.g. dust) or ions. These include dust, smoke, ash, ammonium salts from vehicle exhausts, ocean salt, sulfate particles produced by phytoplankton, and organic vapors from forests. Most are less than 0.2 μ in diameter, but some are greater than 1 μ. Condensation nuclei affect the cloud droplet size, which affects both the amount of precipitation and the radiation transfer properties of the cloud.

Clouds form when moist air is cooled below its dew point (*see* dew). This commonly happens as warm moisture-laden air rises and is cooled by expansion as, for instance, when air warmed by the hot ground rises by convection (e.g. in the tropics, where clouds form by midday and rain falls in the afternoon); where a cold front near the ground pushes up a mass of warmer air as it moves forward; or where warm and cold air currents mix.

cloud forest Tropical forest, usually found above 1000 meters elevation, where rainfall is heavy and moisture-laden air is deflected upward by the mountain, forming clouds for most of the day. The high humidity and abundant moisture leads to low tree growth, often 5–8 meters, but supports dense growth of EPIPHYTES, especially mosses, ferns, filmy ferns, clubmosses, and lichens. In the forest clearings, ferns and herbs such as begonias often reach large sizes.

clumped dispersion *See* dispersion.

cluster sampling *See* sampling.

Cnidaria A large phylum of aquatic, mostly marine invertebrates – the most primitive of the multicellular animals. Cnidarians are radially symmetrical and diploblastic, the body wall having two layers separated by a layer of jelly (mesoglea) and enclosing the body cavity (coelenteron). The single opening (mouth) is surrounded by a circle of tentacles, which are used for food capture and defense and may bear stinging cells (cnidoblasts). Two structural forms include the sedentary polyp (e.g. *Hydra*, sea anemones), and the colonial CORALS and the mobile medusa (jellyfish); either or both forms occur in the life cycle. *See also* Anthozoa.

coagulation The clumping of particles into larger particles. It is an important stage in the processing of SEWAGE, in which small light suspended particles that will not otherwise settle out are first clumped together using chemical coagulants such as aluminum sulfate to form lumps called *flocs*. This process is called *flocculation*. After coagulation, the suspension is transferred to a *sedimentation tank*, and the (now larger) particles will settle out.

coal A carbonaceous deposit formed from the remains of fossil plants. Most of the world's coal deposits formed about 300 million years ago during the Carboniferous (Pennsylvanian) period, under warm tropical or subtropical climates when vast swamp forests flourished. The stages in the formation of so-called humic or woody coals pass through partially decomposed vegetable matter such as PEAT, through lignite, subbituminous coal, semianthracite, to anthracite. During this process, the percentage of carbon increases and volatile components and moisture are gradually eliminated. Sapropelic coals are derived from algae, spores, and finely divided plant material. *See* Carboniferous; fossil fuel.

COD *See* chemical oxygen demand.

codon A group of three nucleotide bases in a messenger RNA (mRNA) molecule that codes for a specific amino acid or signals the beginning or end of the message (start and stop codons). Since four different bases are found in nucleic acids there are 64 ($4 \times 4 \times 4$) possible triplet combinations. The arrangement of codons along the mRNA molecule constitutes the *genetic code*. When synthesis of a given protein is necessary the segment of DNA with the appropriate base sequences is transcribed into messenger RNA (*see* transcription).

When the mRNA migrates to the ribosomes, its string of codons is paired with the anticodons of transfer RNA molecules, each of which is carrying one of the amino acids necessary to make up the protein. The term codon is sometimes also used for the triplets on DNA, which are complementary to those on messenger RNA but contain the base uracil instead of hymine.

coefficient of haze (COH)   A measure of the amount of dust and smoke in the atmosphere. COH units are defined as the quantity of particulate matter that produces an optical density of 0.01 on a paper filter. A COH value of less than 1.0 is considered to represent relatively clean air. Values greater than 2.0 occur where there is a high proportion of combustion-generated particles, such as those from vehicle exhausts, forest fires, or the burning of fossil fuels. The COH is a measure of the light-absorbing capacity of the particles, which is due mainly to carbon produced by combustion. To determine the COH, ambient air is drawn through a paper filter, and a photometer is used to detect changes in the amount of light transmitted through the paper, producing an electrical signal proportional to the optical density.

coevolution   The complementary evolution of two or more ecologically interdependent species, such as a flowering plant and its pollinators. An adaptation in one species triggers the evolution of a complementary adaptation (*coadaptation*) in the other(s). For example, the evolution of long tubelike flowers favors the evolution of birds with long beaks and tongues to reach the nectar deep inside. These birds are more efficient pollinators, since they are more likely to visit similar flowers, while they benefit from reduced competition with birds with shorter beaks. *See also* mutualism; pollination; predator–prey relationships.

coexistence   The living together of two or more species or organisms in the same habitat, such that none is eliminated by any of the others, i.e. there is no competition between them. For example, this may occur when predators or parasites limit the competitive power of certain species that might otherwise be more effective competitors. *See* competitive exclusion principle; niche; predator–prey relationships.

COH   *See* coefficient of haze.

cohort   A group of individuals that began life at the same time in the same population.

cohort generation time   *See* generation.

cohort life table   *See* life table.

cold acclimation   *See* cold tolerance.

cold desert   *See* boreal forest; desert.

cold forest   *See* boreal forest.

cold hardening   *See* cold tolerance.

cold tolerance   (cold acclimation; cold hardening; hardening; frost tolerance)   The development of an ability to tolerate very low temperatures, especially temperatures below freezing, in advance of those temperatures actually occurring. Most plants and animals accumulate free amino acids (mainly plants), polyhydroxyalcohols or polyols (mainly animals), or other organic compounds (e.g. oligosaccharides) in their cells when exposed to a period of low (but not damaging) temperatures, helping prevent damage to cell membranes when temperatures fall still further. In some insects, most of the animal's glycogen stores are converted to polyols as temperatures fall, and these allow the formation of ice between cells, but protect the membranes from damage due to dehydration as water is withdrawn from the cells to form the ice. The degree of cold tolerance shown by a species depends on the normal range of temperatures of the habitat in which it has evolved. The most cold-tolerant species are certain mosses and lichens of polar regions, which can tolerate temperatures of −70°C. *See* vernalization.

coliform bacteria Gram-negative rod-shaped bacteria that can obtain energy aerobically or anaerobically by fermenting sugars to produce acid or acid and gas. They live in soil, water, or as pathogens of plants, but most are found in the intestines of warm-blooded animals (*fecal coliforms*), e.g. *Escherichia coli*. Many, such as *Salmonella*, are pathogenic to humans. The presence of fecal coliforms in water indicates that the water is contaminated with feces, so may also contain other pathogenic organisms. The *Coliform Index* gives a rating of water purity based on the fecal bacteria count.

Coliform Index *See* coliform bacteria.

colloid A dispersion of microscopic finely divided insoluble solid particles in a liquid. They remain suspended as a result of their small size and electrical charge: the particles have identical electrical charge, so they repel each other and cannot form clumps that would settle out. Colloid particles range from 10^{-7} to 10^{-3} cm in diameter.

colonization The movement and subsequent spread of a species into a habitat, location, or population from which it was absent.

combined heat and power plant (CHP) The generation of heat and power (electricity) from waste. This may involve incinerating the waste, or using the methane that forms in landfills. The heat from the combustion process can be used to provide hot water for houses in the vicinity, and/or electricity for local use or for the national grid. While not a highly efficient energy production process, CHP can substitute for energy derived from FOSSIL FUELS, so reducing undesirable emissions.

combined sewer A sewer system that carries both SEWAGE and storm-water runoff. Under normal conditions, its entire flow passes through a waste treatment plant, but during heavy rain storms the extra volume of water may cause some of this sewage and storm water mixture to overflow, bypassing the treatment system.

combustion chamber The compartment in which waste is burned in an incinerator.

commensalism An association between two organisms in which one, the *commensal*, benefits while the other (the *host*) remains unaffected either way, e.g. the saprophytic bacteria in animal intestines and EPIPHYTES on the branch of a tree. *Compare* amensalism; mutualism; parasitism; symbiosis.

commercial waste Solid waste produced by businesses such as shops, offices, restaurants, and places of entertainment.

community A group of POPULATIONS of different species living together in a certain environment and interacting with each other. *See* association; consociation.

community character displacement *See* character displacement.

community structure The pattern of species ABUNDANCE and population interactions within the community. This is determined by the interactions between the different species and between the species and the ABIOTIC environment. Community structure encompasses the number of species present (*see* species richness), their relative abundance, feeding relationships, and the way the resources of the local environment are partitioned between the species. When considering a community on a larger regional scale encompassing diverse habitats, interactions between populations and patterns of habitat selection are also important. Longer-term community structure will also be affected by SPECIATION, which will affect species richness. Thus community structure depends on both gradients of environmental conditions and phylogenetic history, as well as on the local abundance of particular predators, herbivores, and parasites. Where data on community structure is plotted against gradients of environmental conditions (e.g.

moisture, temperature, light intensity, salinity) or topography, the boundaries between different communities may be sharp (*closed communities*) or blurred (*open communities*). In open communities there is a gradient of species abundance and some population interaction across community boundaries.

comparative risk assessment (CRA) The assessment or ranking of proposed projects, especially environmental management strategies, according to predictions of their potential environmental effects.

compensation depth   *See* photic zone.

compensation level   *See* sublittoral.

compensation point (light compensation point)   The light intensity at which the rate of PHOTOSYNTHESIS is exactly balanced by the combined rates of RESPIRATION and PHOTORESPIRATION, so that there is no net exchange of oxygen and carbon dioxide and the rate of synthesis of organic material equals the rate of breakdown by respiration. At normal daylight light intensities the rate of photosynthesis exceeds the rate of respiration. Shade plants tend to reach their compensation points faster than sun plants, as they are better able to make use of dim light. The point at which photosynthesis ceases to increase with increase in light intensity is termed the *light saturation point*. This point occurs at much higher light intensities in C_3 PLANTS than in C_4 PLANTS. Low compensation points indicate high photosynthetic efficiency.

competition   The utilization of the same resources by one or more organisms of the same species (*intraspecific competition*) or between different species living together in a community (*interspecific competition*), when the resources are not sufficient to fill the needs of all the organisms. The closer the requirements of two species, the less likely it is that they can live in the same community, unless they differ in behavioral ways, such as periods of activity or feeding patterns. The presence of one or more competing species will depress a species' birth and/or growth rate and/or increase its death rate.

Competition may be aggressive, e.g. by excluding others from resources or by the use of toxins (e.g. some plants and bacteria), which is termed *direct competition*, or it may simply exclude others by consuming so much of the resource that it reduces their availability to others – *indirect competition*.

Competition may be simply competition for a renewable resource (*consumptive competition*), for space in which to establish (*preemptive competition*), or for territory in which to feed or to compete for females (*territorial competition*). One plant species may outcompete another by growing on or over it, so depriving it of light or of water containing nutrients (*overgrowth competition*), or it may produce a toxic chemical that prevents seedling establishment or larval settlement (*chemical competition*), as happens with some plants, which produce toxic root exudates.

Where competition involves access to a resource, such as food, rather than the level of the resource itself, it is called *contest competition*: animals compete with each other for access to the resource, limiting the number that can utilize it at any one time and using up energy. The converse is *scramble competition* or *exploitation*, in which competition is directly for the resource itself. *See also* niche.

competitive exclusion principle   The principle that when members of different species compete for the same resources (i.e. occupy the same ecological NICHE) in a stable environment, it is likely that one species will be better adapted and outcompete the others, which will consequently decline in numbers and eventually die out. *See* competition.

competitive release   The expansion of the NICHE of a species due to the absence of competition with other species. For example, it may occur on an island where competitors are absent, or a recent environmental event such as a hurricane may have removed other species (e.g. trees).

competitive strategy A life strategy adopted by organisms that live in habitats where resources are abundant and disturbance rare. The main issue for survival is competition for resources as population numbers increase and the habitat becomes crowded. *See* Grime's habitat classification.

complementary resources *See* resource.

compliance monitoring The collection and evaluation of data on POLLUTANT concentrations and loads in permitted discharges in order to ascertain whether or not they fall with within the limits allowed by the discharge permit.

compost An organic fertilizer made from the breakdown by bacteria and other microorganisms of garden rubbish, kitchen vegetable waste, and other biodegradable material. Bacteria in soils are mixed with the waste and the mixture turned regularly to admit air. The organic material breaks down to a relatively stable humus-like material.

concentrate and contain (containment) A method of waste management that involves concentrating potential pollutants and enclosing them in escape-proof containers. For example, the sealing of radioactive substances and their burial underground, or the depositing of compressed waste in sealed and covered landfill sites.

concentration The relative amount of a substance in a given area or volume. For example, the mass of a solute per unit volume of solution, or the concentration of a gas in parts per million by volume.

condensation The conversion of a gas or vapor into a liquid or solid by cooling.

condensation nuclei *See* cloud; fog.

cone of depression A depression in the WATER TABLE that forms around a well as water is pumped out of it. It is usually cone-shaped, and its cross-section indicates the scope of the influence of the well on the water table.

confidence interval In statistical data, a range of values for a given variable, such that the range has a specific probability of including the true value of the variable. The upper and lower values of this range are called the *confidence limits*. *See* probability.

confidence limits *See* confidence interval.

confined aquifer *See* aquifer.

confounding factor In epidemiological studies, the level of exposure to a pathogen or disease may appear to be correlated with the incidence of the disease, when in fact both are related to a third factor – a *confounding factor*. For example, a population living in an industrial area may have a high level of respiratory disease that appears to be related to levels of pollution, when it is due to the poor living conditions, which may also contribute to the incidence or severity of the disease.

conglomerate A coarse-grained sedimentary rock made up of more or less rounded fragments larger than 2 mm in diameter, e.g. pebbles or boulders.

Coniferophyta The largest phylum of GYMNOSPERMS, comprising about 560 species of evergreen trees and shrubs, with many important species, e.g. *Pinus* (pine), *Picea* (spruce), *Taxus* (yew), and *Abies* (fir). They dominate the vast boreal forests of the northern hemisphere. Most are evergreen, but *Larix* (larches) and *Taxodium* (swamp cypresses) are deciduous. The wood lacks xylem vessels. Conifers have simple linear or scale-like leaves, and the male and female reproductive structures are borne on the undersides of scales arranged in cones. Conifers produce pollen grains and after fertilization seeds develop.

coniferous forest *See* boreal forest.

conifers *See* Coniferophyta.

connectance The total number of links in a food web divided by the number of possible links.

connectedness *See* food web.

conservation The management of wild plants and animals or other natural resources to ensure their survival for use by future generations. This may involve the maintenance of particular natural habitats and the control of environmental quality. Modern conservation has to consider people as part of natural ecosystems, and to balance the needs of the wild fauna and flora against the social and economic needs of local people to promote the sustainable use of resources. Conservation usually aims to preserve BIODIVERSITY, but may on occasion focus on the survival of a particular ENDANGERED SPECIES, including such measures as captive breeding or seed and gene banks. Where the habitat being conserved is a subclimax (*see* climax community), it may be necessary to intervene to prevent the natural succession running its course. For example, reedbeds may be cut back to maintain areas of open water. *See* biosphere reserve; CITES.

conservation biology The application of population ecology and genetics to problems of decreasing biodiversity.

consistency The absence of BIAS in experimental procedures or data analysis. Consistency can be improved by the use of properly calibrated instruments and by the use of blind trials, where the identity of individual samples is not known to the experimenter.

consociation A CLIMAX COMMUNITY of natural vegetation dominated by one particular species, such as oakwood, dominated by the oak tree, or *Calluna* heathland dominated by the heather, *Calluna vulgaris*. Many consociations together may form an ASSOCIATION, for example oak-wood, beechwood, and ashwood consociations together make up a deciduous forest association. Consociations are usually relatively small climax communities.

conspecific Describing individuals that belong to the same SPECIES.

constant (constant species) A species found in a particular ASSOCIATION or COMMUNITY but not confined to it, as compared to a *faithful* species, which is seldom found outside a particular association or community. *See* phytosociology.

consumer An organism that feeds upon another organism, e.g. animals, parasitic and insectivorous plants, and many heterotrophic protoctists. A primary consumer feeds on plants or other primary producers. A secondary consumer feeds on primary consumers, and a tertiary consumer feeds on secondary consumers, and so on. *See* food chain; trophic level. *Compare* decomposer; producer.

consumption efficiency The percentage of the available energy that is actually consumed at a given TROPHIC LEVEL. For a primary consumer (herbivore) it is the percentage of the net PRIMARY PRODUCTIVITY that is ingested, the remainder being left to decompose. For higher level CONSUMERS, it is the percentage of the productivity of the prey that is eaten by a carnivore. *See* productivity.

consumptive competition *See* competition.

contact pesticide A herbicide or insecticide that kills a plant or insect on contact rather than needing to be absorbed or ingested, e.g. paraquat (herbicide), pyrethrin (insecticide). *Compare* systemic.

contagious dispersion *See* dispersion.

containment *See* concentrate and contain.

contest competition *See* competition.

continental drift   The theory that present-day continents have arisen by the breaking up and drifting apart of a previously existing ancient land mass (Pangaea). There is much evidence to support the theory, and it serves to explain the distribution of contemporary and fossil plants and animals. Continental drift is now believed to reflect the movement over geological time of underlying plates in the Earth's crust – the theory of PLATE TECTONICS.

continental crust   *See* crust.

continentality   A measure of the effect of isolation from marine influences upon climate. Different temperatures prevailing in January and July are generally used as the indicator. With increasing distance from the oceans, seasonal temperature differences tend to become more extreme.

continental shelf   The shallow, gently sloping submarine plain around a continental margin, formed by the submergence of the edge of the continent. The shelf is normally covered in mud, silt, or sand. It is seldom deeper than 100–200 m, and usually ends abruptly in a steep slope to the oceanic abyss, the *continental slope*. Continental shelves vary in width, but are usually about 65 km wide. For example, on the western side of the USA the shelf is only about 32 km wide, but along the eastern coast it is in places more than 120 km wide.

continuous sampling   The continuous monitoring of the composition of a substance, such as an effluent, by removing samples for analysis. In an industrial or treatment plan, for example, part of the effluent stream may be diverted to a point where samples are taken. This may happen at different points in the processing, providing information about the effluent at all stages of processing.

continuous variation   *See* quantitative variation.

continuum   The gradual change in species composition of overlapping POPU-LATIONS in a large COMMUNITY along an environmental gradient, such as rainfall. This can be quantified as a *continuum index*: the populations are assigned arbitrary values according to how closely their species composition approaches either end of the continuum. Increasing values of the continuum index correspond to seral stages leading to the mature CLIMAX COMMUNITY. The lower values may also be regarded as representing local climax communities for particular topographic or soil conditions, rather than as populations in a larger community. The concept was originally based on work in southern Wisconsin, where the continuum was of woodland along a moisture gradient from dry sites dominated by oak and aspen to moist sites dominated by sugar maple, basswood, and ironwood. *See* sere.

contour line   A line drawn on a map linking points of equal elevation/altitude.

contour plowing   A method of farming in which plowing and planting are carried out along the contours, i.e. at right angles to the slope rather than up and down it. This reduces soil erosion.

contour strip farming   A method of farming in which strips of crops are alternated with strips of close-growing forage crops such as grass, hay, or forage grains, to further reduce soil erosion. *See* contour plowing.

control   In experimental technique, the use of an untreated sample – the control – as a standard against which to compare treated samples. For example, the control may remain free of chemical treatment, addition of nutrients, or physical changes in its environment.

control action threshold (CAT)   The level of pest density and the density of the pest's natural enemies above which it is necessary to intervene to prevent the pest population rising above acceptable levels.

convenience sampling   *See* sampling.

Convention on Biological Diversity (CBD) An international convention, signed at the United Nations Rio Summit in 1992, that aims to conserve BIODIVERSITY, the sustainable use of natural resources, and the equitable sharing of the benefits arising from exploitation of genetic resources. It covers all aspects of biodiversity from genetic resources to ecosystems, and aims to conserve resources while promoting sustainable development, and to encourage the sharing of costs and benefits between developing and developed countries. The United Nations Environment Program (UNEP) responded by commissioning the Global Biodiversity Assessment, to collect and review data, theories, and opinion relating to current issues worldwide, which provided valuable information on the rapid rate at which environmental change is taking place and natural ecosystems are being modified.

Convention on International Trade in Endangered Species *See* CITES.

Convention on Wetlands of International Importance (Ramsar Convention) An international convention between member states of the United Nations that aims to coordinate international action to conserve wetlands of international importance. It does so by restraining the removal of wetlands by drainage, cultivation, and human settlement, and recognizing the role of wetlands as regulators of the hydrological regime and as habitat for plants and animals, as well as their scientific, economic, cultural, and recreational value. A list of Wetlands of International Importance (*Ramsar sites*) is maintained, and the World Conservation Union holds a register of the wildlife and management information relating to these sites. The convention was signed at a meeting in Ramsar, Iran, in 1971, and amended in 1982 and 1987. Signatures may or may not be followed by ratification with UNESCO (United Nations Educational, Scientific and Cultural Organization). *See* World Conservation Union.

convergence The occurrence of SPECIES with similar morphology and adaptations in similar habitats in different parts of the world. For example, the similarity of cacti and euphorbias in the deserts of North America and Africa.

convergent evolution The independent development of similar (analogous) structures in unrelated organisms as a result of living in similar ecological conditions and selection pressures. An example is the evolution of similar succulent forms in the Cactaceae and the Euphorbiaceae, in which the leaves are reduced to pale spines. The wings of vertebrates and insects are an example in which quite distinct groups of animals have independently adapted in a similar way to life in the air.

cooling tower In electricity-generating plants, including coal-, oil-, and nuclear-powered plants, a large structure that provides a large internal surface area across which the heat from the exhaust steam (from turbines) is used to preheat water for the boilers.

cool temperate rainforest *See* temperate deciduous forest.

cooperative breeding The involvement of individuals other than the mating pair in the breeding of the species. For example, in social insects, the workers feed, clean, and care for the young, and in some species of birds and mammals, nonbreeding relatives help rear the young. In social insects the workers share 75% of their parents' genes, whereas their own offspring would have only 50% of their genes, so cooperation helps them perpetuate their genes.

copper (Cu) A metal trace element (*see* micronutrient), essential for plant growth. Copper deficiency in plants results in chlorosis, suggesting that copper may be involved in the synthesis of chlorophyll. It has a role in certain metabolic activities in animals. Dietary sources of copper include liver, seafood, fish, nuts, whole grains, and legumes. Copper sulfate is used as a fungicide in Bordeaux mixture. Copper compounds are often toxic to living organisms.

coppice (copse) A woodland managed for wood production by cutting trees back to ground level at regular intervals (usually 10–15 years) and allowing adventitious shoots to grow up from the base. The young shoots, which are often very straight, are used for fencing, charcoal burning, and firewood. This labor-intensive practice has largely died out, and today coppicing is done mainly for conservation purposes.

coprophagy The habit of feeding on dung. Examples of coprophagous organisms include dung beetles and fungi such as *Pilobolus*. *Compare* ruminants.

copse *See* coppice.

coral bleaching *See* corals.

coral reef *See* corals.

corals Colonial animals (members of the phylum CNIDARIA), class ANTHOZOA that consist of *polyps* embedded in a gelatinous, horny, or calcareous matrix. They are suspension feeders, trapping tiny organisms and organic detritus with their tentacles. Many corals contain symbiotic *zooxanthellae* (*see* dinoflagellates), whose pigments give corals their brilliant colors. These symbionts are photosynthetic, and contribute to the corals' nutrition. In the absence of zooxanthellae, the rate of calcium carbonate deposition is much slower, with serious implications of the maintenance of coral reefs. Temperatures greater than 30°C cause corals to expel their zooxanthellae and die, causing *coral bleaching*. Pollution by sediment, such as that carried into coastal areas by untreated sewage or by rivers passing through areas of deforestation and soil erosion, reduces light penetration and prevents the zooxanthellae from photosynthesizing.

Many coral species show asexual reproduction by budding. Buds remain attached to the parent polyp, so building up a colony. In time, the older polyps die, but the matrix surrounding them remains, building up a *coral reef*. Accumulations of colonial corals with calcareous skeletons

form substantial reefs in warm shallow waters. Solitary corals may also contribute to the bulk of the reef, some reaching diameters of 25 cm, as do red algae that are encrusted with calcium carbonate. Coral reefs have a complex topography, and provide many ecological niches for other animals and algae; their biodiversity has been likened to that of rainforests.

core The central part of the Earth, lying below the crust and mantle at a depth below 2900 km. It is thought to be composed mainly of molten nickel and iron.

corer A device use to extract a relatively undisturbed sample of a vertical section of a sediment or of ice. It usually comprises a rotating cutter that isolates a column of sediment.

Coriolis force In a rotating object, a force of inertia that acts to the left of the direction of movement in clockwise rotation, and to the left in a counterclockwise rotation. It causes an apparent deflection of an object from its path when moving in a rotating system, especially when moving in a longitudinal direction, an effect called the *Coriolis effect*. Thus an object moving in a longitudinal direction will appear to be deflected to the left in the southern hemisphere and to the right in the northern hemisphere because the Earth rotates eastward, and because the velocity of the Earth's rotation is greater at the equator than at the poles.

The Coriolis force affects the direction of prevailing winds, such as the Westerlies in the northern hemisphere and the easterly trade winds in the southern hemisphere. It also affects the rotation of ocean currents, which circulate clockwise in the northern hemisphere and counterclockwise in the southern hemisphere.

corm *See* perennating organ.

correlated extinction *See* extinction.

correlated response A response of the PHENOTYPE of one character to selection acting upon another character. For example, selection for increased body weight in chickens results in increased egg size but decreased rates of laying.

correlation The relationship between two or more qualitative or quantitative variables. On a graph, the drawing of a straight line through a scattering of points so as to minimize the STANDARD DEVIATION of all the points from the line (a *line of best fit*), thus allowing quantification of the way in which the two variables are related to each other. The closer the points fit the line, the stronger the correlation between the two variables. This may be done subjectively, or it may be done statistically, giving confidence limits for the correlation. *See* correlation coefficient.

correlation coefficient The strength of the linear relationship between two variables can be expressed as a correlation coefficient, which equals the covariance divided by the square root of the product of the variances of the two variables. The values of the correlation coefficient vary from −1 to +1, a value of +1 indicating a positive relationship, a value of −1 a negative relationship, and a coefficient of 0 indicating that there is no linear relationship between them. The closer the correlation coefficient is to +1 or −1, the less the scatter of points about a line of best fit (*see* correlation). However, the correlation coefficient does not imply any dependency of one variable upon the other, or give any information about the nature of the relationship between the two.

corridors Connections between fragments of similar HABITAT in a patchy landscape. For example, a thin strip of woodland may link two much greater areas of woodland, allowing mixing of species and individuals between the two communities. In the conservation of BIODIVERSITY, such corridors are vital to preserve a good level of OUTBREEDING and genetic diversity (and hence breeding success and the ability to adapt to environmental change) in populations or communities that are becoming fragmented (and therefore smaller) because of habitat loss due to

human activity or climate change. In ecosystems where certain species need to migrate seasonally, as for example in the African savannas, where grazing herbivores need to migrate to areas where recent rainfall has led to new grass growth, or where a species has distinct breeding areas away from its normal range, corridors allow this to happen.

cost-benefit analysis A quantitative evaluation of the economic, social, and environmental costs that would be incurred in implementing a project or regulation versus the social, economic and, sometimes, environmental benefits.

courtship The specialized patterns of behavior that are preliminary to mating and reproduction. Its function is to synchronize precisely the activities of male and female so that copulation can occur. It is also important in enabling the partners to identify each other as potential mates – i.e. as of the same species and opposite sex. In this, courtship also serves as an isolating mechanism, preventing different but closely related species from interbreeding. *See* reproductive isolation.

covariance A statistical measure of correlation of the fluctuation in two variables (i.e. the deviation from their means), based on the original units of measurement of the variables. Covariance differs from correlation in that it is based on the actual units of the original measurements of the variables, not on standardized units. *Compare* correlation.

cover The proportion of the ground occupied by plants of a given species, as determined by a perpendicular projection on to the ground of their aerial parts. It is usually expressed as a percentage of the total area, and gives an estimate of plant abundance. It may be estimated visually using scales such as the Domin scale (*see* Braun-Blanquet scale). Visual assessment may be aided by the use of a wire grid, or sampling may be made more objective by taking a number of sample points using a point quadrat (a quadrat with pins hanging down at intervals; in which each hit by a pin is recorded). Where leaves overlap, this can give cover values greater than 100%. For this reason, visual estimates are often more meaningful in dense vegetation. *See* abundance; frequency.

cover crop A crop grown to protect delicate seedlings of a later crop, or to provide protection against soil erosion between crop plantings or in certain seasons.

C$_3$ plant A plant in which the first product of PHOTOSYNTHESIS is a 3-carbon acid, glycerate 3-phosphate (phosphoglyceric acid, PGA). Most plants are C$_3$ plants. They are characterized by high carbon dioxide COMPENSATION POINTS owing to PHOTORESPIRATION. *Compare* C$_4$ plant.

C$_4$ plant A plant in which the first product of PHOTOSYNTHESIS is a 4-carbon dicarboxylic acid, oxaloacetic acid. C$_4$ plants have evolved from C$_3$ plants by a modification in carbon dioxide fixation, leading to more efficient carbon dioxide uptake. The modified pathway is called the *Hatch-Slack pathway* or C$_4$ pathway. In the leaves, the mesophyll cells surrounding the vascular bundles (*bundle sheath cells*) contain the carbon dioxide-fixing enzyme phosphoenolpyruvate carboxylase (PEP carboxylase) in their cytoplasm. This has a higher affinity for carbon dioxide than does ribulose bisphosphate (RuBP) carboxylase. The product of carbon dioxide fixation is oxaloacetate, which is rapidly reduced to the C$_4$ acid malate or transaminated to the C$_4$ acid aspartate. These C$_4$ acids are transported to the bundle sheath cells encircling the leaf veins, where they are decarboxylated, releasing CO$_2$, which is then refixed as in C$_3$ plants.

C$_4$ plants are mainly tropical or subtropical, including many tropical grasses (e.g. maize, sugar cane, and sorghum). They are more efficient than C$_3$ plants, producing more glucose per unit leaf area. They have up to double the maximum rate of photosynthesis of C$_3$ plants and lose less water by transpiration and have lower compensation points. *See* photorespiration. *Compare* C$_3$ plant.

CPUE *See* catch per unit effort.

CRA *See* comparative risk assessment.

crab *See* Crustacea.

Craniata *See* Chordata.

Crassulacean acid metabolism (CAM) A form of photosynthesis in which the light and dark reactions are separated in time: carbon dioxide taken in through the stomata at night is converted into an organic acid for storage, then converted back to carbon dioxide for use in photosynthesis during the day. Instead of initially combining with ribulose bisphosphate, the carbon dioxide combines with the 3-carbon phosphoenolpyruvate (PEP) to form oxaloacetic acid, which is then converted to malic acid for storage in the vacuole. The following day, when the stomata are closed, the malic acid is broken down and the carbon dioxide is released to take part in the light reactions of photosynthesis. CAM allows plants of arid environments to keep their stomata closed to conserve water during the day, opening them at night to exchange gases. When stressed by drought, CAM plants can last for days or even weeks without opening their stomata, conserving water and living on stored resources. CAM was first discovered in the family Crassulaceae and has since been found in many other succulent plants, such as *cacti*. Its disadvantage is that it is a slow form of photosynthesis, resulting in rather slow growth. *See* photosynthesis. *Compare* C$_4$ plant; C$_3$ plant.

creep The slow movement of unconsolidated weathered material, including rock fragments, mineral grains, and soil, down slopes as a result of gravity. It may be aided by alternate freezing and thawing, freezing causing a raising of the material followed by a down-slope movement on thawing; expansion and contraction caused by changes in temperature; the burrowing of animals; and the lubrication by water (*see* solifluction). *Compare* rain-wash.

Cretaceous The most recent period of

the Mesozoic era, 145–66 million years ago. It was marked by continued domination of land and sea by dinosaurs until a rapid extinction toward the end of the period. The marine ammonites and aquatic reptiles also became extinct in this period. Primitive mammals were present, but were relatively insignificant in number, size, and variety until the Cenozoic era, which followed. Birds and fishes evolved into structurally modern forms during the Cretaceous. The dominant vegetation on land in the early Cretaceous consisted of forests of cycads, conifers, ginkgoes, and ferns.

Most modern-type ferns, gymnosperms, and angiosperms arose during the Cretaceous, and by the end of the period the flowering plants had replaced the gymnosperms as the dominant terrestrial vegetation, forming vast broad-leaved forests with magnolias, figs, poplars, sycamores, willows, and herbaceous plants and a great variety of insect pollinators. The Cretaceous is named for the large amounts of chalk (fossilized plankton) found in rocks of the period. *See also* geological time scale.

crista   *See* mitochondrion; respiration.

criteria pollutants   In the United States, certain pollutants known to be hazardous to human health for which the Environmental Protection Agency is obliged under the 1970 Clean Air Act to set National Ambient Air Quality Standards.

critical day length   *See* photoperiodism.

critical load   A quantitative estimate of the level of exposure to pollutants below which significant harmful effects on specified elements of the natural environment do not occur.

cropping-off   The harvesting (by humans or by natural predator or grazing) of a particular species, trophic level in a FOOD CHAIN, or link in a FOOD WEB. For example, in fish farming, nitrogenous waste may be removed by allowing algae to grow on it, then adding fish that will remove the algae by eating them (i.e. cropping them off).

crop rotation   The practice of planting different crops on the same land in successive years. Many plant diseases and pests are species-specific, and this avoids the perpetuation of sources of infection from year to year, reducing the build-up of pest populations. It also avoids the removal of exactly the same balance of minerals from the soil year after year, as different crops have different requirements. It is common to include a leguminous crop in the rotation to replenish the nitrates in the soil, since legumes contain nitrogen-fixing bacteria in nodules on their roots. The crop may be may plowed into the soil to fertilize it.

cross   1. The act of cross-fertilization. *See* backcross; pollination; plant breeding. 2. An organism resulting from cross-fertilization.

cross-contamination   1. The movement of underground contaminants such as toxins, pollutants, or pathogen-infested soil from one level or area to another as a result of burrowing or other activity that mixes the soil or invades or permeates the rocks. 2. The transfer of pathogens or toxins from one substance to another as a result of bad storage procedures. For example, the infection of an animal by picking up bacteria or other pathogens from ground, straw, or other bedding soiled by an infected animal.

crossing over   The exchange of genetic material between homologous chromatids by the formation of chiasmata during MEIOSIS, which results in the new combinations of alleles on the daughter chromosomes. This is the main source of genetic variation during sexual reproduction.

cross-pollination   *See* pollination.

crown-of-thorns starfish   A large star fish (*Acanthaster*) that preys on certain species of CORAL. In the 1960s and 1980s population explosions of this starfish in the Pacific Ocean, and especially on Australia's Great Barrier Reef, killed up to 90% of the corals in some areas. The coral skeletons became overgrown with algae, and reef fish and other residents disappeared.

Once the starfish numbers declined, the corals recovered. Opinion differs as to whether this is a natural phenomenon or one triggered by human disturbance. Crown-of-thorns starfish on reefs off the Pacific coast of Central America particularly consume a species of coral, *Pocillopora*, that tends to crowd out other species, so helping maintain species diversity on the reef.

crumb structure The texture of a SOIL in terms of the size of the soil particles in the surface horizons and how they interact, often determined simply by rubbing the soil with fingers. Some soils, such as sandy soils, have a loose crumb structure with good drainage whereas in clay soils the particles are small and stick together to give a dense crumb structure and poor drainage.

crust The outermost layer of the LITHO-SPHERE, which consists of relatively light rock and comprises less than 1% of the total mass of the Earth. On average, it extends 35 km below the surface. There are two types: *continental crust* consists mainly of granitic rock, with a veneer of sedimentary rocks near the surface, thickest in high mountain belts; *oceanic crust* is mainly formed of basalt, which forms from lava that wells up along mid-oceanic ridges. In places, a thick layer of sedimentary rock is building up on the deep ocean floor. The crust comprises a series of distinct plates (*see* plate tectonics) that move relative to each other as a result of convection forces in the semiliquid mantle below.

Crustacea A large group of arthropods containing the mostly aquatic gill-breathing crawfish, crabs, lobsters, barnacles, water fleas, etc., and the terrestrial woodlice and sowbugs. The body is divided into a head, thorax, and abdomen. The head bears compound eyes, two pairs of antennae, and mouthparts composed of a pair of mandibles and two pairs of maxillae. The thorax is often covered with a dorsal carapace (crust-like shell). The appendages are typically forked and specialized for different functions. The sexes are

usually separate and development is indirect, via one or more larval stages. In the FIVE KINGDOMS CLASSIFICATION the Crustacea are given the rank of phylum, but in some other classification systems they constitute a class of the phylum ARTHROPODA.

Crutzen, Paul (1933–) Dutch meteorologist. In 1970 Crutzen argued that nitrous oxide, arising from the use of nitrogen-rich fertilizers and the combustion of fossil fuels, could cause depletion of the OZONE LAYER. Soon after, Crutzen's warnings were overshadowed by the greater threat from the chlorofluorocarbons, which was first identified in 1974 by ROWLAND and MOLINA, with whom Crutzen shared the 1995 Nobel Prize for chemistry. Crutzen was also one of the first scientists to warn (1982) of the dangers of a NUCLEAR WINTER.

crypsis *See* camouflage.

cryptophyte *See* Raunkiaer's life-form classification.

cultivar Any agricultural or horticultural variety of a species. The term is derived from the words *culti*vated *var*iety. *See* variety.

cumulative exposure The total exposure of an organism to a chemical or form of radiation over a given period of time: the sum of repeated exposure events, expressed in quantitative terms; for example, the number of millisieverts of radiation.

cumulative frequency The frequency with which an observed variable takes on a value equal to or less than a specified value. Such frequencies are often plotted as histograms to show cumulative frequency distributions.

current A directional movement of water whose speed is measured in volume passing a particular point in a particular time, for example, liters or gallons per second. Oceanic curents are large, relatively permanent movements of water due to differences in temperature and/or salinity, both of which affect the density of the water. The direction of these currents is determined by the ocean bed topography, the positions of the continents and continental shelves, and the CORIOLIS FORCE. Changes in the strength and position of these currents can have dramatic effects on global climate (*see* El Niño; La Niña).

Cyanobacteria A phylum of EUBACTERIA containing the blue-green bacteria (formerly called blue-green algae) and the green bacteria (chloroxybacteria). Both groups convert carbon dioxide into organic compounds using photosynthesis, generally using water as a hydrogen donor to yield oxygen, like green plants. Under certain circumstances some use hydrogen sulfide instead of water, yielding sulfur. The main storage product is starch or starch-like compounds. These bacteria are spherical or form long microscopic filaments of individual cells. The cyanobacteria are an ancient group, and their fossils (*stromatolites*) have been dated at up to 3500 million years old. They reproduce asexually by binary fission, or by releasing sporelike propagules or filament fragments. Today, most species are found in soil and fresh water. Many species, e.g. *Nostoc* and *Oscillatoria*, have certain cells that are non-photosynthetic and fix nitrogen. They are important sources of nitrates in rice paddies. Some nitrogen-fixing species form symbioses with plants and lichens. It is thought that certain ancient cyanobacteria became permanent symbionts of ancestral algae and green plants, taking up residence in their cells as photosynthetic organelles (plastids), which eventually evolved into chloroplasts.

Cycadophyta (cycads) A phylum of tropical and subtropical cone-bearing GYMNOSPERMS with palmlike compound leaves and special coralloid roots at or near the ground surface, which contain symbiotic nitrogen-fixing CYANOBACTERIA. The stems are thick and woody and the leaves are arranged in a dense rosette at the base. The bases of dead leaves provide important mechanical support. Cycads are slow-growing and are pollinated by wind and beetles. The female cones and ovules are unusually large, except in the genus *Cycas*.

cyclic climax A community in which there is a cyclical pattern of dominant species and associated biota. For example, where heaths growing in exposed locations reach a certain size, they tend to suffer wind damage and split open, allowing the wind access to other parts of the vegetation. The damage spreads through the area, as regeneration or recolonization begins on the parts that were damaged first.

cyclic succession *See* succession.

cyclone *See* depression.

cytoplasm The living contents of a CELL, excluding the nucleus and large vacuoles, in which many metabolic activities occur. It is a colorless substance enclosed within the plasma membrane and contains organelles and various inclusions (e.g. crystals and insoluble food reserves). The cytoplasm is about 90% water. It is a true solution of ions (e.g. potassium, sodium, and chloride) and small molecules (e.g. sugars, amino acids, and ATP); and a colloidal solution of larger molecules (e.g. proteins, lipids, and nucleic acids). It can be gel-like, usually in its outer regions (*exoplasm*), or sol-like (*endoplasm*).

damped oscillation A fluctuation in the number of individuals of a population that decreases in size as the population approaches its equilibrium.

dark bottle A bottle covered to exclude light, used to estimate the respiration rates of aquatic phytoplankton in PRODUCTIVITY experiments. A sample of water is placed in the bottle and suspended at the depth from which the sample was obtained. Changes in oxygen concentration inside the dark bottle are due to the respiration of microorganisms and phytoplankton. When paired with a *light bottle*, in which photosynthesis can proceed, gross PRIMARY PRODUCTIVITY can be estimated (assuming that the respiration rates are the same in both light and dark bottles) by adding the oxygen consumed by respiration in the dark bottle to the change in oxygen concentration in the light bottle. The change in oxygen in the light bottle alone represents the net productivity of the aquatic organisms – the balance between oxygen uptake due to photosynthesis and oxygen depletion due to respiration. This is called the *oxygen method*.

dark reaction *See* photosynthesis.

darwin A measure of the rate of evolution, expressed as units of change per unit time.

Darwinism Darwin's explanation of the mechanism of evolutionary change, which basically states that in any varied POPULATION of organisms only the individuals best adapted to that environment will tend to survive and reproduce. Less well-adapted individuals will tend to perish without reproducing. Hence the unfavorable characteristics, possessed by the less well-adapted individuals, will tend to disappear from a species, and the favorable characteristics will become more common. Over time the characteristics of a species will therefore change, eventually resulting in the formation of new species. Darwin called this process of selective birth and death *natural selection*. It derived from his observations that the individuals of a species show variation, and that although more offspring are produced that are required to replace their parents, not all survive and population numbers tend to remain fairly stable, which suggests competition for survival. The main weakness of Darwin's theory was that he could not explain how the variation, upon which natural selection acts, is generated, because at the time it was believed that the characteristics of the parents become blended in the offspring. This weakness was overcome with the discovery of Mendel's work and its description of particulate inheritance, the later discovery of genes and chromosomes, and the modern science of genetics, which led to a modification of Darwin's theory known as NEO-DARWINISM.

day degrees The number of degrees by which the average daily temperature of a given location differs from a standard, such as the minimum temperature needed for growth or flowering. It can be used to compare growing seasons in different years or in different places, for example.

day-neutral plant *See* photoperiodism.

DDT (dichlorodiphenyltrichloroethane) A synthetic ORGANOCHLORINE insecticide introduced in the late 1930s and subsequently widely used to control insect carriers of diseases, such as malaria, typhus, and yellow fever. It is highly toxic to insects at low concentrations, both on contact and when ingested, affecting the nervous system. However, many species of insects evolve resistance to DDT. Its persistence in the environment (it has a half-life of 15 years) led to widespread poisoning of certain animals, especially fish, birds, and mammals, which accumulated high concentrations of DDT in their fatty tissues. DDT thus accumulates in the food chain, affecting especially top predators. The breakdown product *DDE* causes thinning of eggshells in birds. DDT is widely present in human milk, and can cause premature shortening of the lactation period in nursing mothers. It is also suspected of being a carcinogen and possibly harming the immune system. Its sale and use is now banned in many countries.

death rate   The number of deaths in a population per unit time.

decay constant (λ)   The probability that an atom of a RADIOACTIVE ISOTOPE will decay within a particular period of time (t). The rate of disintegration is given by:
$$(dN/dt) = \lambda N$$
where N is the number of radioactive atoms present. The total lifetime of a radioactive parent atom is theoretically infinite, but its persistence in the environment is usually expressed as its *half-life* (T): the time it will take for half the quantity of the parent material to decay. T is related to the decay constant by $T = 0.6993/\lambda$.

decay products   The chemical elements or compounds that result from the radioactive disintegration of a substance. For example, the isotopes argon-40 and calcium-40 are produced by decay of the radioactive isotope potassium-40.

deciduous   Describing plants that seasonally shed all their leaves, for example before the winter or a dry season. It is an adaptation to prevent excessive water loss by transpiration when water is scarce. *Compare* evergreen.

decomposer   An organism that feeds upon dead organisms, breaking them down into simpler substances. Decomposers recycle nutrients, making them available to producer organisms. Bacteria and fungi are important decomposers in most ecosystems. DETRITIVORES, by breaking down partially decomposed material still further, may also be considered to be decomposers. *See also* food chain.

decomposition   The gradual breakdown of dead organic material to simple inorganic substances of lower energy content. This may be entire carcasses, shed parts of bodies (e.g. leaves, root caps, old coral polyps), or feces. Decomposition takes place by both chemical and physical processes; eventually organisms such as bacteria, fungi, and DETRITIVORES attack and respire the last products of disintegration, releasing inorganic nutrients, water, and carbon dioxide. Decomposition involves both the release of energy and the demineralization of the nutrients – their conversion from an organic form to an inorganic form. This makes possible the recycling of nutrients in the ecosystem, as nutrients are released back into soil or water for uptake by other organisms. Under aerobic conditions, early stage decomposers respire aerobically, releasing carbon dioxide. However, if oxygen is in short supply, fermentation (anaerobic respiration) takes place, producing by-products such as alcohol and organic acids, increasing the acidity of the local environment and affecting the succession of decomposers. In terrestrial habitats, fungi dominate in acidic conditions, but in the ocean sediments specialist bacteria such as denitrifying bacteria, sulfate-reducing bacteria, and METHANOGENS are important. *See also* decomposer.

decontamination   The removal of unwanted bacterial, chemical, radioactive, or otherwise harmful materials or pests from an area.

deep scattering layer A layer of water in the oceans that reflects/scatters sound waves, e.g. from an echo sounder. It is a stratified layer, up to 200 m thick, containing dense concentrations of fish and zooplankton.

deep-well injection The disposal of hazardous waste by pumping into deep wells, where it remains trapped in the pores of permeable rocks.

deficiency disease A disease caused by deficiency of a particular essential nutrient, such as a VITAMIN or MICRONUTRIENT, usually with a characteristic set of symptoms. For example, lack of vitamin C in mammals causes scurvy, with a number of symptoms, including bleeding gums, loose teeth, painful joints, internal bleeding, anemia, and poor wound healing.

deflected climax A CLIMAX COMMUNITY that is maintained by the activity of various organisms, e.g. by grazing, browsing, or human interference.

deflocculating agent A substance added to a suspension to prevent settling. See clarification; coagulation.

defluoridation The removal of excess fluoride from water to prevent the discoloration of teeth.

defoliation The removal of leaves from plants. Certain kinds of HERBICIDES, called defoliants, act by causing leaf ABSCISSION.

deforestation The clearing of forests. The consequences of the removal of large tracts of forest leads to an increase in soil erosion by winds and/or heavy floods and results in a decrease of BIODIVERSITY. For example, the tropical RAINFORESTS, which are the most diverse BIOMES in the world, are in great danger as a result of deforestation. See also Dust Bowl.

degenerate phase In the cyclical pattern of succession in grassland and heathland communities, the stage in which lichens move into the mature hummocks and ero-

sion of the hummocks begins. In grasslands this phase starts with the colonization of grass hummocks by lichens; in heathlands it begins when 20–30-year-old bushes of ling (*Calluna vulgaris*) become full of gaps and older central stems are colonized by lichens and mosses. The phase ends when new grass or heath plants invade the gap. See hummock and hollow cycle.

degradative succession (heterotrophic succession) A short-lived SUCCESSION of DETRITIVORE communities on dead organic remains, each releasing further nutrients to the environment for the next stage to utilize.

deimatic behavior Animal behavior that warns off potential predators. For example, the hissing of a snake before it strikes, or the puffing up of the body in toads. See also aposematism.

deionized water See water.

deme A subpopulation of a SPECIES: a discrete interbreeding group of organisms with recognizable cytological or genetic characters that is spatially distinct from other such groups, although it may be adjacent to them. Within the deme there is an equal chance of all possible male and female pairings, but cross-breeding with members of other demes is rare.

demersal Living on or near the seabed, but being able to swim actively.

demography The statistical study of human populations. It encompasses population size and density, distribution, birth and death rates, immigration and emigration, marriage and divorce rates, and the interaction of populations with external factors such as the urban environment and the economy.

dendrochronology A method of archaeological dating by the ANNUAL RINGS of trees, used when the lifespans of living and fossil trees in an area overlap. Exact dates for sites can be calculated and the method is more accurate than radioactive dating techniques (*see* isotopic dating). Some extremely long-lived trees, such as bristlecone pines, which can live for up to 5000 years, have been used in this work. For trees that are not so old, ring patterns can be matched between samples from trees whose ages overlap.

In addition to dating, the study of annual rings gives a record of past climate, since ring width correlates with availability of water and with the current weather. Traces of pollutants such as lead may also be detected in the rings.

dendroclimatology The study of the record of past climates found in the ANNUAL RINGS of trees. The radial growth of the wood is affected by the weather during a particular year, and is especially sensitive to seasonal changes in water availability, which affect the date at which growth commences in that year, the rate of growth, and the size of the cells. Climatic stress may give rise to incomplete or missing rings, or to distortions at certain points on the circumference or at different heights up the tree. Drought may give rise to an extra ring ('false ring') that does not show the usual gradation of increasing cell wall thickness and decreasing cell diameter. The number of cells in each ring, the diameter of their lumens, and differences in the thickness and composition of the cell walls all give information about previous climatic conditions.

dendrogram A branched diagram used in taxonomy to demonstrate relationships between species, families, or other taxa, or to show the relationships between individuals, as in family trees. *See also* cladistics.

dendrograph An instrument that continuously records the circumference of a tree. It is used, for example, to record daily fluctuations in girth caused by net changes in water content.

denitrification The loss of nitrogen or nitrogen compounds, especially the chemical reduction of nitrates or nitrites by bacteria, usually with the release of gaseous nitrogen. Denitrifying bacteria such as *Pseudomonas*, *Micrococcus*, and *Clostridium* use nitrate as the terminal electron acceptor in anaerobic respiration, so denitrification occurs mainly in environments that have low levels of oxygen, such as waterlogged soils and marine and freshwater sediments. The process is important in terms of soil fertility because some of the products of denitrification (e.g. nitrites and gaseous nitrogen) cannot be used by plants as a nitrogen source. *See* nitrogen cycle. *Compare* nitrification.

denitrifying bacteria *See* denitrification.

density In ecology, the number of a specific type of organism per unit area. Density is usually measured by counting individuals in a series of randomly placed QUADRATS. It gives a measure of ABUNDANCE, and allows comparison of different species or different areas. *Relative density* is sometimes used to give a measure of the numerical importance of one species relative to others of similar habit in a community, and is often expressed as a percentage.

density dependence The regulation of population size by mechanisms that are themselves controlled by the density of the population: the rate of growth in numbers of the population depends on the number already present in the population. It can be seen clearly in a population in a new environment, which may show a typical S-shaped growth curve, with growth proceeding slowly at first, then increasing rapidly, before declining to a more or less stable peak, which represents the carrying capacity of the environment for that organism. The birth or growth rate in the population decreases or the death rate increases as the population density increases. It is thought that most populations experience density-dependent regulation at some stage.

density independence The tendency for the birth, growth, or death rates in a population to remain unaffected as the population density increases. The population size is being regulated by mechanisms that are independent of the population size or density until the final population crash. In a new environment, such populations typically increase rapidly then stop abruptly as factors such as the end of the breeding seasons or changing seasons limit growth, producing a J-shaped growth curve. Algal blooms are an example.

deoxyribonucleic acid   *See* DNA.

dependent variable   *See* variable.

depensation   The increase in parasitism and predation that occurs when the numbers in a population fall below a critical threshold. Depensation further reduces numbers, making it unlikely that the population will recover, e.g. for some species of commercially overfished stock such as the Atlantic cod.

deposit feeder   An animal that feeds on the sediment of the seabed, ingesting mud or sand and extracting organic material from it.

deposit gauge   A device for collecting and measuring solid and liquid atmospheric POLLUTANTS, especially dust.

depression (cyclone)   A circulating body of air in which the atmospheric pressure decreases to a minimum at the center. In the northern hemisphere, the winds in a depression circulate in a counter-clockwise direction; in the southern hemisphere they circulate in a clockwise direction. Depressions are usually about 1000 to 4000 km in diameter. The winds are generally stronger than those associated with anticyclones. Depressions are usually associated with stormy, unsettled weather, and occur mainly in the mid-latitudes. They tend to travel mainly across the oceans, because the water surface offers less resistance to their strong winds than does land. They therefore affect particularly the maritime parts of continents. Their occurrence is affected by large land masses and especially by mountain ranges. Cyclones that form in latitudes 10° to 15°, called *tropical cyclones*, are of much smaller diameter, only

100–500 km across, but are much more violent, with wind speeds up to 60 meters per second or 216 kph (almost 134 mph) – twice the speed of mid-latitude cyclones. In the Atlantic Ocean they are called *hurricanes*, and in the Pacific Ocean *typhoons*. *Compare* anticyclone.

dermal toxicity   *See* contact pesticide.

desalination   The removal of salt from brine to produce fresh water. Desalination is used to irrigate arid regions in which sea water is available, especially if solar power can be used as an energy source. Several methods are employed, the most common being evaporation of the sea water by heat or by reducing the pressure on it (*flash evaporation*). The vapor is condensed to form relatively pure water. Freezing is another technique; pure ice forms from brine as it freezes. The method theoretically requires less energy than evaporation but the process is slower and technically more difficult. Reverse OSMOSIS is another method used. Pure water and salt water are contained on either side of a permeable membrane. The pressure of the salt water is raised above the osmotic pressure, causing water to pass from the brine to the pure side. Because the osmotic pressure required is about 25 atmospheres there are difficulties with large-scale application. In *electrodialysis*, the ions are subjected to an electric field instead of increased pressure, the positive and negative ions being filtered off through separate membranes. Another method used for low salinities is ion exchange; in this method the salt ions are chemically removed from the solution. *See illustration overleaf.*

desert   A major BIOME characterized by low rainfall and thus supporting little or no vegetation. A *hot desert* may form in any climate where average annual precipitation is less than 250 mm and intermittent. A *true desert* has no higher plant life. A hot *semidesert*, in which plant and animal life is sparsely distributed and adapted to long periods of drought, has an annual rainfall of less than 400 mm. The rain falls in brief heavy showers and varies from year to

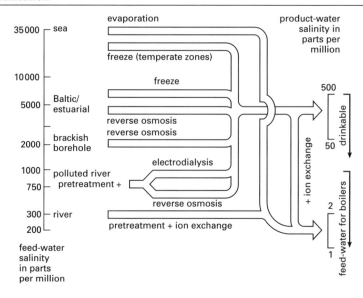

Desalination

year. There are few sparsely distributed shrubs. The perennial plants consist of succulent xerophytic trees, shrubs, and herbs such as euphorbias and cacti. There are also EPHEMERAL plants with seeds that lie dormant until a brief shower of rain that prompts them to germinate, flower, and set seed in a short period of time, as little as two weeks in some cases. Snakes and lizards are common, as are bees, butterflies, and ants. Larger animals face particular problems with scarcity of food as well as water, very cold nights, and difficulties making permanent burrows and shelters in the shifting sand. Camels are well adapted to this harsh environment.

Cold deserts include the TUNDRA and other regions permanently covered with ice and snow.

desertification The process of desert formation, which may be due to climate change or to overgrazing or other human activities such as the removal of vegetation and concomitant soil erosion, overcultivation, or the overextraction of water for irrigation, or for domestic or industrial use. During desertification, the water table falls, and the topsoil and groundwater become increasingly saline as water evapo-

rates from the bare soil, leaving salt behind. The reduction of surface water causes increased soil erosion, and the loss of soil and salinization cause a loss of vegetation. This alters the ALBEDO of the land surface, encouraging the climate to change toward increasing dryness. An example is the southward spread of the Sahara Desert into former areas of steppe grassland and scrub woodland in the Sahel region of Africa as a result of severe drought coupled with overgrazing, overcultivation, and the removal of valuable plant cover for firewood.

desert pavement A thin layer of gravel and stones left on the ground after wind and water have removed the smaller mineral particles from the surface of a desert.

desulfurization The removal of sulfur from FOSSIL FUELS to reduce pollution.

determinate growth *See* growth.

detritivore An organism that feeds on DETRITUS. Examples are various kinds of worms, snails, slugs, insects, mites, bacteria, and fungi. Detritivores are DECOMPOSERS: the larger species break down the detritus by maceration or by grinding it in their intestines into smaller particles with relatively large surface areas that pass out in the feces and are then consumed by the smaller detritivores in the litter and soil. In aquatic ecosystems many detritus feeders filter particles of detritus from the water, or sieve the sediments using tentacles or fine mucus nets.

detritus Finely divided fragments of dead material, such as dead or partially decayed plants and animals, leaf litter, feces, and products of the breakdown of organic material by DECOMPOSERS. In forest ecosystems most of the plant production ends up as detritus, because wood is difficult for herbivores to digest.

developmental response A response to environmental factors that results in the development of alternative forms of an organism according to the particular conditions under which it grows and develops. For example, arrowhead (*Sagittaria sagittifolia*) has linear straplike leaves below the water, which offer little resistance to flowing water, but leaves that develop above the water are arrow-shaped; the buck's horn plantain (*Plantago coronopus)* has hairier leaves when growing in windy situations. European water striders (*Gerris*), small freshwater insects that hunt on the surface film of ponds and lakes, show four different wing lengths according to whether they inhabit large, permanent lakes (short wings or none, because they do not need to disperse and growing wings costs energy), or temporary ponds (long wings so they can disperse to other bodies of water to breed). In some *Gerris* species the late summer generation develop long wings, so they can move to the shelter of woodlands to overwinter.

developmental threshold The minimum body temperature of an organism that is necessary for development to occur.

Devonian The geological period some 405–355 million years ago, between the Silurian and the Carboniferous periods of the Paleozoic era. It was characterized by

an enormous number and variety of fish, most of which have become extinct without leaving any modern relatives. During the late Devonian, primitive amphibians were evolving from crossopterygians (lobe-finned fish). Terrestrial fauna included insects and spiders. During this period the first major invasion of the land by plants occurred. The very first land plants date from late Silurian but during the Devonian many vascular land plants appeared, such as the rhyniophytes (all now extinct), club-mosses (Lycophyta), horsetails (Sphenophyta) and early ferns (Filicinophyta). The period is named after rocks found in Devon, England. *See* geological time scale.

dew Moisture that condenses from the air onto the surface of cool objects, especially at night. It forms on clear calm nights when exposed surfaces lose heat by radiation faster than the air does. The cooled surface cools the adjacent air, and if it cools below its *dew point* (the temperature at which water vapor begins to condense, which depends on the degree of saturation of the air with water vapor), water vapor condenses out onto the surface. If the exposed surface is colder than the freezing point of water, the vapor condenses out as *hoar frost. See also* fog; frost.

dew point *See* dew.

diadromous (amphidromous) Describing fish that migrate regularly between fresh water and salt water. Fish that spend most of their lives in the sea, but migrate to fresh water to spawn (e.g. the Atlantic salmon, *Salmo salar*) are termed *anadromous*, whereas those such as the European eel (*Anguilla anguilla*), which migrate from fresh water to the sea to spawn, are termed *catadromous*.

diapause In some insects and mites, a period of DORMANCY in the life cycle during which growth and development cease and metabolism is greatly decreased. It is often seasonal, occurring, for example, in winter or during a dry season; but in some cases it may last for several years. It enables the insect to survive unfavorable environmental conditions.

diatomaceous earth *See* diatoms.

diatoms A phylum of the Protoctista (formerly called the *Bacillariophyta*) whose members are unicellular algae found in freshwater, the sea, and soil. Much of PLANKTON is composed of diatoms and they are thus important in aquatic FOOD CHAINS. Diatoms have silica cell walls (*frustules*) composed of two overlapping valves ornamented with perforations, which are arranged differently in each species. Diatoms are typically pillbox-shaped (*centric*) or coffin-shaped (*pennate*). The frustules of dead diatoms are highly resistant to decay, and form deep sediments on the floors of oceans and lakes called *diatomaceous earth* or *kieselguhr*.

dicotyledons *See* Angiospermophyta.

dieldrin An organochlorine contact PESTICIDE used to control insects in the soil and to mothproof carpets and other furnishings. It is also formed in the body by the breakdown of the pesticide ALDRIN. It is a powerful carcinogen, and persists in the environment. Dieldrin has been banned by most countries for some time, but there are still remnants in trade and in waste tips.

diesel A fuel derived from PETROLEUM consisting of various alkanes used as a fuel for diesel (compression-ignition) engines. Diesel-powered vehicles are up to 40% more fuel efficient and have lower emissions of carbon dioxide, carbon monoxide, and hydrocarbons but higher emissions of nitrogen oxides and small particulates (PM_{10}) than gasoline-fueled vehicles. Diesel-powered vehicles can be fitted with an oxidation catalyst to reduce emissions of carbon monoxide, hydrocarbons, and particulates (*see* catalytic converter). *Ultralow sulfur diesel* (USLD) is highly refined and has a sulfur content of less than 15 parts per million.

differential resource utilization The use of different resources by different species in the same habitat or the use of the same resource at a different time or place, or in a different manner. The term is generally used in the context of interspecific COMPETITION. For example, in a rainforest, different species feed at different levels of vegetation, and some are nocturnal and some diurnal. Specialized diets may mean that while some insect species feed on the nectar or pollen of flowers, others may feed on the foliage of the same tree.

differentiation A process of change during which cells with generalized form become morphologically and functionally specialized to produce the different cell types that make up the various tissues and organs of an organism.

diffuse competition *See* competition.

diffusion The movement of molecules along a concentration gradient from areas of high concentration to areas of lower concentration as a result of random movement. Gaseous exchange in plants and animals takes place by diffusion. *See* transpiration. *Compare* active transport; osmosis.

diffusion tube A tube, usually about 7 cm × 1 cm in size, sealed at one end, which houses a chemical absorbent, used to monitor gaseous POLLUTANTS, especially nitrogen oxides.

digester A strong closed vessel or tank in which complex organic substances are biologically degraded ('digested') by anaerobic microorganisms. For example, SEWAGE SLUDGE may be broken down to methane, carbon dioxide, and water, with the release of energy. The energy released may be trapped and used to generate electricity, and the methane (BIOGAS) may also be piped off for use in power generation.

The reaction vessel in which wood chips are heated with chemicals to separate the fibers by dissolving LIGNIN during the process of pulp and paper manufacture is also called a digester.

digestion **1.** The process by which en-

zyme-mediated chemical reactions break down ingested food into simpler, soluble substances that can be assimilated by the tissues.
2. The decomposition of organic matter by microorganisms, usually carried out industrially in large chambers or vats called DIGESTERS.

dilute and disperse A method of dealing with POLLUTANTS in which the pollutant is discharged into a large body of water or air that will reduce its concentration and carry it away from its source, further reducing its concentration on the way. Examples are the emission of gases from power stations burning FOSSIL FUELS and the discharge of small quantities of untreated sewage into the sea. *Compare* concentrate and contain.

dilution ratio The ratio of the volume of water entering a stream to the total volume of water in the stream. It gives an indication of the capacity of the stream to dilute and assimilate waste.

dilution series A series of progressively more dilute solutions of known concentration. Dilution series are used to prepare standard solutions and can also be used to determine the efficiency of removal of a substance such as a dissolved POLLUTANT, by comparing the solution after extraction with a dilution series of known concentrations.

dimictic Describing a lake in which there are two seasonal periods (spring and fall) of vertical mixing when water circulates throughout the volume of the lake. In summer, there is thermal STRATIFICATION as the surface waters warm and become less dense, floating on top of and not mixing with the colder layers below. In the fall, the surface waters cool to the temperature of the waters below, and mixing may also be aided by the surface friction generated by fall storms. In winter, when the surface waters cool below 4°C, they expand, again becoming less dense than the water below, giving a reverse stratification. In spring, the surface waters warm, becoming denser than the water below, so they sink and mixing occurs. *Compare* monomictic.

dimorphism The existence of two different forms of an organism. An example is sexual dimorphism in animals. *See* POLYMORPHISM.

dinoflagellate A marine or freshwater protoctist of the phylum Dinoflagellata (Dinomastigota) that swims in a twirling manner by means of two undulipodia (flagella). These lie at right angles to each other in two grooves within the organism's rigid body wall (test). Many possess stinging organelles that they discharge to catch prey; and some produce potent toxins that are capable of killing fish. Roughly half of all known dinoflagellates are capable of photosynthesis.

dioxin Polychlorinated dibenzo-*p*-dioxins, found as contaminants in commercial products and generated by certain industrial processes, such as the manufacture of herbicides and disinfectants. They are known to be highly toxic and persistent POLLUTANTS, with effects on reproduction, development, and the immune system; they may also be carcinogenic. Dioxins were accidentally released by an explosion at a chemical plant in Seveso, Italy in 1976, causing widespread mortality in animals, and among humans the disfiguring skin disease chloracne. *See also* Agent Orange.

diploid A cell or organism containing twice the haploid number of chromosomes (i.e. $2n$). In animals the diploid condition is generally found in all but the reproductive cells and the chromosomes exist as homologous pairs, which separate at meiosis, one of each pair going into each gamete. In plants exhibiting an alternation of generations the sporophyte is diploid, while higher plants are normally always diploid. Exceptions are those species in which polyploidy occurs.

direct competition *See* competition.

directional selection A type of selection that acts on the PHENOTYPES for a particular characteristic in a population (and hence on the ALLELES that give rise to them) by moving the mean phenotype toward one extreme of its expression. Directional selection is often a response to a progressive change in the environment (e.g. climate change). *Compare* disruptive selection; stabilizing selection.

dirty dozen A popular name for the twelve toxic substances the United Nations Environment Program (UNEP) has targeted in a legally binding international convention: ALDRIN, chlordane, DIELDRIN, endrin, DDT, dioxins, furans, HEPTACHLOR, hexachlorobenzene, mirex, POLYCHLORINATED BIPHENYLS (PCBs), and toxaphene.

disaggregation The process of breaking up material into smaller pieces or units.

disassortative mating Negative assortative mating. *See* assortative mating.

discontinuous variation *See* variation.

discrete generations A series of generations of an organism in which one generation dies before the next begins. For example, annual plants in a desert flower and set seed after rain, then die as another drought begins. Their seeds do not germinate until the next time rain falls, which may be long after the parent plant has died.

dispersal The tendency of an organism to reproduce in areas some distance away from its breeding site. In plants, fungi, and microorganisms dispersal is passive, and the degree of dispersal depends on the nature of the seeds, fruits, and spores, and the agents that disperse them, which may be animals, wind, or water. Dispersal is important in the COLONIZATION of new areas, and in the recolonization of areas where vegetation has been lost. Many animal species have special behaviors to promote dispersal. For example, at a certain age young male lions are driven away from their pride and wander a wide area until eventually competing for or establishing prides of their own.

dispersal barrier (ecological barrier) An area of unsuitable habitat separating two areas of suitable habitat, of a sufficient size that a particular species is unable to cross it to colonize suitable areas on the other side. *See also* corridors.

dispersal polymorphism The existence of more than one type of dispersal structure in a species or among the progeny of an individual. Dispersal polymorphism is found in both animals and plants. Certain plants produce two kinds of seeds: one that remains near the parent plant, exploiting the conditions that favored the growth of its parent, and the other that disperses farther away, to colonize new habitats or escape any local disaster that may befall the habitat of its parent. Plant hoppers, which are usually wingless, develop winged progeny in crowded conditions when resources are under strain.

dispersant A chemical, typically a detergent, used to break up oil into smaller droplets in clearing oil spills.

dispersion The pattern of distribution of organisms in an environment. There are three main dispersion patterns. In *random dispersion* there is an equal chance of finding an individual at any place at a given time. This happens when individuals are unaffected by the presence of other individuals or by patchiness in the environment. In *regular dispersion* individuals are more localized than would be expected on purely statistical grounds. This may be due to a tendency to avoid other individuals, or to a failure to survive if too close to others. Regular dispersion is also referred to as *even, uniform,* or *spaced dispersion* or as *overdispersion*. In *aggregated dispersion* (*clumped* or *contagious dispersion* or *underdispersion*) individuals are closer together than expected. This may be due to an attraction to other individuals or to certain parts of a patchy environment, or to a pattern of reproduction in which individuals give rise to offspring close to them (as, for example, in many forms of asexual reproduction in plants). *See also* distribution. *Compare* dispersal.

dispersive mutualism   *See* mutualism.

disposables   Objects that are intended to be thrown away after a single use or a few uses.

disruptive coloration   Coloration that disrupts an animal's outline, making it more difficult for a predator to see it. Examples include the stripes of a zebra, which make it hard to distinguish individuals when seen in a large group, as for example, when fleeing from a predator. Many angelfish of coral reefs have striking blocks of color, so do not present the image of a fish shape.

disruptive selection   A type of selection that acts on the phenotypes for a particular characteristic in a population (and hence on the alleles that give rise to them) to progressively move them toward two opposite extremes. In other words, there is selection for two different alleles of the gene. This may happen, for example, where a population is located at the boundary between two distinct habitats, which exert different selection pressures; or where a population adapts to two different pollinators. Disruptive selection can lead to SPECIATION if divergence becomes so pronounced that individuals exhibiting opposite extremes of phenotype become unable to interbreed. *Compare* directional selection; stabilizing selection.

dissimilation   A biochemical transformation that results in the oxidation of organic compounds and their conversion to inorganic compounds or elements, so that they no longer form part of the *biotic* environment. *See* denitrification.

dissolved load   The part of a river's load that is carried in solution. It usually consists mainly of calcium, sodium, bicarbonate, sulfate, and chloride ions.

dissolved oxygen (DO)   The quantity of oxygen that is dissolved in an aqueous solution. It is usually expressed as mgL^{-1} or as a percentage saturation. It represents the oxygen available to fish and other aerobic organisms, and indicates the water's ability to support aquatic life. Dissolved oxygen can be measured by means of a *dissolved oxygen electrode* or a *fiber optic oxygen sensor*.

distilled water   *See* water.

distribution   The way in which a species, population, or other ecological unit. This is distributed over a particular area.

disturbance   An event or change in the environment that changes the composition of a community and may also affect the succession of communities in the area by removing individuals and opening up new spaces for colonization. Examples include hurricanes, fires, or changes in land use caused by agricultural development or urban settlement.

diurnal rhythm   *See* circadian rhythm.

divergence   1. A horizontal flow of water in a different direction away from a particular area, e.g. the spreading of ocean surface water in areas of UPWELLING. 2. The differential diversification and segregation of certain parts of a TAXON to produce new species, genera, families, orders, or classes. For example, the diversification of the ancestors of horses into zebras, horses, and asses. *See* divergent evolution.

divergent evolution   The evolution of different forms from a single basic structure in response to different selection pressures.

diversity   *See* species diversity.

diversity index   A mathematical expression of SPECIES DIVERSITY in a community, taking into account the relative ABUNDANCE of different species. It is used to compare the structure of different communities and to assess the effects of environmental factors.

divide (drainage divide; watershed)   A dividing ridge between adjacent CATCH-

MENT AREAS. Water on one side of the ridge drains into one catchment area (drainage basin), and water on the other side of the ridge drains into the adjacent catchment area.

DNA (deoxyribonucleic acid) A substance mainly found in the chromosomes, that contains the hereditary information of organisms. The molecule is made up of two helical chains coiled around each other to give a *double helix*. Phosphate molecules alternate with deoxyribose sugar molecules along both chains and each sugar molecule is also joined to one of four nitrogenous bases – adenine, guanine, cytosine, or thymine. The two chains are joined to each other by bonding between these bases. The sequence of bases along the chain makes up a code – the genetic code – that determines the precise sequence of amino acids in proteins. The two purine bases (adenine and guanine) always bond with the pyrimidine bases (thymine and cytosine), and the pairing is quite specific: adenine with thymine and guanine with cytosine. DNA is the hereditary material of all organisms with the exception of RNA viruses. Together with RNA and histones it makes up the chromosomes of eukaryotic cells.

DO   *See* dissolved oxygen.

doldrums   The region on either side of the equator from 30°S to 30°N, between the two belts of trade winds, in which intense heating of the land and sea causes warm, most air to rise, creating low pressure, high humidity, and cloudy weather with light, variable winds.

domain   A taxonomic grouping of organisms that ranks higher than kingdom. There are generally considered to be three domains: the ARCHAEA, BACTERIA, and Eukarya (*see* eukaryotes). The Five Kingdoms system of classification recognizes only two 'superkingdoms': the Prokarya (which includes the single kingdom Bacteria comprising both archaea and the bacteria) and the Eukarya.

domestication   The selective breeding of plant and animal species by humans so that they are exploited in some way. Domestication is a form of imposed DIRECTIONAL SELECTION. *See* hybridization; vegetative propagation. *Compare* natural selection.

dominance   In vertebrate societies, a strict hierarchy (*dominance hierarchy*) in which each individual occupies a particular position that is recognized by others in the group. High-ranking individuals are usually aggressive toward subordinates. Establishment of the hierarchy often involves fighting or ritualized contests, or displays. Once the hierarchy is established, subordinate animals usually avoid threatening dominant animals and perform submissive displays to avoid being threatened by them. Disruption of the hierarchy usually occurs only if new animals move into the population or if an animal dies. In some extreme hierarchies, such as wolves (*Lupus*), only one pair of animals in a pack, the dominant pair, breeds. Among domestic poultry, where dominant individuals tend to peck at subordinates, the hierarchy is sometimes termed the *peck order*.

dominant   1. Describing a species that is the most important member of a community. *See* dominance.
2. An allele that, in a heterozygote, prevents the expression of another (recessive) allele at the same locus. Organisms with one dominant and one recessive allele thus appear identical to those with two dominant alleles, the difference in their genotypes only becoming apparent on examination of their progenies. The dominant allele usually controls the normal form of the gene, while mutations are generally recessive.

Domin scale   *See* Braun-Blanquet scale.

dormancy   A period of minimal metabolic activity of an organism or reproductive body. It is a means of surviving a period of adverse environmental conditions, e.g. cold or drought, and occurs in both plants and animals.

dosage   1. The measured amount of a

chemical or form of radiation to which an organism is exposed or which it ingests (its *exposure dose*).

2. The amount of an absorbed substance that actually reaches specific tissues and is available to interact with metabolic processes, for example, to elicit a toxic response (the *target dose*).

dose equivalent (H_T)   The dose of ionizing radiation absorbed by an organism multiplied by a factor that takes into account the difference in the biological effects of different kinds of radioactive particle having the same energy. It is expressed in sieverts (Sv). 1 sievert is the radiation dose delivered in one hour at a distance of one centimeter from a point source of 1 mg of radium enclosed in platinum 0.5 mm thick. For beta-particles, gamma-rays, and x-rays this factor is 1, but for alpha-particles it is 20. *See also* dosage.

downstream drift   The drifting downstream of inorganic and organic material and organisms as a result of the flow of water. It affects especially bottom-dwelling invertebrates such as the larvae of mayflies, caddisflies, and midges in streams and rivers. For some of these species downstream drift is part of a regular cycle of migration, the larvae drifting downstream and the adults flying back upstream to lay their eggs. It allows for the rapid repopulation of parts of the stream after floods or pollution have removed the original inhabitants, and also gives immature insects the chance to avoid overcrowding. Downstream drift is most marked at night, and during spring and summer.

drainage   The movement of water from land, either naturally or artificially, as it flows over the surface and through rocks and soil under the influence of gravity, eventually reaching the sea, an inland lake, or underground reservoir.

drainage basin   *See* catchment area.

dredging   The removal of mud or other sediments to deepen channels or keep waterways and harbors clear. The sediments removed by dredging may be used for land reclamation. Dredging can have deleterious effects on the environment, releasing toxic chemicals and heavy metals from bottom muds, and stirring up silt that kills aquatic life and benthos.

drift edition   *See* map.

drought   A prolonged period when water loss by evaporation and transpiration exceeds precipitation, resulting in depletion of groundwater and soil moisture and reduction of stream flow, and a shortage of water. A *drought cycle* is a temporary but repeated period of drier conditions in a particular environment. For example, in the North American grasslands droughts occur at approximately 22-year intervals.

drought resistance   Physiological mechanisms that enable living organisms to cope with drought conditions. These may range from adaptations to conserve water (e.g. water storage in the tissues of succulents or in the bladders of desert tortoises, extra-thick cuticles on plant surfaces), to adaptations to minimize heat stress in situations where evaporative cooling is no longer possible or much reduced (e.g. the production of heat-stress proteins). In extreme cases the organism may be able to tolerate drying out, as in the resurrection plants of the American deserts, which dry out and blow around the deserts, then take root and turn green again after rain.

dryfall   The settling out of suspended particles from the atmosphere during periods without rain. Dryfall is a source of nutrients for living organisms. *Compare* wetfall.

dry-matter production   PRODUCTIVITY expressed in terms of the dry weight of organic matter produced per unit area per unit time.

dry season   A regularly occurring period without precipitation or with much reduced precipitation that occurs at a particular time of year every year. In

Mediterranean, west-coast, and subtropical climates the dry season occurs in summer, and is exacerbated by high rates or evaporation and transpiration. In the tropics close to the Equator, there are often two dry seasons a year, corresponding to the migration north and south of the equatorial rain belts. In other parts of the tropics, such as many parts of India, the dry season occurs in winter.

dry weather flow (DWF; baseflow) In a stream or river, the flow of water derived from the seepage of groundwater and from the flow of water through the upper soil horizons. During periods of drought, most of the river flow may be due to dry weather flow, but during peak flow periods it is only a very small proportion of the total flow.

dune A ridge or hill of sand typically between 1 and 50 m high deposited there by the wind. Sand dunes usually migrate downwind with time, as sand is picked up by the wind on the windward slope, then dropped as the wind speed decreases when it reaches the downwind side of the dune. Dunes are formed in sandy deserts and along coastlines. Coastal dune systems form important barriers against the sea, protecting land and settlements in the hinterland.

Along sandy coasts the main band of bare sand dunes facing the shore are called *foredunes*. In time grasses such as marram or beachgrass colonize the foredunes, their stems and roots trapping more sand. As vegetation dies and decays, soil starts to build up, and more species of plants move in. The hollows between dunes may be eroded by the wind down to the water table, so they contain pools or lakes, and often support diverse flora. Such wet hollows are called *dune slacks*. Once the vegetation cover is more or less complete, the dunes are called *fixed dunes*, and will not migrate unless the vegetation cover is seriously broken, by off-road vehicles or storm seas, for example. At the shoreward edge of a dune system – and in deserts – small *embryo dunes* form as sand piles up against objects washed up by the tide or small tufts

of vegetation. In some parts of the world, large mature dune systems often have a *dune heath* vegetation dominated by low-growing ericaceous shrubs, relatively drought-tolerant plants.

Dust Bowl An area in the USA, extending across western Kansas, Oklahoma, and Texas, and into Colorado and New Mexico. During the 1930s droughts and over-farming caused topsoil erosion.

Dutch elm disease A serious disease of all species of elm (*Ulmus*) caused by the fungus *Ceratocystis ulmi*. The fungus spreads through the XYLEM vessels, and the plant responds by producing tyloses (bladderlike ingrowths of the adjacent parenchyma cells that penetrate the xylem vessels through the pits) that seal off the affected vessels. Symptoms include yellowing and curling of the leaves, wilting and rapid death of branches and eventually of the whole tree. The fungus is spread by elm-bark beetles (usually *Scolytus scolytus*). In the 1970s an especially virulent strain of the disease was introduced to Britain from Canada on imported timber, and virtually all the hedgerow elms of England (many millions) were killed, dramatically changing the landscape.

dynamic-composite life table *See* life table.

dynamic equilibrium A state of an ECOSYSTEM whereby there is a relatively stable balance between populations of organisms and the environment despite constant small-scale change and disturbance. Such an equilibrium is due to the action of opposing forces (e.g. numbers of births and deaths) that proceed at approximately equal rates.

dynamic life table *See* life table.

dysphotic zone *See* photic zone.

dystrophic Describing lakes whose water contains a high content of organic matter. *Compare* eutrophic; oligotrophic.

earthquake A sudden movement of the Earth (*seismic activity*) due to release of stress in the CRUST, e.g. by movement along a fault line. The movement is propagated through the rocks and soil as a series of shock waves (seismic waves), causing the ground to shake. Earthquakes range in depth from near-surface events to (rarer) movements as deep as 720 km. The stresses causing earthquakes may be associated with volcanism (volcanic eruptions and intrusions of magma into the crust) and/or with movements of crustal plates (*see* plate tectonics). Movements of the ground at the surface and below ground are measured by a *seismograph*. A *seismometer* is an electronic device that detects, amplifies, filters, and records the motions of the earth in a particular direction. The magnitude of an earthquake is usually measured on the *Richter scale*.

Earth Summit A United Nations Conference on Environment and Development in Rio de Janeiro, Brazil, in 1992 to discuss ways of achieving SUSTAINABLE DEVELOPMENT on a global scale. It produced the *Rio Declaration on Environment and Development*, which defined sustainable development as development that 'meets the needs of present generations without compromising the ability of future generations to meet their own needs,' and outlined key policies to help individual countries devise national strategies for sustainable development. It also gave rise to two legally binding conventions: the CONVENTION ON BIOLOGICAL DIVERSITY and the FRAMEWORK CONVENTION ON CLIMATE CHANGE, as well as a Statement of Forest Principles. A second Earth Summit took place in Johannesburg, South Africa, in 2002.

earthworms *See* Annelida.

ebb tide The TIDE when it is going out.

eccritic temperature The preferred body temperature of ECTOTHERMS, usually maintained by 'shuffling' between exposed sunny areas and those providing shade. *Compare* thigmotherm.

ecesis The GERMINATION and establishment of colonizing plants in an area, this being the first stage in a SUCCESSION. The associated verb is *ecize*. *See* colonization; pioneer species.

Echinodermata A phylum of marine and intertidal invertebrates containing the starfish, sea urchins, feather stars, sea cucumbers, and brittlestars. Some are predators, while others feed on detritus. The phylum includes both commercially important edible species and important predators of commercially harvested shellfish. Most echinoderms have a radial symmetry, typically with five rays radiating from a central disk. All have calcareous skeletal plates and most have spines. Part of the coelom (central body cavity) is modified as the *water vascular system*, a kind of hydrostatic skeleton that extends into hydraulic tube feet, used typically in locomotion. The water vascular system has an external opening, the *madreporite*. The nervous system is simple and there are no excretory organs. Echinoderms produce one or more stages of planktonic larvae. *See also* crown-of-thorns starfish.

echolocation A method of navigation used by some animals, such as bats and dolphins. They emit high-pitched sounds, often inaudible to humans, which are reflected back off nearby objects and detected by the ears or other sensory receptors.

echo sounder (sonar device)   A device that sends pressure waves from pulses of sound to the sea bed, then interprets the echoes to give a depth profile, the delay between sending and receiving the waves being a measure of depth. Echo sounding is also used to detect shoals of fish, the position of underwater objects such as wrecks, and the depth of polar ice. The procedure is sometimes called *sonar*, by analogy to radar, which works in air.

ecize   *See* ecesis.

ecocline   A CLINE that is due to a specific environmental factor, such as a gradient of heavy metal concentration in the soil.

ecoengineering   The incorporation of a natural process in a controlled environment to provide a useful 'end-product', for example, the use of reed beds to remove heavy metals from sewage farms.

ecological amplitude   The range of tolerance by a species to a particular environmental condition(s). When mapped, this provides a bell-shaped curve, the actual shape of which can be used to identify ecological indicator species.

ecological diversity   *See* biodiversity; species diversity.

ecological efficiency (biological efficiency)   The efficiency of energy transfer between one TROPHIC LEVEL and the next, usually expressed as the percentage of the energy of the BIOMASS produced by the donor trophic level.

ecological energetics   The study of the energy transformations in ecosystems.

ecological indicator   *See* indicator.

ecological isolation   The separation of POPULATIONS or subpopulations as a result of changes in the environment or in their ecology. Isolation leads to restricted GENE FLOW between the populations, and to further divergence due to NATURAL SELECTION pressures in the local environment. The resulting differences in gene frequencies may eventually lead to SPECIATION. *See* genetic drift. *Compare* geographical isolation; reproductive isolation.

ecological niche   *See* niche.

ecological range   The range of environmental conditions under which individuals of a species can survive.

ecological sustainability   *See* sustainability.

ecology   The study of the relationships of organisms to one another and to their living (BIOTIC) and nonliving (ABIOTIC) environment.

economically optimum yield (EOY)   The best YIELD that can be obtained in a given situation in terms of economic value rather than BIOMASS. It is defined as the difference between the gross value of the harvested crop and the sum of the fixed costs (such as finance, insurance, and administrative overheads), and the variable costs such as fuel and labor, which increase with the effort put into harvesting the crop. *Compare* maximum sustainable yield.

economic injury level (EIL)   The level of abundance of a pest below which putting more effort into controlling it will cost more than is saved by pest control. Above this level, it is cost effective to control the pest.

ECOPATH   An ecosystem MODEL that calculates energy flow and biomass production in various conditions for a steady-state ecosystem containing many species. It is used particularly for aquatic ecosystems, especially for estimating sustainable levels of cropping.

ecophysiology   *See* physiological ecology.

ecoregion A geographic area defined by its natural vegetation, fauna, soil type, physical features, and climate. *Compare* biogeographical region; biome.

ecosphere (biosphere) The global ecosystem; the part of the Earth and its atmosphere that is inhabited by living organisms.

ecosystem A unit made up of all the living and nonliving components of a particular area that interact and exchange materials with each other. The concept of the ecosystem differs from that of the COMMUNITY in that more emphasis is placed on abiotic factors. The term can be applied on various scales, from small ponds to the whole planet.

ecotone A transition zone between two or more different communities, e.g. at the boundary between land and water, or where rocks and soils change. It is sometimes known as an *edge habitat*. Such habitats are usually species-rich. *See* edge effect.

ecotourism Tourism that promotes travel to natural, 'unspoiled' habitats to observe wildlife or indigenous peoples. Ecotourism can be an important source of income to local people as part of a SUSTAINABLE DEVELOPMENT program, but all too often it generates disturbance and pollution, destroying the very qualities it set out to promote.

ecotoxicology The study of the adverse effects of manmade substances on populations and communities, including humans.

ecotype A subpopulation of individuals genetically adapted to the combination of environmental factors in their local habitat, but still able to mate with other ecotypes belonging to the same species and produce fertile offspring. Differences between ecotypes may be physiological or morphological. *Compare* biotype. *See also* adaptive radiation; speciation.

ectomycorrhiza *See* mycorrhiza.

ectoparasite *See* parasitism.

ectotherm An animal that keeps its body temperature within fairly narrow limits by behavioral means, such as basking in the sun or moving into shade, e.g. snakes and lizards. *See* poikilothermy. *Compare* endotherm.

ectotrophic mycorrhiza *See* mycorrhiza.

edaphic factors The physical, chemical, and biological characteristics of the SOIL that together form an important component of the habitat because of their influence on plant distribution. The main edaphic factors are water content, PH, organic matter, and SOIL TEXTURE.

edge effect 1. The sampling errors that occur at the edges of sampling plots/areas. This may be due to items at the edge of the plots experiencing significantly different conditions from those in the center, or because items overlap the periphery. This effect is especially pronounced in small sampling areas.
2. The tendency for a transitional zone between communities (an ecotone) to contain a greater variety of species and more dense populations of species than the communities surrounding it.

edge habitat *See* ecotone.

effective dose equivalent *See* dose equivalent.

effective population size (N_e) The average number of breeding individuals in a population that contribute genes to the next generation. This is often less than the observed number of breeding individuals. There may be a number of causes including imbalances in the numbers of individuals of each sex, a BREEDING SYSTEM in which a higher proportion of one sex may mate, differences in fertility between individuals and, especially in small populations, loss of heterozygosity (*see* heterozygous) and INBREEDING DEPRESSION. The effective population size determines the rate at which genetic VARIATION is lost as a result of GE-NETIC DRIFT.

effluent Waste material that is discharged into the environment, for example, from sewage outfalls and factory chimneys.

EIA *See* Environmental Impact Assessment.

EIL *See* economic injury level.

EIS *See* Environmental Impact Statement.

electrochemical sensor A sensor that uses electrodes specific to particular ions to determine the quantity of substances in solution.

electromagnetic energy (electromagnetic radiation) Energy propagated by vibrating electric and magnetic fields. It can be thought of as being in the form of waves or as streams of photons (units of light). The energy carried depends on the frequency. The frequency and wavelength are related by the equation:

$$\lambda v = c$$

where c is the speed of light, λ the wavelength, and v the frequency. The *electromagnetic spectrum* ranges from low-frequency radio waves to high-frequency gamma rays.

electromagnetic radiation *See* electromagnetic energy.

electromagnetic sense A sense used by some fish (particularly cartilaginous species) to detect subtle changes in magnetic impulse, which they use to locate prey. Some insects, e.g. fire ants, are also attracted to the electromagnetic fields produced by underground cables, which they subsequently attack.

electromagnetic spectrum *See* electromagnetic energy.

electron transport chain *See* photosynthesis.

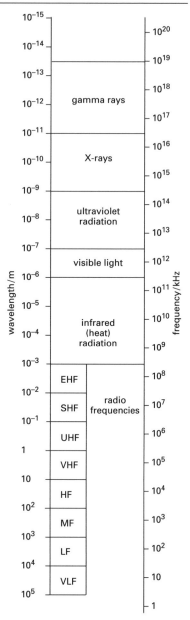

Electromagnetic radiation: the electromagnetic spectrum

electrophoresis The migration of electrically charged particles toward oppositely charged electrodes in solution under an electric field – the positive particles to the cathode and negative particles to the

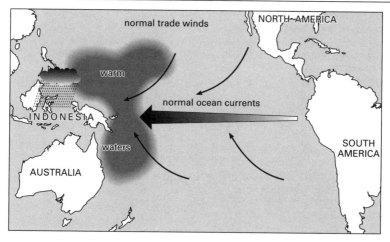

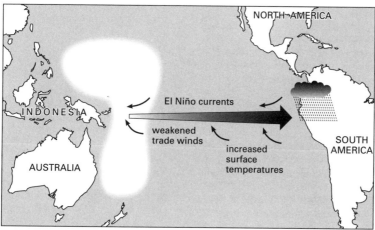

El Niño

anode. The rate of migration varies with molecular size and shape. The technique can be used to separate or analyze mixtures of organic molecules, such as proteins or nucleic acids. The medium used may be wetted filter paper, a gel of starch or a similar polymer (gel electrophoresis), or a similar inert porous medium.

element A substance that cannot be chemically decomposed into simpler substances. The atoms of an element all have the same proton number, and thus the same number of electrons, which determines its chemical activity. *Compare* compound; ion.

elfin forest (Krummholz) Forest dominated by dwarfed and deformed trees and shrubs found between the timberline and the tree line on tropical mountains. It is normally between 1 and 3 m tall. The zone of elfin forest is termed the *Kampfzone*.

El Niño (El Niño-Southern Oscillation Event; ENSO) A warm ocean current that flows southward along the Pacific coast of tropical South America, approximately every seven years. It is associated with a change in the atmospheric circulation known as the SOUTHERN OSCILLATION. El Niño flows during the southern midsummer when the prevailing trade winds weaken, allowing the Equatorial counter-current to strengthen; warm surface waters are no longer driven westward, but flow eastward instead, to overlie the cold water of the northward-flowing PERU CURRENT. Severe El Niño events prevent the up-welling of nutrient-rich cold water along the Pacific coast and cause widespread death of plankton and declines in fish. The climatic effects of El Niño are felt through-out the Pacific and beyond, causing in-creased rainfall in desert areas fringing the Pacific coast of South America, but droughts in parts of Africa. *See also* La Niña.

El Niño Southern Oscillation Event *See* El Niño.

Elton, Charles Sutherland (1900–91) British ecologist. Elton joined an Oxford University expedition to Spitzbergen (1921), where he carried out ecological studies of the region's animal life. Further Arctic expeditions were made in 1923, 1924, and 1930, leading to his appoint-ment as biological consultant to the Hud-son's Bay Company, for which he carried out investigations into variations in the numbers of fur-bearing animals using trap-per's records dating back to 1736. In 1932 Elton helped establish the Bureau of Ani-mal Population at Oxford – an institution that subsequently became an international center for information on and research into animal numbers and their ecology. In the same year he became editor of the new *Journal of Animal Ecology*, launched by the British Ecological Society. Elton was one of the first biologists to study animals in relation to their environment and other animals and plants. His demonstration of the nature of food chains and cycles, as well as such topics as the reasons for dif-ferences in animal numbers, were discussed in *Animal Ecology* (1927). In 1930 he pub-lished *Animal Ecology and Evolution*, in which he advanced the idea that animals were not invariably at the mercy of their environment but commonly, perhaps through migration, practiced environmen-tal selection by changing their habitats.

EMAP (Environmental Monitoring and Assessment Program) A research pro-gram developed by the United States ENVIRONMENTAL PROTECTION AGENCY to provide tools for monitoring and assessing the status and trends of national ecological resources. It aims to further scientific un-derstanding to assist the use of monitoring data in assessing current ecological status and forecasting future risks to natural re-sources.

embryo dune *See* dune.

emergent 1. An individual tree or group of trees that rises significantly higher than the continuous CANOPY of tropical RAIN-FORESTS.
2. An aquatic plant that has most of its veg-etative parts above water.

emergent poverty Poverty that has only recently arisen in a particular area. For example, the migration from the coun-tryside to the cities in many parts of the world has created new urban poor; redis-tribution of land or the introduction of mechanization may lead to poverty in rural areas as people lose their jobs or their source of free food.

emigration The movement of individu-als out of a POPULATION or from one area to another. *Compare* immigration.

emission A discharge of gaseous, partic-ulate, or soluble waste material or other POLLUTANTS into air or water. Vehicle ex-haust gases are an example. Emissions that are trapped by emission control systems are called *primary emissions*. *Secondary emissions* are those not catered for by such controls, for example emissions through leaks in pipes.

emission credit *See* emissions trading.

emissions trading The control of total emissions by allowing overproducers of emissions to purchase *emission credits* from companies that produce less than their quota of emissions, or by allowing companies that reduce their emissions substantially to save credits against future emissions.

emulsifier A device or chemical that aids the suspension of one liquid in another, for example an appropriate organic chemical in an aqueous solution.

emulsion The dispersion of one liquid in a second immiscible liquid.

encounter competition *See* competition.

endangered species A species that is in danger of EXTINCTION. *See* CITES; Endangered Species Act; Red Data Book.

Endangered Species Act A federal act passed in 1973 to protect the United States' endangered and threatened species, *endangered species* being those in imminent danger of extinction throughout their range, e.g. the black-footed ferret, the blue whale, and the California condor, and *threatened species* being likely to become endangered in the foreseeable future, e.g. the northern spotted owl. Many states also recognize *rare species*, with low or declining numbers, shrinking habitats, and so on. Endangered species and their habitats are protected by law, and states are required to set up recovery plans.

endemic Describing a population or species that is restricted geographically. Some endemics have evolved in geographical isolation on islands or mountaintops; others may be relics of once widespread species that have become restricted owing to climate change, geological change, or human activity.

endobiont (endosymbiont) A symbiotic organism that lives within the body of its host. *See also* endosymbiont theory.

endocrine disrupter *See* endocrine toxicity.

endocrine toxicity Toxicity that affects the endocrine systems of animals – their hormones – often by blocking receptors for certain hormones. Such endocrine-disrupting chemicals are thought to be involved in fertility problems in humans as well as other animals, and especially in fish and other aquatic organisms. They are also implicated in some cancers, such as breast and testicular cancer. Natural estrogens and synthetic estrogens used in hormone treatments and contraceptives are found in SEWAGE effluent. Other *endocrine disrupters* include tributyl tin oxide, which causes female dog whelks to become masculinized, and certain industrial detergents, which cause male trout to produce vitellogenin, an egg yolk protein normally produced only by females.

end-of-the-pipe Pollution or waste treatment processes or devices located immediately at the outlet of a collecting system, e.g. at a sewage outfall. The pollutants are treated before they are discharged to the environment.

endogenous rhythm Cyclical, physical, and biochemical processes that occur in an organism in response to internal stimuli, e.g. the opening and closing of stomata, which usually coincides with dawn and dusk respectively, but which continues even if kept in continuous darkness. *See* circadian rhythm.

endomycorrhiza *See* mycorrhiza.

endoparasite *See* parasitism.

endoparasitoid *See* parasitoid.

endoplasmic reticulum *See* nucleus.

endosperm The nutritive tissue that surrounds the embryo in angiosperms. Many endospermic seeds (e.g. cereals and oil seeds) are cultivated for their food reserves.

endosymbiont *See* endobiont.

endosymbiont theory The theory that eukaryotic organisms evolved from symbiotic associations between bacteria. It proposes that integration of photosynthetic bacteria, for example purple bacteria and

Cyanobacteria, into larger bacterial cells led to their permanent incorporation as forerunners of the plastids (e.g. CHLORO-PLASTS) seen in modern eukaryotes. Similarly, other symbiotic aerobic bacteria gave rise to the MITOCHONDRIA. Chloroplasts and mitochondria contain some circular DNA and small ribosomes, both more similar to those of bacteria than those of eukaryotes, and a double membrane envelope. The chloroplasts of some groups of algae have a triple envelope, suggesting more than one symbiotic event. *See* Bacteria; eukaryote; symbiosis.

endotherm An animal that can generate heat internally to raise its body temperature, e.g. birds and mammals. *See* homoiothermy. *Compare* ectotherm.

energetic efficiency The ratio of energy stored or useful work done to energy intake.

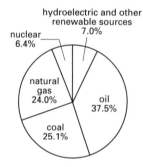

hydroelectric and other renewable sources
7.0%
nuclear 6.4%
natural gas 24.0%
oil 37.5%
coal 25.1%

Energy: primary sources (World 1993)

energy The capacity for doing work. Energy has several forms, some of which are convertible; they include chemical energy (energy stored in chemical compounds), ELECTROMAGNETIC ENERGY, potential energy (the capability of doing work), kinetic energy (movement), mechanical energy, radiant energy (light), and thermal energy (heat). Energy is measured in JOULES.

energy budget The comparison between the energy entering and the energy leaving the body of an animal, a particular TROPHIC LEVEL, or an industrial process. It

is an indication of the efficiency of energy use. *See also* production efficiency.

energy conservation The strategy for reducing the energy usage by individuals, households, industrial units, or modes of transport. For example, the reduction in energy use for heating by installing insulation, double-glazing, and energy-efficient boilers and the use of low-energy lighting. The strategy can also be applied to the unnecessary use and discarding of objects that use energy in their manufacture, such as plastic bags.

energy efficiency The use of processes and fuel-consuming devices designed to ensure that they make efficient use of energy: that the energy input is low in relation to energy output. For example, the introduction of condensing boilers for heating, COMBINED HEAT AND POWER PLANTS that burn waste, low-energy light bulbs, and the development of vehicles that make efficient use of fuel. *See also* energy budget.

energy flow web *See* food web.

energy recovery The conversion of organic wastes to usable energy. For example, microbes may be used to digest organic wastes and generate methane gas, refuse may be burned to produce steam power, or heated to produce oil. *See* combined heat and power plant.

energy subsidy The use of energy that is not derived from an organism's own metabolism.

energy transformation The transformation of one kind of energy into another. For example, during photosynthesis, light energy is transformed into chemical energy, in respiration chemical energy is transformed into heat energy, and in locomotion chemical energy is converted to mechanical energy.

enhanced treatment process *See* digestion (def. 2).

enriched uranium   *See* enrichment (def. 3).

enrichment   1. The planting of young trees in a forest, e.g. after timber extraction. 2. The addition of nutrients, for example, from agricultural runoff or sewage effluent, to surface water. It greatly increases the growth of algae and aquatic plants. *See* eutrophication. 3. The processing of uranium to increase the proportion of the fissionable form U-235 in relation to U-238, or to add plutonium P-239 to natural uranium, for use in nuclear reactors or nuclear weapons. Enriched uranium contains up to 3% U-235, as compared with 0.7% in natural uranium. *See* radioactive isotope.

ENSO   *See* El Niño.

entropy   Symbol: *S*. In a given system, the availability of energy to do work. In any system that undergoes a reversible change, the change of entropy is defined as the heat absorbed divided by the thermodynamic temperature:
$$dS = dQ/T$$
The concept of entropy has been widened to take in the general idea of disorder – the higher the entropy, the more disordered the system. For example, a chemical reaction involving polymerization may exhibit a decrease in entropy because there is a change to a more ordered system.

environment   The complete range of external conditions under which an organism lives, including physical, chemical, and biological factors, such as temperature, light, the availability of food and water, and the effects of other organisms.

environmental biology   *See* binomics.

environmental hazard   One or more factors in the environment that pose a threat to air, water or soil quality, flora and fauna, or human health and safety, for example spills of oil or toxic chemicals.

Environmental Impact Assessment

(EIA)   An attempt to predict the effects of proposed industrial and other building developments, land-use changes or legislation on the natural and human environments, and on human health and wellbeing. The United States Environmental Protection Agency requires a written EIA prior to federal activities such as the building of highways or airports and, if it is still in doubt about the impact of the proposed activity, it will request an ENVIRONMENTAL IMPACT STATEMENT.

Environmental Impact Statement (EIS or ES)   A written analysis prepared for or by the United States Environmental Protection Agency that identifies and analyzes in detail the environmental impacts, both positive and negative, of a proposed action. *See also* Environmental Impact Assessment.

environmentalism   A social movement that attempts to understand the impact of human activity on the environment.

environmental load   The concentration (or other measure of quantity) of one or more chemicals (e.g. waste gases from a factory chimney) or physical fluxes (e.g. heat or sound) emitted by a source into the environment.

environmental management   Procedures and controls aimed at conserving the status of an environment. It may involve active intervention to maintain a particular habitat, for example, a particular stage in plant succession. Alternatively it may aim to balance the conservation of a particular natural resource with the needs of local human communities to promote sustainable development.

environmental monitoring   Continuous or periodic measuring of environmental parameters and/or the amounts of particular chemical or physical pollutants to prevent deterioration of the environment. Monitoring may also include attempts to predict changes to the ecosystem and take preventive action. *See* EMAP.

Environmental Monitoring and Assessment Program *See* EMAP.

environmental noise Background signals that derive from outside the environment, but which interfere with the measurement or detection of environmental processes or biotic factors.

environmental polymorphism *See* polymorphism.

environmental potential *See* land-use capability.

environmental potential maps *See* map.

Environmental Protection Agency (EPA) The United States Government body responsible for the monitoring and control of pollution, such as hazardous wastes, pesticides and other toxic substances, radioactive materials, and the quality of air, water, and drinking water. It has the power to enforce its regulations. The EPA also carries out research in the effects of pollution and human activities on the environment.

environmental resistance *See* biotic potential.

environmental stochasticity Random variation in the ABIOTIC environment.

enzootic Describing a disease constrained by geographic boundaries.

Eocene The second oldest epoch of the TERTIARY PERIOD, 55–38 mya. It was characterized by predominance of early hoofed mammals. Many other mammals (e.g. carnivores, bats, and whales) and birds were also present. It was a period of widespread temperate and subtropical forests. The extensive grasslands of the late Tertiary had not yet developed, but large hoofed grazing mammals, such as horses and elephants, were evolving rapidly. *See also* geological time scale.

eolian Caused or deposited by wind.

EOY *See* economically optimum yield.

EPA *See* Environmental Protection Agency.

ephemeral A plant that has a very short life cycle, and may complete more than one life cycle within a year. Examples are shepherd's purse (*Capsella bursa-pastoris*) and certain desert plants that grow, flower, and set seed in brief periods of rain. *Compare* annual; biennial; perennial.

epidemic A relatively sudden and widespread outbreak of a disease, affecting either a wide area or a large number of people or animals in a small area.

epidemiology The study of the incidence, distribution, and control of diseases affecting large numbers within a population. These include both epidemics of infectious diseases and diseases associated with environmental factors and dietary habits (e.g. lung cancer, some forms of heart disease, etc.).

epifauna Organisms living on the surface of a substrate.

epilimnion In a stratified lake (a lake in which there are distinct layers of different temperature and/or density), the upper layer of warm water. It usually has a fairly uniform temperature, because it undergoes mixing by wind action. *See* stratification.

epipelagic The oceanic zone extending from the surface to about 200 m, where enough light penetrates to allow photosynthesis.

epiphyte (air plant) Any plant growing upon or attached to another plant or object merely for physical support.

epoch *See* geological time scale.

Equator An imaginary line at latitude 0° that girdles the Earth and divides it into the northern hemisphere and southern hemisphere.

equatorial forest   *See* rainforest.

equilibrium   In population ecology, the state in which the numbers of births and deaths are equal, so there is no fluctuation in population numbers. *See also* dynamic equilibrium; equilibrium theory.

equilibrium hypothesis   *See* equilibrium theory.

equilibrium isocline   (zero net growth isocline)   On a graph that shows the effect of the availability of two potentially limiting resources on population numbers, a line that indicates zero net population growth rate. It represents the balance between the consumption and renewal of the resources. In a situation where two species compete for the same resources, by plotting the equilibrium isocline for both populations, the species whose isocline lies closest to the resource axis will eventually competitively exclude the other species from occupying the same NICHE. *See* competitive exclusion principle.

equilibrium species   A species whose main survival strategy is competitive ability rather than reproductive rate or dispersal success. Such species are typical of stable environments. In unstable environments, such species rely on surviving unfavorable conditions by saving energy, entering a sate of DORMANCY or ESTIVATION with minimal metabolic demands, rather than relying on, in plants, for example, producing seeds that will germinate when conditions improve. *Compare* fugitive species.

equilibrium theory   (equilibrium hypothesis) The theory that a COMMUNITY in a defined area is in a state of DYNAMIC EQUILIBRIUM, the number of species at a given time being determined by the balance between IMMIGRATION and EXTINCTION. Following a disturbance such as a natural CATASTROPHE, the community will tend to return to equilibrium. This theory is based on the assumption the organisms occupying the same NICHE compete for resources, and that when resources are limiting, one

species will inevitably exclude another (*see* competitive exclusion principle). *Nonequilibrium theory*, by contrast, is concerned with the temporary behavior of a system when it moves away from the equilibrium point, and in particular its variation over time.

equinox   The period of the year when the Sun is overhead at local solar time at the Equator. It happens twice a year and day and night are equal time periods of twelve hours. In the northern hemisphere the *vernal* or *spring equinox* occurs on 20 or 21 March and the *fall* or *autumnal equinox* on 22 or 23 September. This is reversed in the southern hemisphere.

equitability   A measure of the evenness with which species are distributed in a community. For example, if out of a community of 100 organisms, 95% were of one species, and the other 5% of three other species, this would be an uneven distribution.

equivalent dose   *See* dose equivalent.

era   *See* geological time scale.

erg   A large expanse of sand (*sand sea*) in a hot desert. The sand usually accumulates in an enclosed basin, where sediments derived from the surrounding rocky desert were deposited by rivers or in a lake.

ericaceous   Of, relating to, or being a heath or of the heath family of plants, which are mostly shrubby, dicotyledonous, and often evergreen plants that thrive on open, barren soil that is usually acidic and poorly drained.

erosion   1. The wearing away of rocks, soil, or buildings by physical or chemical processes. Physical processes include the effects of PRECIPITATION, rivers, RUNOFF, frost, changes in temperature, and wind. Chemical processes include corrosion by water containing dissolved substances, including ACID RAIN, and by gases such as those containing pollutants. These processes are often exacerbated by human ac-

tivities such as land-clearance, logging, industrial pollution, or diversion of waterways. The products of erosion are often transported by wind or water and deposited elsewhere, e.g. in the sea. *See* soil erosion; weathering. *See also* solifluction; runoff.
2. (genetic erosion) The depletion of genetic diversity in domestic crops and livestock.

essential amino acid   *See* amino acid.

essential resource   A RESOURCE that limits the numbers of a consumer population independently of other resources.

estivation   A period of inactivity seen in some animals during the summer or dry hot season. For example, lungfish respond to the drying up of water by burying themselves in the mud bottom. They re-emerge at the start of the rainy season. *See also* HIBERNATION.

estuary   A complex ECOSYSTEM that occurs where fresh water draining off the land meets salt water, for example, at the mouths of rivers, SALT MARSHES, bays, and LAGOONS. The boundary of tidal estuary occurs where the fresh water starts, or at the point in a river to which the highest spring tides penetrate. Estuarine habitats are often subject to cyclical fluctuations in salinity.

ethanol   A CLEAN FUEL (used as an alternative fuel for motor vehicles) that can be produced from grain crops, and from paper and wood wastes. Because it is derived from recently grown crops, it does not contribute to carbon dioxide accumulation as do FOSSIL FUELS. Ethanol and ethanol/gasoline blends (GASOHOL) have been used for many years. *Methanol*, another alcohol produced from wood and coal, has similar benefits, and provides a much better performance for motor vehicles than ethanol.

ethene   *See* ethylene.

Ethiopian   One of the six zoogeographi-cal regions of the earth. It encompasses Africa south of the Sahara Desert, the southern half of Arabia, and (according to certain authorities) Madagascar. The animals characteristic of this region are the gorilla, chimpanzee, African elephant, rhinoceros, lion, hippopotamus, giraffe, certain antelopes, ostrich, guinea fowl, secretary bird, and, in Madagascar, the lemur.

ethnobotany   The study of the use of plants by humans.

ethology   The study of the behavior of animals in their natural environment.

ethylene   (ethene)   A gaseous hydrocarbon produced in varying amounts by many plants, which functions as a plant hormone. Ethylene is involved in the control of GERMINATION, cell growth, fruit ripening, ABSCISSION, and SENESCENCE; it inhibits longitudinal growth and promotes radial expansion. In ripening fruit a rapid rise in ethylene production precedes respiration to reach the climacteric ripeness; production of ethylene stimulates further production of ethylene. Thus ripe fruits stimulate other fruits to ripen quickly; this process can be controlled to some extent for fruit storage and transport.

ethylene dibromide   A colorless volatile organic liquid used as a fuel additive to remove LEAD.

Eubacteria   A DOMAIN or a major subkingdom of the BACTERIA, containing a large, diverse, and widespread group of bacteria, principally distinguishable from the other major subkingdom, the ARCHAEA, by differences in the RNA sequences of their RIBOSOMES, which are thought to reflect the very early evolutionary divergence between the two groups. Most are single-celled organisms that divide by binary fission. The group includes photosynthetic forms, some utilizing water and carbon dioxide, as do plants, but others deriving energy from a range of inorganic substrates. Heterotrophic types include both aerobic and anaerobic forms, and many important DECOMPOSERS and nitrogen fixers. Some phyla, such as the ACTINOBACTERIA, are important sources of antibiotics. *See* nitrogen fixation. *Compare* Archaea.

eucalypts Australian hardwood trees of the genus *Eucalyptus* (family Myrtaceae). Widely cultivated as fast-growing timber trees, they produce a strong-smelling resin that is highly flammable, so are susceptible to fire. Many eucalypts have characteristic peeling bark. They include jarrah (*E. marginata*) and karri (*E. diversicolor*), and the mountain ash (*E. regnatus*), probably the tallest tree in the world.

eucaryote *See* eukaryote.

Eukarya *See* eukaryote.

Eukaryota *See* superkingdom.

eukaryote Any member of all the living kingdoms except the Bacteria (Archaea and Eubacteria), which are the prokaryotes. Eukaryotes are defined by the presence of a much more elaborate cell than the prokaryotes. Features not found in PROKARYOTES include the genetic material packaged in chromosomes within a membrane-bound nucleus; possession of mitochondria (*see* mitochondrion) and, in photosynthetic eukaryotes, chloroplasts; a quite different and much more elaborated membrane structure including internal membranes such as the endoplasmic reticulum and Golgi body; different-sized ribosomes and complex undulipodia composed of arrays of microtubules. Some of these features probably arose through endosymbiosis of prokaryotes. In the FIVE KINGDOMS CLASSIFICATION, the eukaryotes constitute a superkingdom or DOMAIN, the *Eukarya*, containing the kingdoms Animalia, Fungi, Plantae, and Protoctista. *See* endosymbiont theory.

Eulerian measurements *See* waterflow measurements.

Eulerian models *See* Lagrangian models.

euphotic zone *See* photic zone.

euryecious Describing an organism that is able to live in a wide range of habitats.

euryhaline Describing organisms that are able to tolerate wide variations of salt concentrations (and hence osmotic pressure) in the environment. For example, the eel (*Anguilla*) can live in both fresh water and salt water. *Compare* stenohaline.

eusociality The complex social organization of certain groups of insects, especially termites, ants, bees, and wasps, in which only a few members of the community breed, while the other CASTES are sterile or non-breeding workers, and where there is an overlap of generations, so that the offspring can help rear their younger siblings.

eutrophic 1. Describing lakes or ponds that are rich in nutrients and consequently are able to support a dense population of plankton and littoral vegetation. *See* eutrophication. *Compare* oligotrophic. 2. Describing fen peats that contain high concentrations of minerals and bases.

eutrophication The process that results when an excess of nutrients enters a lake, for example, as sewage or from water draining off land treated with fertilizers. The nutrients stimulate the growth of the algal population, giving a great concentration or BLOOM of such plants. When these die they are decomposed by bacteria, which use up the oxygen dissolved in the water, so that aquatic animals such as fish are deprived of oxygen and die from suffocation.

evaporation The changing of a liquid to a vapor by the expenditure of heat energy, which is drawn from the surroundings, thus cooling them. *Compare* transpiration. *See also* evapotranspiration.

evaporimeter An instrument for measuring the rate of evaporation.

evaporites *See* salt pan.

evapotranspiration The loss of water vapor by EVAPORATION and TRANSPIRATION.

evergreen Describing plants that retain their leaves throughout the year or through several years. *Compare* deciduous.

evolution The process of genetic change that occurs in populations of organisms over a period of time. It manifests itself as new characteristics in a species, and may eventually lead to the formation of new species. *See* Darwinism; Lamarckism; natural selection.

evolutionary tree A branching diagram showing the phylogenetic relationships between different TAXA.

excreted energy The part of the energy assimilated by an animal that is eliminated in the form of nitrogenous waste.

excretion The process by which excess, waste, or harmful materials, resulting from the chemical reactions that occur within the cells of living organisms, are eliminated from the body. The main excretory products in animals are water, carbon dioxide, salts, and nitrogenous compounds.

exon A segment of a GENE that is both transcribed and translated and hence carries part of the code for the gene product. Most eukaryotic genes consist of exons interrupted by noncoding sequences (*introns*). Both exons and introns are transcribed initially, but the introns are then removed, leaving mRNA, which has only the essential sequences and is translated into the protein. Bacteria do not have introns.

exotic A non-native species, often introduced into surroundings that it finds hospitable and lacking in population-control mechanisms such as disease or predators. *Compare* indigenous.

expectation of further life (e_x) The average remaining lifetime of an individual of age x.

experiment A procedure carried out under controlled conditions in order to establish or verify a hypothesis or law. Experiments involve exploring or testing causal relationships between variables, and usually involve a control procedure against which the others are compared, randomization of design, and repeats to ensure validity or assign significance. *See* control; null hypothesis; sampling.

exploitation 1. Direct competition for a resource.
2. The removal of BIOMASS, including individuals, from a population by consumers.

exponential growth A type of GROWTH in which the rate of increase in numbers at a given time is proportional to the number of individuals present. Thus, when the population is small multiplication is slow, but as the population gets larger, the rate of multiplication also increases. An exponential growth curve starts off slowly and increases faster and faster as time goes by, so becoming J-shaped (when plotted on a logarithmic scale, it is linear). However, at some point factors such as lack of nutrients, accumulated wastes, etc. limit further increase, when the curve of number against time begins to level off and the curve becomes *sigmoid* (S-shaped).

exposure indicator An environmental characteristic that can be measured to provide evidence of exposure to, and sometimes also the magnitude of, a chemical or biological stress.

exposure limit The threshold concentration in workplace air of a chemical above which it is deemed to be harmful to workers.

exposure meter *See* light meter.

expression The occurrence in the PHENOTYPE of characteristics related to a particular ALLELE or alleles. *See* genotype.

extinction The complete disappearance of a species from the planet; or from the whole or a particular part of its natural range (*compare* extirpation).

The natural, slow replacement of species that occurs as ecosystems change is termed *backround extinction*. *Anthropogenic extinctions* are due to human activity, e.g. the dodo (a flightless bird that lived on the island of Mauritius in the Indian Ocean) was rendered extinct by human hunters and predation by and competition with introduced animals. *Secondary extinction* occurs where one species becomes extinct as a consequence of another becoming extinct. For example, the dodo was the sole distributor of the seeds of a species of the tree *Sideroxylon* (formerly *Calvaria major*), consuming the fruits and passing out the seeds in its droppings. Following the demise of the dodo, the tree also declined to extinction. Sometimes extinction is due to a situation where population density is correlated with environmental factors (*correlated extinction*). For example, if the numbers of individuals in adjacent populations decline as a result of environmental factors, such as drought, and if emigration rates are related to population density, the density may become too low to trigger emigration, so both populations may become extinct.

At intervals in the past there have been extinctions of large numbers of species as a result of natural catastrophes – *mass extinctions*. Some authorities consider that a similar mass extinction is happening at the present time as a result of human activity and unsustainable consumption of natural resources. The current rate of extinctions is estimated to be several thousand species every year. Such mass extinctions have occurred at least three times in the past 750 million years, one at the end of the Permian period, associated with large movements of the tectonic plates bearing the continental land masses, resulted in the loss of some 90% of all the species on Earth, while over 50% were lost in the *Cretaceous mass extinction* some 65 million years ago, when the dinosaurs were among those life forms that vanished. The effect on global climate of a large asteroid impacting the Earth is thought to have been the cause.

extinction vortex   A cascade of extinctions that spreads from an initial species through the community of which it was a part. This is especially likely to happen if a KEYSTONE SPECIES becomes extinct. *See* extinction.

extirpation   The loss of a species from certain areas but not from the whole of its range. *Compare* extinction.

extrapolation   The use of data derived by observation to deduce estimated values for variables or conditions that have not been observed. For example, the continuation of a line of best fit to the origin of a GRAPH, even though there are no values plotted in that part of the graph.

extremophile   A microorganism that lives in extreme environments, such as hot springs or highly saline water. *See* halophile; methanogen; thermophile.

extrinsic factors   Factors acting from outside the organism – the physical and chemical impacts of the environment and the effects of other organisms.

facilitation   1. In a community SUCCESSION, the facilitating of the establishment of later species in a succession by the action of earlier species, which change the local environment, making it more favorable for later species to colonize. For example, in the colonization of newly exposed ground, early pioneers often include species such as alder (*Alnus*), which possess root nodules containing symbiotic bacteria that fix nitrogen. Decomposition of these species releases soluble nitrogen compounds into the soil, increasing its fertility. 2. The reinforcement of a particular behavior due to the presence of other individuals of the same species. For example, increased numbers of foraging birds in a flock may lead to increased rates of discovery of new food sources or, by improving the chances of detecting predators, may increase the time available to each individual for feeding.

factorial experiment   An experiment in which the material is divided into a number of groups, such that every possible combination of treatments can be tested separately on at least one group.

facultative   Describing an organism that can utilize certain conditions but is not dependent on them; i.e. an organism that can adopt an alternative mode of living. For example, a *facultative anaerobe* is an organism that can grow under anaerobic conditions but is also able to survive in aerobic conditions. A *facultative annual* may complete its life cycle in a single year but may contain a small proportion of individuals that do not breed until their second summer, while the rest of the population breeds in the first summer. *Compare* obligate.

facultative mutualism   *See* mutualism.

fairy ring   A ring of dark green grass caused by the presence of the mycelia and fruiting bodies of certain FUNGI. The mycelium grows outward from the center, depleting the soil as it grows. When the mycelium dies, it rots and releases nutrients into the soil, so the vegetation behind it may show increased growth and a darker green color. Rings may continue growing for hundreds of years, and can reach a diameter of over 200 m.

faithful   *See* constant.

fall bloom   A BLOOM of algae that occurs in the fall (autumn) when a decrease of air temperature and wind activity stirs up the water of a lake, bringing up nutrients from deeper waters where they have been trapped all summer by STRATIFICATION. This vertical mixing of layers of water is called *fall overturn*.

fallout   Particulate matter that is transported in the atmosphere by turbulence, but eventually reaches the land surface in rain or settles out as dust. The term is used especially for radioactive material released by nuclear explosions.

family   In classification, a TAXON that comprises a collection of similar genera. Families may be subdivided into subfamilies, tribes, and subtribes. Plant family names generally end in *aceae* whereas animal family names usually end in *idae*. Similar families are grouped into orders. *See* taxonomy.

Farman, Joe   (1930–) British atmospheric chemist. Farman had been engaged in the study of the Antarctic atmosphere since 1957. In 1982, investigating atmospheric ozone, he first noted a very low reading in the ozone level. Initially he suspected that his readings were caused by instrumental or other errors, but, when similar low levels were recorded in the following season, Farman investigated further. An examination of past records revealed that there had been a decline in ozone levels since 1977. By the end of the 1984 season Farman was convinced that he had detected experimental evidence for a persistent seasonal fall in Antarctic ozone levels of about 40%. Farman was aware of the work of ROWLAND and began to suspect that ozone depletion was connected with atmospheric CFCs. He published his results in May 1985.

fast breeder reactor   A nuclear reactor that uses fast neutrons to convert uranium to plutonium. The process is up to 120 times more efficient than energy generation by other types of nuclear reactor, but it produces excess plutonium, which can be recycled, and RADIOACTIVE WASTE. *See* nuclear energy.

fats   Triglycerides of long-chain carboxylic acids (fatty acids) that are solid below 20°C. They commonly serve as energy storage material in higher animals and some plants. *See also* lipid.

fatty acid   *See* lipid.

fault   A fracture in a body of rock along which there has been some displacement of one part relative to the other. Faults may bring one rock type adjacent to another, causing changes in soil type, fauna, and flora. Where a permeable rock is brought up against an impermeable one, a spring line may result. Movements along fault lines are common causes of EARTHQUAKES. The stresses that cause such movements are often related to tectonic movements. *See* plate tectonics.

fauna   The animal life of a region or a geological period. *See* biogeographical realms.

fecundity   The capacity of an organism to produce offspring. Many organisms show such enormous fecundity that the size of their population would rapidly increase if all of them survived. In practice most offspring do not survive. *See also* natural selection; fertility.

fecundity table   A table showing the number of offspring (for example, the number of bird eggs) produced by individuals of different ages in a population. Such tables are used to compare fecundity from year to year, or in different seasons, as well as showing the effects of maturity or aging on fecundity.

feedback mechanism   A mechanism in which the product of a process affects the rate of the process. For example, an increase in the number of herbivores in an ECOSYSTEM leads to overgrazing and reduction of the herbivore population by starvation or increased susceptibility to disease, which allows the plants to grow back, and so on – the population of both species fluctuates about a mean. This is an example of *negative feedback*, opposing the change in the system. *Positive feedback* has the opposite effect, reinforcing a change in the system.

fell   An open mountainside with short vegetation.

female choice   A form of SEXUAL SELECTION in which the female chooses her mate. It may lead to exaggerated male characteristics, such as the peacock's tail, with females choosing the male with the most impressive display. It is thought that since the growth of such features uses a lot of energy, this is a way of selection for FITNESS. In such species, the males are usually more showy than the females.

fen   An area of wet PEAT that has resulted from the silting up of an open stretch of water. Fen is a subclimax community in a HYDROSERE, prevented from developing to the CLIMAX COMMUNITY by the wet conditions, which prevent the growth of trees. *See* sere; succession.

feral Describing formerly domesticated plants or animals that have escaped into the wild, or their descendants.

fermentation The breakdown of organic substances, particularly carbohydrates, under anaerobic conditions. It is a form of anaerobic respiration and is seen in certain bacteria and in yeasts. The incompletely oxidized products of alcoholic fermentation – ethanol and carbon dioxide – are important in the brewing and baking industries.

ferns *See* Filicinophyta.

fertility 1. The degree to which a SOIL possesses sufficient quantities of the elements necessary for vigorous plant growth. *See* macronutrient; micronutrient. 2. The number of fertilized eggs (ova) produced by an organism. In plants fertility is also measured by the number of seeds produced.

fertilization The fusion of a male GAMETE with a female gamete to form a ZYGOTE, the essential process of SEXUAL REPRODUCTION. *External fertilization* occurs when gametes are expelled into a watery substrate from the parental bodies before fusion; it is typical of aquatic animals and spore-bearing plants. *Internal fertilization* takes place within the body of the female and complex mechanisms exist to place the male gametes into position. Internal fertilization is an adaptation to life in a terrestrial environment, although it is retained in secondarily aquatic organisms, such as pondweeds or sea turtles.

Internal fertilization allows terrestrial organisms to become independent of external water for reproduction. It also allows a considerable degree of nutrition and protection of the early embryo, which is seen in both mammals and seed plants.

fertilizer A substance applied to land to increase soil fertility and hence plant growth. Fertilizers include 'organic' products such as MANURE, COMPOST, bone meal, fishmeal, and GUANO, and inorganic chemicals such as nitrates, phosphates, and potash.

fetch The length of water surface over which wind blows, causing waves. The length of fetch dictates the degree of wave activity.

field capacity The point at which the soil contains all the water it can hold by capillary and chemical attraction. Any more water added to soil at field capacity would drain away by gravity. Soil under these conditions has a high water potential (*see* osmosis). Water content at field capacity is usually expressed as a percentage of the weight or volume of oven-dry soil. This is affected by the SOIL TEXTURE, CLAY SOIL having a high field capacity and SAND a low field capacity. *See* capillarity.

field layer The herb and shrub layer of a plant community. *Compare* canopy; ground layer.

field work (field studies) Work carried out in a particular habitat, i.e. 'IN THE FIELD.'

FIFRA The Federal Insecticide, Fungicide and Rodenticide Act, which requires certain pesticide ingredients to be registered, and regulates the manufacture, usage, and labeling of pesticides in the United States.

Filicinophyta (ferns) The largest phylum of spore-bearing VASCULAR PLANTS, comprising some 12 000 species of ferns. There are many fossil genera, some dating back to the DEVONIAN period. Ferns have large spirally arranged leaflike fronds bearing spores on their margins or undersurfaces. The spores give rise to heart-shaped haploid prothalli, delicate plates of cells that bear archegonia and antheridia. Motile male gametes from the antheridia fertilize the ova in the archegonia. The fertilized ovum develops into the diploid fern plant while still embedded in the prothallus, deriving nutriment from it for a while. Ferns do not show secondary thickening, but in large ferns such as tree ferns bands of sclerenchyma (tissue containing fibers reinforced with lignin) and the overlapping bases of the fronds help support the plant. Many ferns spread by means of rhizomes. Most grow on the ground, but many species are epiphytes, especially in the tropics. *See also* alternation of generations.

filter feeding The method of feeding of some aquatic animals, especially invertebrates, in which small suspended food particles are strained from the surrounding water. Many animals simply allow water to flow over or through them, but others actively produce a current, often by means of cilia. The type of filter used varies from minute hairlike cilia to the fringes of bristles on the legs of shrimps, the sticky nets of radiolarians, the serrated bills of flamingoes, and the large horny baleen plates of certain whales. Small filter feeders are DETRITIVORES, whereas the larger ones are CARNIVORES. *See* ciliates.

filter strip An area of vegetation used to remove organic matter, pollutants, or sediments from runoff and waste water. *See* bioremediation.

filtration The removal of suspended particles from a fluid by passing or forcing it through a porous material (the *filter*). The fluid that passes through the filter is the *filtrate*. In the treatment of water, sand or filters are used to remove solid particles that may contain pathogens.

fire ant *Solenopsis invicta*, a pest introduced to the southern United States. They are small red or yellowish ants that have an extremely painful sting. They cause damage to grain crops and attack poultry.

fish *See* Pisces.

fission 1. *See* asexual reproduction. 2. *See* nuclear energy.

fitness In an evolutionary context, the ability of an organism to produce a large number of offspring that survive to a reproductive age. This is Darwinian fitness, a measure of evolutionary success – the ability to pass on genes to the next generation. POPULATION GENETICS is concerned with *relative fitness* – the relative contribution of the individual organism to the GENE POOL of the next generation. This may be expressed as a *fitness set* – a graph plotting the different phenotypes in a population and their values for various components of fitness, such as reproductive rate or survivorship. The sum of an organism's Darwinian fitness and that of its relatives is called *inclusive fitness*. Fitness is thus a measure of the degree to which an organism has successfully adapted to its environment. The phrase '*survival of the fittest*' summarizes the principles of the theory of natural selection. *See* Darwinism; natural selection; population genetics.

Five Kingdoms classification A classification system that recognizes five kingdoms containing at least 96 phyla. It comprises two superkingdoms, the Prokarya (PROKARYOTES) and Eukarya (EUKARYOTES). The Prokarya contains a single kingdom, the BACTERIA, which is divided into two subkingdoms, the ARCHAEA and the EUBACTERIA. The Eukarya contains four kingdoms, the ANIMALIA, PLANTAE, FUNGI, and PROTOCTISTA.

The Protoctista is not a natural phylogenetic grouping, but contains those organisms that do not fit into the other four kingdoms, including numerous unicellular groups (e.g. algae, oomycetes, and slime molds). The recognition of a single superkingdom, the Bacteria, is disputed, and most current classification systems recognize three DOMAINS (taxa above the rank of kingdom): the Bacteria, Archaea, and Eukarya.

fixation The incorporation of inorganic materials such as carbon and nitrogen into living tissues by living organisms. An example is the incorporation of carbon from carbon dioxide into new organic compounds during PHOTOSYNTHESIS. *See also* nitrogen fixation. *Compare* assimilation.

fixed dune *See* dune.

fixed quota A harvesting strategy in

which a fixed number of individuals are removed from a population during a given period of time. The balance between the fixed quota and the RECRUITMENT rate determines whether such harvesting is sustainable.

flagellate An organism that uses UNDULIPODIA for locomotion. Several phyla of the PROTOCTISTA contain flagellates, especially the Zoomastigota.

flammable Describing any material that ignites easily and burns rapidly.

flash point The lowest temperature at which a flammable liquid produces sufficient vapor to form an ignitable mixture with air.

flavonoids A common group of plant compounds and an important source of nonphotosynthetic pigments in plants, which include the yellow and ivory flavones and flavonols and the red, blue, and purple anthocyanins. They are widely used in CHEMOTAXONOMY to distinguish between certain plant groups. Some flavonoids have a defensive role as PHYTOALEXINS.

floc *See* coagulation.

flocculation 1. The aggregation of soil particles into crumbs. Compacted structureless clay soils can be flocculated by the addition of neutral salts, particularly of calcium. The addition of lime to saturated clay soils (*liming*) is a common agricultural practice, improving soil structure by encouraging crumb formation and making heavy soils more workable. *See also* soil texture.
2. *See* coagulation.

flocking In certain species of birds, social attractions between individuals that keep groups together and may lead to a recognizable social structure. For example, many species of fruit- and seed-eaters form flocks outside the breeding season to forage for food. Foraging flocks may contain several different species, e.g. the mixed

winter flocks of tits (family Paridae) of the northern hemisphere, and the antbirds (family Formicariidae) and other species that follow columns of army ants in South American rainforests. Flocking at some stage in the life cycle is found in almost half of all bird species.

flood-plain The flat or nearly flat part of a river valley that is covered by water during periodic floods. It consists mainly of coarse unconsolidated sediment left behind as the river changes course and migrates across the flood-plain, and finer sediment deposited when the river overflows its banks.

flora All the plant species in a given area, e.g. the Cape flora. The term is also used with reference to the plant species in a particular period of geological time, e.g. the Cretaceous flora.

floral province One of a number of regions into which the world is divided on the basis of plant distribution. Each province has one or more CENTERS OF DIVERSITY and each contains unique families and genera of plants. The main provinces are the *Boreal*, which covers all of the north temperate zone; the *Neotropical*, covering tropical Central and South America; the *Paleotropical*, covering tropical Africa and Asia; the *Australian*; the *Cape Province*, covering a small area around the Cape of Good Hope; and the *Antarctic*, including New Zealand and temperate South America. Oceanic islands, particularly in the Pacific, form minor provinces. There are floral affinities and connections between the provinces, particularly near their boundaries.

floristic Relating to all the plant species present in a particular area, region, or community.

floristic analysis The identification and listing of all the plant species present in a particular area. The resulting check list gives the *floristic composition* of the region or community. *See also* phytosociology.

flotation The separation of solids and liquids according to their specific gravities. Flotation is used to separate metal wastes in suspension from the liquid phase of SEWAGE and industrial effluents. Metal hydroxides and other suspended particles are subjected to streams of gas bubbles, which adhere to the particles, causing them to float to the surface, where they can be skimmed off.

flowering plants *See* Angiospermophyta.

flowmeter An instrument for measuring the velocity of a moving fluid (liquid or gas).

flow web *See* food web.

flue gas The mixture of gases leaving a chimney venting a combustion chamber. It is a mixture of water vapor, oxides of carbon, nitrogen and sulfur, and other POLLUTANTS. *See also* fly ash.

fluidized Describing a mass of solid particles that has been injected with water or gas so that it flows like a liquid.

fluidized bed incinerator An incinerator in which a hot granular material such as sand is FLUIDIZED by the injection of waste, to which it transfers its heat. Such incinerators are used to destroy municipal SEWAGE SLUDGE. *See* incineration.

flume 1. A channel that diverts water. 2. An experimental channel used to study the relationships between flow conditions and the movements of sediments, for example to determine the erosive force of the water.

fluoridation The addition of small quantities of FLUORIDE compounds to drinking water in order to reduce the incidence of tooth decay in children. *See also* fluoride.

fluoride A compound containing fluorine. Inorganic fluorides contain the fluoride ion, F⁻. Excessive amounts of fluorides

in food or drink can lead to *fluorosis* (mottling and softening of teeth). Some toothpastes contain fluoride to help reduce tooth decay. *See also* fluoridation.

fluorosis *See* fluoride.

fluvial Relating to rivers or streams and their actions.

flux The rate of flow of mass, volume, or energy expressed per unit cross-section perpendicular to the direction of flow.

fly ash The noncombustible particles carried by flue gases, which settle out as a fine ash.

fog A suspension of fine droplets of liquid in a gas, e.g. water in air, which reduces visibility to below 1 km. Fog may form as a result of the cooling of a body of air due to contact with cold ground or cold water, or where a body of warm air meets a cold front. Fog formation is accelerated by the presence of *condensation nuclei* such as smoke particles, which may lead to fog formation at humidities lower than the condensation point.

fogging The application of a pesticide in the form of a FOG generated by rapidly heating the liquid chemical. Fogging is commonly used to destroy such insects as mosquitoes and blackfly.

folivore An animal that feeds mainly on foliage, for example, a tapir (*Tapirus*).

Food and Drug Administration (FDA) The branch of the US federal government responsible for enforcing the Food, Drug, and Cosmetic Act and related consumer safety issues.

food begging Behavior that solicits food from a parent or other adult, or from a mate. For example, the bright orange or yellow lining of the beaks of some juvenile birds, especially passerines, is exposed when the birds open their beaks wide to beg, and the sight of this induces the adult to feed them.

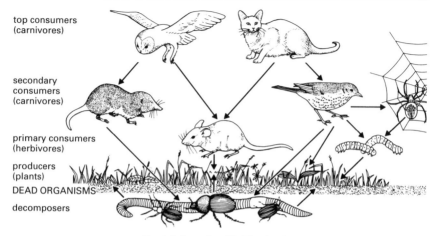

top consumers
(carnivores)

secondary
consumers
(carnivores)

primary consumers
(herbivores)

producers
(plants)

DEAD ORGANISMS

decomposers

Food chain: a simplified food web

food chain The chain of organisms existing in any natural COMMUNITY, through which food energy is transferred. Each link in the food chain obtains nutrients and energy by eating the one preceding it and is in turn eaten by the organisms in the link following it. At each transfer a large proportion (80–90%) of the potential energy is lost as heat, and a smaller proportion as mechanical energy (e.g. for locomotion) and, at higher levels of the food chain, as electrical energy for nerve transmission, for example. For these reasons the number of links in a sequence is limited, usually to four or five. The shorter the food chain, the greater the available energy, so total energy can be increased by cutting out a step in the food chain, for example if people were to consume cereal grains instead of consuming animals that eat cereal grains.

Food chains are of two basic types: the *grazing* or *herbivore-based food chain*, which goes from green plants to grazing herbivores and finally to carnivores; and the *detritus-based food chain*, which goes from dead organic matter to microorganisms and then to detritus-feeding organisms. The food chains in a community are interconnected, because most organisms consume more than one type of food, and the interlocking pattern is referred to as a FOOD WEB.

In most food chains and food webs, the greatest diversity of PRODUCERS occurs when population regulation by CONSUMERS is at its most intense. This is termed a *trophic cascade*. For example, it occurs in a lake in summer, when the pressures of high numbers of predatory zooplankton lead to high diversity in the phytoplankton (*see* plankton). Trophic cascades occur typically where the producers are relatively short-lived and fast-growing, so they can respond to the impact of the consumers. Such food webs show *top-down control*, where productivity is limited by the rate at which nutrients can be recycled, and hence by the number and activity of consumers. *See also* bottom-up control; ecological efficiency; trophic level; biomanipulation.

food web The feeding relationships by which energy and nutrients are transferred from one species to another (*see* food chain). Food webs whose links emphasize feeding relationships between organisms are sometimes called *topological food webs* or *static food webs*. They show who eats what, but give no information about the strength of the interaction – the amount of nutrients or energy taken in at each TROPHIC LEVEL. Food webs that show the energy flux between trophic levels are called *energy flow webs*, *flow webs*, or *bioenergetic webs*. A final category of food web, the *functional food web* or *interaction food web*, shows the importance of each population in the community in influencing the growth rates of the other populations. Experimentation is needed to determine the structure of functional food webs.

Species in a food web that have a similar trophic habit, i.e. have similar predators and prey or other food, are termed *trophospecies*. They form *nodes* in the food web from which *links* (which represent interactions) to other species radiate. The number of links between the *basal species* (the producers or, in the case of detritus-based food webs, the primary consumers) and the TOP PREDATOR is called the *chain length*. The average number of links per species is called the *linkage density* of the food web. The total number of links in the food web divided by the number of possible links is its *connectance* or *connectedness*.

foot-and-mouth disease A disease of livestock caused by a wind-borne virus, causing fever and characteristic eruptions on the mucous membranes and skin, especially in the mouth and in the clefts of the hoofs. In countries where it is endemic, it is controlled by vaccination, but in countries subject to sporadic outbreaks it is controlled by slaughter of affected animals and adjacent herds.

foraging Behavior associated with searching for, collecting or capturing, and consuming food.

foraging strategy FORAGING behavior that has evolved to enable an organism to achieve a high rate of net energy intake, taking into account the energy costs of obtaining that energy/food. The *optimal foraging theory* proposes that foraging strategy involves decisions that maximize the net energy/food intake or some other nutritional goal such as the acquisition of essential elements, and uses mathematical models to attempt to predict foraging behavior. It aims to explain why organisms do not always make use of the full range of their diet, preferring to select certain items above others. Such choices are likely to be related to competition for resources in

their present environment, but may also be a result of NATURAL SELECTION in the past.

Foraminifera (forams) A class of PROTOCTISTA that contains small marine organisms from 10 mm to several centimeters in diameter enclosed in pore-studded shells. Free-swimming planktonic forms are important components of marine FOOD CHAINS. Many other foraminifera live on the seabed in mud or sand or attached to rocks, algae, or other organisms. Cytoplasmic projections protrude through the pores of the shell, forming nets to trap food. Forams feed on small particles of dead or living organic matter. Their shells sink to the seabed at death, forming an important component of SEDIMENTARY ROCKS such as LIMESTONES. The delicately sculpted shells are preserved as FOSSILS (microfossils), and are used to distinguish different rock strata of the same age, and are especially important in exploration for petroleum deposits.

forbs Herbaceous plants that are not grasses (Poaceae), sedges (Cyperaceae), or rushes (Juncaceae).

forcing A horticultural practice that causes a plant to flower or fruit earlier than it would under normal environmental conditions, for example, by growing it in a greenhouse or by subjecting it to an artificial daylength.

foredune *See* dune.

foreshore The part of the shore that lies between the normal high- and low-water marks.

forest 1. A plant COMMUNITY dominated by trees whose crowns touch, forming a continuous CANOPY. *Compare* woodland. 2. A major plant community (biome) in which the dominant plants are trees. *See* biome; boreal forest; temperate deciduous forest; tropical forest.

forestry 1. The study of trees and timber production systems. 2. The growing and managing of trees for commercial timber production. This includes the commercial management of existing forests as well as the planting of native or exotic species, and the selective breeding or other genetic modification of trees for specific uses.

form   The lowest taxonomic group, ranking below the variety, normally referring to the recognizably different morphology found in several members within a plant species. *See* taxonomy.

formalin   A mixture of about 40% formaldehyde, 8% methyl alcohol, and 52% water (the methyl alcohol is present to prevent polymerization of the formaldehyde). Formalin is a powerful reducing agent and is used as a disinfectant, germicide, and fungicide and also as a general preserving solution, especially for biological specimens.

formation   A unit of classification of vegetation that is based on the form and structure of the plant COMMUNITY rather than its species composition. Different classification schemes define formations in different ways. The term *formation type* is generally used to refer to a world vegetation type (BIOME) of relatively uniform appearance and life forms, e.g. tropical rainforest or savanna. A formation type is equivalent to a CLIMATIC CLIMAX vegetation type. *See* life form.

fossil   The remains of, or impressions left in rocks by, long dead animals and plants. The fossil assemblage in a particular stratum gives an indication of the environmental conditions that prevailed at that time.

fossil fuel   Fuel derived from ancient organic remains, e.g. COAL, crude oil (*see* petroleum), NATURAL GAS, and PEAT. These fuels have been formed by the decomposition or partial alteration of organic remains under pressure, usually in anaerobic conditions. Fossil fuels are a nonrenewable resource and burning such fuels releases carbon dioxide that was fixed from the atmosphere millions of years ago, contributing to the GREENHOUSE EFFECT and GLOBAL WARMING. They are, however, the main source of energy used in modern society.

fossil water   Water contained in an underground AQUIFER that was stored there in the distant past and is no longer being replenished from present-day PRECIPITATION. This water is a nonrenewable RESOURCE, and overexploitation will lead to the depletion of the aquifer and, in an *unconfined aquifer*, to the retreat of the WATER TABLE to greater depths, which may have a serious effect on local agriculture and human settlements.

fragmentation   1. *See* asexual reproduction.
2. The separation through human activity of once continuous habitats into fragments that may or may not contain viable populations of the organisms present. *See* habitat islands; island biogeography theory.

Framework Convention on Climate Change   (Global Warming Convention) An international treaty agreed at the Earth Summit in Rio de Janeiro, Brazil, in 1992, that requires signatory nations to reduce their emissions of GREENHOUSE GASES, especially carbon dioxide and methane, believed to be responsible for GLOBAL WARMING.

free radical   *See* radical.

frequency   The number of individuals of a particular species in a given area. *See* quadrat.

frequency distribution   The pattern of distribution of the values of a VARIABLE, i.e. the number of individuals or events with each value. This is usually plotted as a GRAPH or histogram, with number on one axis and value on the other.

frequency table   A table showing the number of times a value appears in a data set. The table may simply show the values and the number of occurrences, or it may show values and frequency and/or RELATIVE FREQUENCY. *See* frequency; tally chart; variable.

fringing forest *See* riverine forest.

fringing reef A coral reef that forms along a coast rather than at some distance from the shore (*see* corals).

frogs *See* Amphibia.

front A boundary between two distinct masses of air. Often the air masses differ in temperature and/or moisture content, and are moving in different directions. Fronts have important effects on weather. Cold air is denser than warm air, so cold fronts tend to push under warm fronts where the two meet, creating instability in the air masses, often giving rise to thunderstorms and causing a change in wind direction. The warm air is cooled, resulting in cloud formation and PRECIPITATION.

frost The formation of ice crystals on solid objects, such as grass or window panes, when the air temperature falls below the freezing point of water (0°C) and its dew-point temperature is below 0°C. Water vapor condenses out of the atmosphere as ice. *Hoar frost*, by contrast, forms where wet surfaces, e.g. grass on which DEW has formed, is cooled by heat loss due to RADIATION, so the dew freezes. *See* condensation.

frost-free period For any particular species that is susceptible to certain low temperatures, the period between the last FROST of summer and the first frost of autumn. Frost in this context is usually taken to mean a temperature below –2°C, but the actual degree of frost required to prevent growth varies with the species. The frost-free period represents the maximum length of the growing season. The term is also used in relation to the survival of insect pests, for many of which frost is lethal to the active stage. In this context, the relevant degree of frost is that needed to kill the insect.

frost hardiness *See* cold tolerance.

frugivore An animal that feeds mainly on fruit.

fruit *See* Angiospermophyta.

fruticose *See* lichens.

fucoids Brown ALGAE (phylum Phaeophyta) of the order Fucales, which include many common medium to large brown seaweeds, such as the wracks (*Fucus* spp.). These are flattened, often branched, seaweeds attached to rocks by means of holdfasts at their base.

fuel cell A device in which an electrochemical reaction occurs between a fuel (e.g. hydrogen) and oxygen to produce electrical energy directly.

fugitive emission A gas, vapor, mist, fog, fume, liquid, or solid that escapes from a process, product, or piece of equipment and is not trapped by a system designed to capture potential POLLUTANTS.

fugitive species An opportunist SPECIES that is capable of rapid colonization of a newly exposed area, but unable to survive competition from later arrivals. Such species are typical of unstable environments such as deserts and temporary ponds. They usually have short life cycles and good long-distance dispersal mechanisms. *See also* ephemeral.

fumigant *See* fumigation.

fumigation The killing of pests using fumes of volatile pesticides (*fumigants*), e.g. carbon disulphide and formaldehyde. This is an especially useful method in greenhouses, buildings, soil, or compost.

functional response A change in the rate of predation by an individual predator in response to a change in the density of its prey. *Compare* numerical response.

fundamental net reproductive rate *See* reproductive rate.

fundamental niche The ecological NICHE that an organism could potentially occupy if it were not limited by environmental constraints or competition from other species. *See* realized niche.

Fungi A kingdom of nonphotosynthetic mainly terrestrial organisms that lack CHLOROPHYLL and are quite distinct from plants and animals. They are characterized by having cell walls made chiefly of CHITIN, not the CELLULOSE of plant cell walls, and they all develop directly from SPORES without an embryo stage. UNDULIPODIA are never found in any stage of their life cycles. Fungi are generally saprophytic (*see* saprobe) or parasitic (*see* parasitism). They may be unicellular or composed of filaments (termed *hyphae*) that together comprise the fungal body or *mycelium*. Hyphae may grow loosely or form a compacted mass, giving well-defined structures, as in mushrooms. The phylum includes mushrooms, toadstools, yeasts, molds, and mildews.

Fungi are major DECOMPOSERS in soil and fresh water. Some form symbiotic associations with algae or CYANOBACTERIA to form LICHENS, while others form extremely important MYCORRHIZAE with the roots of many plants, including most forest trees. Some parasitic fungi cause diseases in plants and animals, whereas others are sources of ANTIBIOTICS. Yeasts and other fungi are used in commercial FERMENTATION processes such as brewing. Some fungi are pathogens; others stabilize sewage and digest composted waste. *See* symbiosis.

fungicide A chemical that inhibits fungal growth. Fungicides are used to prevent fungal infection as well as to deal with existing infections (chemicals called *fungistats* prevent fungi from growing). There are two main categories: inorganic chemicals such as copper-based fungicides and sulfur dust (which are surface treatments), and organic chemicals such as dithiocarbamate, many of which are SYSTEMIC.

fungistat *See* fungicide.

fusion *See* nuclear energy.

fynbos *See* maquis.

G

Gaia hypothesis   The theory that the Earth functions as a single homeostatic system analogous to a single living organism. The activities of organisms in response to changes in their physical and chemical environment in turn modify the physical and geochemical cycles, including climate, so that the planet continues to support life. The Gaia hypothesis was proposed by the British scientist James LOVELOCK.

gallery forest   *See* riverine forest.

gamete   A cell capable of fusing with another cell to produce a ZYGOTE, from which a new individual organism can develop. Gametes are usually HAPLOID, so fusion of gametes results in the nucleus of the zygote having the DIPLOID number of chromosomes. The typical male gamete is small, motile (by means of undulipodia), and produced in large numbers. The typical female gamete is large because of the food reserves it contains, and is produced in smaller numbers than the male gametes. In all except the most primitive organisms, the female gamete is nonmotile. *See also* sexual reproduction.

game theory   A method of analyzing mathematically social interactions and decision-making between individuals in terms of gains and losses among opposing players. It involves calculating the probabilities and values of the outcomes of various choices made by the decision-makers.

gametophyte   *See* alternation of generations.

gamma diversity   *See* species diversity.

gamma radiation   A form of high-energy electromagnetic radiation emitted by changes in the nuclei of atoms. Gamma rays have very high frequency, about 10^{-10} to 10^{-14} m in wavelength, shorter than x-rays. Gamma rays penetrate tissues to a greater degree than other forms of radiation. *See* radioactive decay.

gap analysis   A method of analyzing species distribution in order to detect which ecosystems are most in need of protection. The ranges of rare and endangered species are mapped; the maps are laid on top of each other and overlaid by a map of existing reserves and protected areas. This reveals gaps that represent areas in need of greater protection.

garrigue   A form of Mediterranean scrub woodland that has developed as a result of burning and grazing of the original broad-leaved evergreen forest of former times. It is found in limestone areas on thin dry soils. It is dominated by aromatic herbs and prickly dwarf shrubs, including species such as lavender (*Lavandula vera*), sage (*Salvia officinalis*), thyme (*Thymus*), and mints (*Mentha*). It is an impoverished flora compared with the MAQUIS that grows on better soils, which contains taller shrubs and scattered small trees such as olives (*Olea europaea*) and figs (*Ficus*).

gasification   The generation of fuel gas from solid material. Examples are producing gas from coal and the production of gas containing carbon dioxide, hydrogen, and methane from SEWAGE SLUDGE.

gasohol   Ethyl alcohol (ethanol) used as a substitute for gasoline or as an additive to gasoline. It can be produced by fermentation of grain, potatoes, sugar-cane residue, etc.

gasoline *See* petroleum.

gauging station *See* water-flow measurements.

Geiger counter An instrument for detecting the presence and quantity of IONIZING RADIATION. It is less sensitive than a scintillation counter.

gene A unit of hereditary material located on a chromosome that, by itself or with other genes, determines a characteristic in an organism. It corresponds to a segment of the genetic material, usually DNA (the genes of some viruses consist of RNA). Genes may exist in a number of forms, termed ALLELES. For example, a gene controlling the characteristic height in peas may have two alleles, one for 'tall' and another for 'short'. In a normal diploid cell only two alleles can be present together, one on each of a pair of HOMOLOGOUS CHROMOSOMES: the alleles may both be of the same type, or they may be different. Genes can occasionally undergo changes, called MUTATIONS, to new allelic forms.

A gene may control the synthesis of a single POLYPEPTIDE or a RNA molecule whose GENETIC CODE is contained within the gene. There are three main types of genes: *structural genes* code for the polypeptide that make up enzymes and other proteins, *RNA genes* code for ribosomal RNA and transfer RNA used in the assembly of polypeptides, and *regulator genes* regulate the expression of other genes. *See* mutation.

gene bank 1. A place where plant material is stored in a viable condition in order to preserve plants that are in danger of becoming extinct in the wild or cultivars that are being lost from cultivation. The material also provides a source of genes for breeding new varieties. Seeds are usually stored at –20°C after being dehydrated to about 4% of their normal water content. They may remain viable for up to 20 years, often much more, depending on the

species, after which they must be germinated and a new crop of seeds harvested. For species that do not store well, material may be kept in tissue culture. Pollen may also be stored, but has a shorter lifespan. **2.** *See* gene library.

gene center *See* center of diversity.

genecology The study of the genetics of populations in relation to their environment.

gene flow The movement of ALLELES within and between populations of the same species through interbreeding. *See also* GENETIC DRIFT.

gene-for-gene effects The evolution of virulence and RESISTANCE of pests and crops, respectively, on a gene-by-gene basis. For example, the pest may have several genes involved in virulence for a particular crop. The crop in turn may have several different GENES involved in resistance, certain alleles of each conferring resistance against particular pest virulence alleles.

gene frequency The proportion of an ALLELE in a population in relation to other alleles of the same GENE.

gene library A collection of cloned DNA fragments derived from the entire GENOME of an organism. Such fragments are usually stored in plasmids inside a suitable host, such as the bacterium *Escherichia coli*. *See also* genomic library.

gene pool The total number and variety of GENES existing within a breeding population or species at a given point in time.

generalist A species with wide food or HABITAT preferences. *Compare* specialist.

generation (cohort generation time) The average interval between the birth of parents and the birth of their offspring.

generation length (T) The average age of parents of the current COHORT (i.e. newborns). Generation length is related to the TURNOVER rate of breeding individuals in a population. Since offspring may give birth to their own offspring during the life of their own parents, generation length so defined is less than the *cohort generation time*.

generation time 1. The mean time between new generations of individuals. 2. In unicellular organisms, the interval between successive cell divisions.

gene sequencing Determination of the order of BASES (def. 2) making up a GENE on a DNA molecule. Sequencing requires multiple cloned copies of the gene (*see* gene library). Long DNA sequences are cut into more manageable lengths using restriction enzymes. These cleave DNA at specific base sequences, and it is possible to reconstitute the overall sequence once the constituent fragments have been analyzed individually.

genet The organism developed from a ZYGOTE, or a CLONE of it. *Compare* ramet. *See also* modular organism.

genetically modified crops (GM crops) Crops that have been modified by GENETIC ENGINEERING.

genetic code The sequence of bases in either DNA or RNA (especially messenger RNA) that conveys genetic instructions to the cell. The basic unit of the code consists of a group of three consecutive bases, the base triplet or *codon*, which specifies instructions for a particular AMINO ACID in a POLYPEPTIDE, or acts as a start or stop signal for translation of the message into a polypeptide assembly. For example, the DNA triplet CAA (which is transcribed as GUU in mRNA) codes for the amino acid valine. There are 64 different triplet combinations but only 20 amino acids; thus many amino acids can be coded for by two or more triplets. Certain triplets code not for an amino acid, but for the beginning and end of a polypeptide chain.

genetic drift The random fluctuation of ALLELE frequencies in a small population due entirely to chance. If the number of matings is small, then the gene frequencies of the offspring may not be a perfect representation of the gene frequencies of the parental generation. *See also* bottleneck; fixation.

genetic engineering The direct introduction of foreign GENES (from other individuals or artificially synthesized) into an organism's genetic material by micromanipulation at the cell level. Genetic engineering techniques by-pass crossbreeding barriers between species to enable gene transfer between widely differing organisms. The commonest method of genetic engineering is RECOMBINANT DNA TECHNOLOGY.

Genes may be taken from living tissues by isolating the messenger RNA produced when the gene is active, then recreating the DNA sequence of the gene using the enzyme reverse transcriptase. Gene transfer is achieved by various methods, many of which use a replicating infective agent, such as a VIRUS, PLASMID, phage (*see* bacteriophage), cosmid, or YAC (yeast artificial chromosome). Other methods include microinjection of DNA into cell nuclei and direct uptake of DNA through the cell membrane.

Modified microorganisms are grown in large culture vessels and the gene product harvested from the culture medium. TRANSGENIC mammals, including mice, sheep, and pigs, have been produced by microinjection of genes into the early embryo, and it is also now possible to clone certain mammals from adult body cells (see CLONING). Dicotyledonous plants, including tobacco and potato, have been transfected using the natural plasmid vector of the soil bacterium *Agrobacterium tumefaciens*. Genes have been introduced to crop plants for various reasons, for instance to reduce damage during harvest or to make them resistant to the herbicides used in controlling weeds. Genetically modified tomatoes and soya beans are now widely available. Human proteins have been successfully transferred to bacteria, and a wide range of therapeutic substances is produced commercially from genetically engineered bacteria. *Pharming* (the production of pharmaceuticals by genetically engineered animals) is increasing, with pharmaceuticals and products such as clotting factors being produced in the milk of sheep or the eggs of chickens. There is hope that in future many genetic diseases will be treatable by manipulating the faulty genes responsible. However, genetic engineering raises many legal and ethical issues, and the introduction of genetically modified organisms into the environment requires strict controls and monitoring.

genetic erosion The loss of genetic VARIATION and shrinking of the GENE POOL of wild plants or old CULTIVARS when new, improved cultivars are grown over wide areas. This can pose problems for the future, for instance if specific diseases arise that target the new cultivars, or if climate change requires drawing on a wider gene pool to develop better-adapted cultivars. It is particularly serious in CENTERS OF DIVERSITY. *See* genetic resources.

genetic fingerprinting *See* DNA.

genetic resources The GENE POOL in natural and cultivated stocks of organisms. The term is used especially in relation to domesticated plants and animals and their wild relatives. A large gene pool increases the opportunity for developing new varieties and the capacity to respond to environmental change and develop resistance to pathogens. Attempts to preserve genetic resources include the establishment of GENE BANKS and GENE LIBRARIES. *Compare* genetic erosion.

genetics The study of inheritance and variation and the factors controlling them.

genetic structure The nature of the genetic VARIATION among the individuals of a POPULATION or COMMUNITY.

genome The set or sets of chromosomes carried by each cell of an organism. Haploid organisms have one set of chromosomes, diploid organisms have two sets,

polyploid organisms have many sets sometimes from the same ancestor (autopolyploids) and sometimes from different ancestors (allopolyploids).

genomic library A GENE LIBRARY containing entire genomes of particular organisms. Such libraries are important resources for comparing the nucleic acid sequences of different organisms, recognizing the genes in newly sequenced organisms, and in studies of gene and protein function.

genotype The genetic makeup of an organism. The actual appearance of an individual (the PHENOTYPE) depends on the dominance relationships between alleles in the genotype and the interaction between genotype and environment.

genus A taxonomic category involving a collection of similar SPECIES. Genera may be subdivided into subgenera and, especially in plant taxonomy, into sections, subsections, series, and subseries. The scientific name of a species always includes the genus name (or its abbreviation) as the first word of the binomial. Similar genera are grouped into families. *See also* classification.

Geographical Information System

(GIS) A system of computer hardware and software that allows various kinds of data to be mapped to a standard survey map showing geographic position, often overlaying several layers of information into one image. For example, data may include vegetation type, topography, geology, soil type, water quality, pollution levels, population density, land use, or land subject to planning controls. Individual observations can be spatially referenced in relation to each other. GIS is especially useful to planners and conservationists and is used in the modeling of processes and for environmental risk assessment.

geographical isolation The separation of two populations of the same or closely related species by a physical barrier such as a large river or mountain range. Random

GENETIC DRIFT and different SELECTION PRESSURES in the two populations may lead to the gene pool diverging, perhaps to morphological differences and sexual incompatibility. *Compare* ecological isolation; reproductive isolation.

geological time scale A system of measuring the history of the Earth by studying the rocks of the Earth's crust. Since new rocks are generally deposited on top of existing material, those lower down are oldest, although this is often disrupted, for example by volcanic activity, tectonic movements (*see* plate tectonics), and EROSION. The strata of rock are classified according to their age, and a time scale corresponding to this can be constructed. The main divisions (*eras*) are the Paleozoic, Mesozoic, and Cenozoic. These are further subdivided into *periods* and *epochs*.

geomagnetic poles *See* geomagnetism.

geomagnetism The force of the Earth's magnetic field, which varies from the poles to the equator, and also through time. The *geomagnetic poles* do not coincide with the geographic poles, but are currently located at 73°N 100°W and 68°S 143°E. The positions of the geomagnetic poles fluctuate over both short periods of time and on a geological time scale (millions of years). The direction of the Earth's magnetic field is reflected in the direction of the magnetism of minerals that crystallize from molten lavas. Reversals of the field occur from time to time, and are used in determining the age and sequence of magnetic rocks, for example on either side of the mid-oceanic ridges.

geometric growth Population growth in which breeding is confined to a particular season, and the rate of increase in numbers is proportional to the number of individuals in the population at the start of the breeding season. *See also* geometric series.

geometric mean *See* mean.

geometric series A series of numbers in which each number is obtained by multiplying the previous one by a constant factor. The numbers of individuals in a population of a pioneer or ephemeral species tend to represent a geometric series over time, as the more successful species restrict other species to the remaining NICHES.

geomorphology The study of the landforms of the Earth's surface and the processes that have shaped them.

geophyte *See* Raunkiaer's life-form classification.

geosphere The inorganic part of the Earth, which does not contain any living organisms. It comprises the ATMOSPHERE, HYDROSPHERE, and LITHOSPHERE.

geothermal power Power generated by using the heat of rocks in the Earth's crust to heat water to drive steam turbines.

germination The first outward sign of growth of a reproductive body, such as a spore or pollen grain. The term is most commonly applied to seeds, in which germination involves the mobilization of food reserves, followed by the emergence of the first root and shoot through the seed coat.

GIS *See* Geographical Information System.

giving-up time The time after which a foraging animal leaves a patch of food to seek another one. According to OPTIMAL FORAGING THEORY an animal attempts to maximize its energy intake over the whole period of foraging, rather than just while actually feeding. The giving-up time depends on the quality of the food being eaten and the distance between patches of food. As a patch becomes depleted, more energy is expended in feeding, which becomes less cost-effective. Optimal foraging theory suggests that a forager will leave all patches at the same extraction rate (i.e. when the cost of extracting the energy from the food increases above a critical level), regardless of how rich they are.

glacial drift Material deposited on the land surface by glaciers and ice sheets.

glacial period A period during which ice covers large areas of the Earth's surface and glaciers advance or remain stationary. The term may be applied to the cold parts of an ice age, or to the entire ice age. *Compare* interglacial period.

glacial retreat The melting of glaciers, when they retreat higher up the valleys.

glaciation The formation of GLACIERS and creeping ice sheets and their effects on the land. The erosive power of glaciers, resulting mainly from the rocks and gravel that become embedded in its base and sides, scours away the surrounding rocks as they move, creating *U-shaped valleys* and *hanging valleys* (tributary valleys whose floors are well above the floor of the main valley, which has been scoured deeper by the larger glacier it contains). The creep of large ice sheets, such as those of Antarctica, levels the underlying surface, forming (after the ice sheets retreat) a landscape of rounded hills and plains, all at the same level (a *peneplain*).

glacier A large mass of ice that moves slowly across the land, usually moving down a valley or spreading outward over the land from its source. It is formed by snow in the surrounding area becoming compacted, eventually forming ice. As the weight of ice accumulates, the pressure of the overlying ice causes the ice at the bottom to melt, lubricating the creep of the glacier downhill. This creep may be permanent or temporary.

glade An open space in a forest.

gley A waterlogged SOIL lacking in oxygen, in which raw HUMUS accumulates as a result of lack of decomposition by bacteria. The defining feature is the *gley horizon* below the humus, blue-gray clay whose color is due to ferrous iron compounds that have been reduced by microorganisms, flecked with localized areas of rust-colored oxidized ferric compounds. The formation of a gley is known as *gleying*. Gley soils are typical of TUNDRA, meadows, and boggy areas.

global positioning system (GPS) A satellite-based coordinate positioning tool and navigation system that can rapidly and accurately determine the latitude, longitude, and altitude of a point on or above the Earth's surface. It is based on a constellation of 24 satellites orbiting the Earth at a very high altitude and uses a form of triangulation based on the known positions and distances of three satellites relative to the surface of the Earth. First developed by the US Department of Defense to provide the military with a state-of-the-art positioning system, GPS receivers are now small enough and economical enough to be used by the general public.

global stability The tendency for a community to return to its original state even after suffering a large perturbation. *See also* resistance. *Compare* local stability.

global warming A warming of the Earth's surface temperature, which is responsible for changes in global climate patterns. While the term may be used broadly to encompass past warming events, such as interglacials, it is more commonly used to refer to recent global warming due to the GREENHOUSE EFFECT and anthropogenic influences. The quantities of many GREENHOUSE GASES within the atmosphere are rising, especially of carbon dioxide. During the past 200 years it has been estimated that carbon dioxide levels in the atmosphere have risen by 25–30% as a result primarily of changes in land use (e.g. deforestation) and the burning of fossil fuels (e.g. coal, oil, and natural gas). In addition, levels of methane, another greenhouse gas, doubled in the 100 years up to the year 2000. Since the late 19th century the globally averaged temperature of air at the Earth's surface has risen by 0.3–0.6°C (0.5–1°F).

Complex computer models have been developed to try to predict the changes to climate that may occur as a result of increased greenhouse gas emissions and anthropogenic influences. One prediction is that if greenhouse gas emissions continue on a business-as-usual basis, the estimated average global temperature rise will be between 1° and 3.5°C (2° and 6°F) by the end of the 21st century. Some of the consequences envisaged by such warming include a rise of global sea levels of about 0.50 m (1.6 ft) during the 21st century, boundary shifts in the world's vegetation zones, extension of desertification, and a reduction in extent of polar ice and glaciers.

Global Warming Convention *See* Framework Convention on Climate Change.

global warming potential (GWP) An index that measures the relative potential of various gases to contribute to greenhouse warming, and avoids the necessity to directly calculate the changes in atmospheric concentrations. The reference gas is carbon dioxide, and the global warming potential of other gases is given as a ratio of the radiative forcing that would result from the emission of that gas over a certain time period. It is calculated as the ratio of the radiative forcing that would result from the emission of 1 kilogram of a greenhouse gas to that from emission of 1 kilogram of carbon dioxide over a fixed time period (usually 100 years).

glyphosate A broad-spectrum herbicide, soluble in water.

GM crop Genetically modified crops. *See* genetic engineering.

Gondwanaland A former *supercontinent* (very large landmass) formed about 180 million years ago by the break-up of an even larger supercontinent, Pangaea. Gondwana later broke up to form the southern land masses of South America, Australia, Antarctica, New Zealand, India, Africa, Madagascar, and some smaller islands. Many species that today have *disjunct distributions* (*see* distribution) evolved on the single landmass of Gond-

wana, e.g. the monkey-puzzle family (Araucariaceae), which is found in South America and Australia, and the Proteaceae, found in Australia and Africa. *See also* continental drift; plate tectonics.

goodness of fit The degree to which observed data fit the values predicted by a model. There are various statistical methods for measuring goodness of fit, including the CHI-SQUARED TEST for frequency data.

GPP *See* gross primary production.

GPS *See* Global Positioning System.

grab sample 1. A single sample of soil, air, or water taken at one particular place at one time.
2. A sample taken using a *grab*, i.e. a sampler with a scoop or bucket that is lowered to a stream bed. Grabs tend to disturb the sediments as they sample.

graft The transplantation of an organ or tissue in plants and animals. In plants, grafting is an important horticultural technique in which part (the *scion*) of one individual is united with another of the same or a different species (the *stock*). Usually the shoot or bud of the scion is grafted onto the lower part of the stock. Incompatibility between species is much less common in plants than in animals, where ANTIBODY reactions may cause rejection of the foreign tissue.

grain A dry indehiscent fruit, typical of the grass family (Poaceae), which includes cereal crops such as wheat, corn, and barley.

granivore An animal that feeds mainly on grain and seeds.

granular activated carbon treatment A filtering system in which water is passed through granules of activated carbon to remove organic material. It is used to treat small water systems and in domestic water and waste systems. It is also highly effective in reducing levels of radon in water.

graph A pictorial representation of the relationship between the values of two or more different variables. Usually there are two axes at right angles to each other, a horizontal x-axis, and a vertical y-axis. The x-axis usually represents an independent variable, the y-axis the dependent variable. The horizontal coordinate of a point on a graph, as measured parallel to the x-axis, is called the *abscissa*, whereas the vertical coordinate, measured parallel to the y-axis, is the *ordinate*.

Data are usually plotted on the graph as points, which may or may not be joined by a line. A simple *plotted curve*, where points are joined by curving lines, is particularly suitable where a variable shows a continuous distribution. Where the points of a frequency distribution are joined by straight lines, it is called a *frequency polygon*. If the graph shows simply scattered points, it is called a *scatter diagram*. If the points take the form of icons, it is called a *pictograph*. Alternatively, the points may be clustered around a *line of best fit*, a line that best represents the distribution of the data on the assumption that a sample of infinite size would average a linear relationship. A *contour diagram* is a representation of the topography of an area using the contours to 'build' the composition of the landscape.

Sometimes a rectangular bar may be drawn from the location centered around each value of the variable on one axis to a point parallel to the location of the corresponding variable on the other axis. The length of each bar represents the number of samples having that value. When bars are drawn horizontally from the y-axis, this is called a *bar chart*; where the bars rise vertically from the x-axis, it is a *histogram*. A *line chart* is a similar kind of graph, but with the bars represent by simple single lines.

When illustrating the relative abundance of certain values or properties in a population of individuals or samples, a circular graph called a *pie chart* may be used, the relative proportions of individuals with each property being shown as a sector of proportionate size.

grassed waterway A belt of vegetation planted where water flows to filter out unwanted particles or to prevent RUNOFF water causing erosion. *See also* bioremediation.

grassland A major BIOME dominated by grasses (Poaceae). Grasslands occur in many temperate regions where annual rainfall is 250–500 mm and in tropical regions with 750–1500 mm. They are widespread in the interiors of continents, where there is a distinct rainy season in spring and summer. Many grasslands, especially in the wetter climates, are naturally maintained by grazing mammals and, to some extent, fire. Grasslands have become much more extensive as a result of the removal of forest cover by humans in the last 5000 years and by deliberate burning and the grazing of domestic livestock. Temperate grasslands contain other herbaceous plants in addition to grasses, though trees are found only along streams and rivers. They include the North American PRAIRIES, the STEPPE of southwest Russia, the grasslands of Mongolia, and the South American PAMPAS. Tropical grassland is known as SAVANNA and is found in large areas of South America, East and South Africa, Southeast Asia, and northern Australia. Grasslands such as the prairies, steppes, and savannas in times past supported vast herds of grazing mammals. Today these areas are also important for growing cereal grasses, such as barley. *See also* chalk grassland.

gravimetric analysis The chemical analysis of substances by separating their constituents and weighing precipitates or residues.

gravitational water *See* soil moisture.

grazer **1.** An animal that feeds on herbs, especially grasses. The term also includes fish and aquatic invertebrates that suck or scrape algae from underwater surfaces. Grazers are larger than the plants or algae they consume.
2. A consumer that feeds on large prey, but takes only a small part of each individual so that, although injured, the prey survives. For example, certain fish feed on the scales of others.

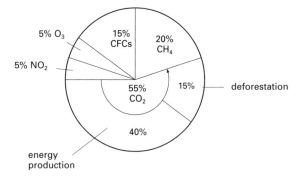

Greenhouse effect: the extent to which various greenhouse gases contribute to global warming

greenhouse effect   A rise in temperature of the Earth's atmosphere due to the presence of certain gases that allow light to penetrate the atmosphere, but do not let heat out, in much the same way as the glass of a greenhouse does. Solar (shortwave) radiation passes easily through the atmosphere (or glass in a greenhouse) and is absorbed by the Earth's surface. It is re-emitted in the form of infrared (longwave) radiation, which is absorbed by water vapor, carbon dioxide, methane, chlorofluorocarbons (CFCS), and other so-called GREENHOUSE GASES in the atmosphere with a consequent increase in the atmospheric temperature. Some scientists believe that increasing atmospheric pollution by carbon dioxide, mainly due to the burning of

FOSSIL FUELS, and many other gases is leading to a rise in global temperatures (GLOBAL WARMING), which will eventually affect other aspects of climate and have profoundly damaging effects on natural ecosystems and agriculture, as well as raising sea level.

greenhouse gas   A gas that is present in the atmosphere and absorbs infrared radiation, causing the GREENHOUSE EFFECT.

green manure   A crop that is grown to be plowed in while still green to increase the organic matter in the soil. Examples are lupines (*Lupinus*) and clovers (*Trifolium*), which increase the nitrogen content of the soil. Such crops are usually grown toward

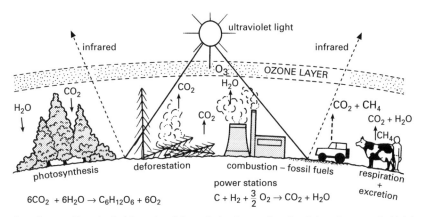

Greenhouse effect: the incident ultraviolet radiation is re-radiated as infrared, some of which is absorbed by greenhouse gases such as carbon dioxide and methane

the end of the growing season, when there is not enough time to grow crops that need to set seed.

Greenpeace An high-profile international conservation organization that employs direct confrontation with government agencies and polluting companies in an attempt to protect endangered species, prevent damage to the natural environment, and increase public awareness of environmental issues.

Green Revolution The widespread development and adoption of high-yield strains of wheat, corn, and rice during the 1960s and early 1970s in an attempt to alleviate world hunger (officially called the Indicative World Plan for Agricultural Development). It followed the World Food Congress, held by the Food and Agricultural Organization in 1963, and the subsequent 'Freedom from Hunger' campaign.

grid lines A pattern of vertical and horizontal straight lines at right angles to each other, running north to south and west to east, that make a series of squares, or grid, on a map, chart, or aerial photograph. A given point on the map is denoted by reference to its position in terms of both the vertical and horizontal axes.

grid north The location of north on a map containing grid lines, by convention in the direction of the top edge of the map.

grike *See* limestone pavement.

Grime's habitat classification A classification scheme devised by the ecologist J. P. Grime, which classes habitats in terms of the concentration of resources and the level of disturbance or stability they offer plants, and the effects on plant life strategies. Habitats that offer reasonable resources and are seldom disturbed tend to be occupied by competitors, large long-lived plants that take up considerable space both above and below ground. Habitats that are cold, dry, or lacking in resources favor stress tolerators, long-lived slow-growing plants that reproduce intermit-

tently when conditions are favorable. Highly disturbed habitats favor the growth of ruderals (weeds), fast-growing plants, often ephemerals, that produce large numbers of seeds.

gross primary production (GPP) The total fixation of energy by an autotroph (phototrophic or chemotrophic) or by the autotrophs in a population or community. It includes the fixation of energy that is later lost in respiration. *See* primary production. *Compare* net primary production.

gross production efficiency The percentage of the food that is ingested by an organism that is used in its growth and reproduction. *Compare* net production efficiency.

ground cover 1. The proportion of the ground occupied by plants of a given species, as determined by a perpendicular projection onto the ground of their aerial parts. It is usually expressed as a percentage of the total area, and gives an estimate of plant abundance. *See* cover.
2. Plants grown to prevent soil erosion.

ground layer The lowermost layer of a plant COMMUNITY, usually comprising mosses, lichens, fungi, and low-growing herbs, especially rosette plants and creeping prostrate forms. *Compare* canopy; field layer.

groundwater Water lying below the Earth's surface, in the part of the rocks or soil that is saturated with water. *See* field capacity; water table.

growing season In an area with a seasonal climate, the season of rapid plant growth. The definition of the growing season varies with geographical location. In the USA it is defined as the period between the last killing frost (temperature low enough to kill a plant or prevent its reproduction) in spring and the first killing frost of the fall. In the UK it is loosely correlated with the period when the mean temperature exceeds 6°C.

growth   An irreversible increase in size and/or dry weight. It excludes certain developmental processes that involve no size change, such as cleavage and uptake of water by seeds (imbibition). Growth involves cell division and cell expansion through synthesis of new materials, and is closely related to subsequent developmental processes. If some measure of growth of an organism, such as height or weight, is plotted throughout its life, a characteristic S-shaped (sigmoid) *growth curve* is obtained for most organisms. In some organisms, including many plants, growth never stops entirely, though it may become extremely slow (*indeterminate* or *indefinite growth*). Limits to growth are genetically set and vary with the species, but actual growth is highly dependent on environmental factors.

growth curve   *See* growth.

growth inhibitor   A substance that slows the growth of a plant, for example the plant hormones abscisic acid and ETHYLENE (ethene), which act at very low concentrations. *Compare* growth retardant.

growth retardant   A synthetic compound used to reduce stem elongation in plants, for example to produce more compact plants or to prevent the stems of cereal crops becoming too long and thin. *Compare* growth inhibitor.

growth ring   *See* annual rings.

guano   The accumulated droppings of birds, bats, or seals, which are collected and used as fertilizer. Guano is rich in nutrients, especially calcium phosphate, and in places is mined on an industrial scale. It tends to accumulate at the sites of dense breeding colonies, such as seabird colonies on offshore islands or bat colonies in caves.

guild   A group of species that live in the same environment and exploit the habitat and its resources in a similar way.

Gulf Stream   A warm ocean current that flows north from Florida to Newfoundland, with an eastern extension, the North Atlantic Drift, which curves east to northwest Europe. Its temperature and salinity are relatively constant. The warming effect of the North Atlantic Drift on northwest Europe is considerable, ameliorating the winters. There is concern that if GLOBAL WARMING continues, increased meltwater from the Arctic may disrupt the Gulf Stream, and especially the North Atlantic Drift, causing climate change.

gully erosion   Severe erosion of land in which water flowing over slight depression erodes them into trenches more than 30 cm deep. It is typical of sloping land in areas of high rainfall or sporadic fierce storms where the natural vegetation has been removed for agriculture or firewood or reduced by overgrazing, e.g. many parts of the uplands of Haiti.

guttation   Loss of water as liquid from the surface of a plant. Water is normally lost as vapor during TRANSPIRATION but, if the atmosphere is very humid, water may also be forced from the leaves through hydathodes (special secretory structures on leaves) as a result of root pressure (a form of hydrostatic pressure).

GWP   *See* global warming potential.

gymnosperms   A general term for any seed plants except the angiosperms (*see* Angiospermophyta), characterized by bearing naked seeds, i.e. seeds not enclosed in an ovary (fruit). Gymnosperms lack ENDOSPERM and many also have advanced tracheids with a structure reminiscent of angiosperm vessels, whereas in other gymnosperms the XYLEM is composed solely of more primitive tracheids.

gyre   A circular or spiral ocean current, like a giant eddy. There is a major gyre in each of the main ocean basins about 30° on either side of the Equator, caused mainly by surface winds. Gyres flow clockwise in the northern hemisphere and counterclockwise in the southern hemisphere as a result of the CORIOLIS FORCE.

H

Haber process An important industrial process for NITROGEN FIXATION. It involves reaction of atmospheric nitrogen with hydrogen to produce ammonia. The reaction takes place at high pressure using an iron catalyst.

habit The typical form or shape of a plant, such as erect, prostrate, woody, etc.

habitat The place where a particular organism lives, described in terms of its climatic, vegetative, topographic, and other relevant factors. Habitats vary in both space and time. For example, in a forest the conditions at ground level are very different from those in the leaf canopy. Conditions also vary between seasons. However the conditions found in a specific habitat at a given time are unique to that habitat even though they may resemble conditions found in other similar habitats. The term *microhabitat* describes a small area, perhaps only a few square millimeters or centimeters in size, e.g. the undersurface of a stone.

habitat fragmentation The breaking up of a habitat into discrete patches separated by other types of habitat that may not be suited to the species from the original habitat. For example, the clearing of forest for urban development.

habitat islands Isolated patches of a particular habitat. *See also* island biogeography theory.

habitat patch 1. A fairly homogeneous area of habitat that is capable of supporting a population that is separated from similar patches by areas of unsuitable habitat.

2. A part of a habitat containing a high abundance of prey or food. *See* foraging strategy.

hadal Describing the part of the ocean below the general level of the deep-ocean floor (600 m). It consists mainly of deep trenches. *Compare* abyssal zone.

Hadean *See* Precambrian.

hail Precipitation in the form of balls or irregularly shaped pieces of ice. It is formed by the coalescence and freezing of supercooled water droplets.

half-life The time required for the amount of a chemical or physical agent to be reduced by one half. The term is most commonly use to indicate the time it takes for half of the atoms of a radioactive isotope to decay (*see* radioactive decay). In theory the time taken for all the parent atoms to decay is infinite, but the half-life can be measured.

When a radioactive substance enters the body, its activity/concentration in the body will depend partly on the balance between ingestion and excretion of the substance. Thus the effective half-life, t_{eff}, the time required to reduce the activity in the body by half, depends in both the radioactive half-life, t_{rad}, and the biological half-life, t_{biol}:

$$t_{eff} = t_{rad}t_{biol}/(t_{rad} + t_{biol})$$

t_{biol} varies from species to species, and also depends on an individual's metabolic rate, physical condition, age, and sex.

halons (bromofluorocarbons) Bromine-containing compounds derived from methane or ethane and used in fire extinguishers. They cause depletion of the OZONE LAYER when they break down in the stratosphere, and have long atmospheric lifetimes.

halophile An EXTREMOPHILE bacterium of the domain ARCHAEA that lives in an extremely saline environment. *See also* methanogen; thermophile.

halophilic Describing an organism that thrives in or grows best in a saline environment.

halophyte A plant that grows in soils with a high concentration of salt, as found in salt marshes and other littoral habitats. Examples are species of *Spartina*. Halophytes are adapted to obtain water from soil water with a higher osmotic pressure than normal soil water, so they need to be able to accumulate a high concentration of salts in their root cells. Some halophytes, such as species of *Limonium*, are also able to excrete excess salt through special glands on their leaves. Many have a succulent growth form with swollen stems or leaves.

halosere A series of successional stages leading to a climax, originating in a saline area, such as the edge of the sea or of a saline lake. The halosere extends from the pioneer plants at the edge of the bare mud to salt marshes as mud and silt are trapped by the plants and raise the land surface. *See also* sere; succession.

HAP *See* hazardous air pollutant.

haplobiontic *See* life cycle.

haploid Describing an organism, cell, or nucleus containing only one representative from each of the pairs of HOMOLOGOUS CHROMOSOMES found in the normal DIPLOID cell. Haploid chromosomes are thus unpaired and the haploid chromosome number (n) is half the diploid number ($2n$). MEIOSIS, which usually precedes spore or, sometimes, gamete formation, halves the diploid chromosome number to produce haploid GAMETES. The diploid condition is restored when the nuclei of two gametes fuse to give the ZYGOTE. Gametes may de-

velop without fertilization, or meiosis may substantially precede gamete formation, leading to the formation of haploid organisms, or haploid phases in the life cycles of organisms. In a haploid cell all the genes are in the homozygous condition (only one kind of allele is present in each individual); heterozygosity (in which an individual may contain more than one kind of allele for the same gene) is impossible. *See also* alternation of generations.

hardening *See* cold tolerance.

hardpan 1. A hardened SOIL horizon, usually in the middle or lower part of a soil profile. A hardpan is typical of PODZOLS, where iron compounds leached from the upper layers accumulate in the B horizon, staining it red-brown and forming an impermeable layer (*iron pan*) that can prevent drainage, resulting in the waterlogging of higher horizons.
2. Also called *calcrete*, a hardened layer in or on a SOIL formed on calcareous substrata, formed in arid and semiarid regions as a result of fluctuating climatic conditions by calcite that precipitates out of solution as water evaporates at the surface. It may be redissolved by rainwater, penetrating the soil interstices and again precipitating to form an impermeable crust. *See* podzol.

hard water Water containing dissolved minerals, especially calcium and magnesium salts, which make it difficult to get soaps to form suds. Hard water deposits calcium carbonate, which accumulates in pipes, boilers, and kettles.

Hardy-Weinberg law In large populations in which there is random mating, no selection pressures, no mutation, and no immigration or emigration, the frequency of alleles and genotypes remains constant down the generations, i.e. there is no GENETIC DRIFT.

harvesting The removal of plants or animals from a population.

Hatch–Slack pathway *See* C_4 plant.

hazard indicator A quantitative measurement of a chemical or physical hazard. It may be based on various factors, such as toxicity, persistence in the environment, or potential exposure. *See also* hazardous ranking system.

hazardous air pollutant (HAP) A pollutant that poses a serious risk to health or the environment and that is not covered by existing legally enforceable ambient air quality standards, for example asbestos and mercury vapor.

hazardous chemical Any of a class of chemicals for which the USA ENVIRONMENTAL PROTECTION AGENCY requires manufacturers, importers, and employers to provide labels and a detailed information sheet (*material safety date sheet* or *MSDS*) covering such information as health effects, safety hazards, precautions and protective equipment required, emergency and first aid measures, procedures for clearing up spills, and recommended or statutory exposure limits.

hazardous ranking system A screening system used by the USA ENVIRONMENTAL PROTECTION AGENCY to evaluate the risks posed to public health and the environment by uncontrolled hazardous waste. Wastes are scored according to their potential for spreading out from the site in air or water, and the proximity and density of human populations.

hazardous waste Solid waste that is a potential threat to public health, other living organisms, or the environment. There are usually national and international laws regulating the handling and disposal procedures for such wastes. The USA ENVIRONMENTAL PROTECTION AGENCY also has a specific and detailed definition of such wastes for legal purposes, which it uses to determine which sites to clean up first.

haze Dust, condensed water vapor, and other material suspended in the atmosphere that reduces visibility. A *heat haze* is caused by warm air rising from hot ground.

heat Energy being transferred between a system and its surroundings as a result of temperature differences. It is equal to the total kinetic energy (*see* energy) of the atoms or molecules of a system. Heat flows spontaneously from hot matter to colder matter.

heath 1. An area dominated by dwarf shrubs of the heath family (Ericaceae). Legumes (family Fabaceae) may also be common components of heath floras, e.g. gorse (*Ulex* spp.). Heaths normally develop on poor sandy well-drained soils in temperate climates, and are usually subclimax communities maintained by grazing, burning, and sometimes cutting. The term is also used, particularly in the UK, for heather-dominated communities on acidic PODZOLS, with heather (*Calluna vulgaris*) mixed with grasses and lichens, and for communities on chalk soils, with heather and chalk grassland species. In Australia heath communities are dominated by the related family Epacridaceae. *See also* chalk grassland; climax community.
2. A dwarf shrub of the heath family (Ericaeae).

heath forest (keranga) FOREST growing usually on sandy nutrient-deficient PODZOLS, dominated by short trees and shrubs with thick bark and leathery leaves, many of them thorny. Heath forests are found in southeast Asia, the coast of Gabon (central Africa), Guyana, and the Amazon basin.

heat island effect Elevated temperatures in urban areas caused by heat fluxes due to the presence of buildings, roads, and so on, and to emissions of pollutants.

heavy metal A metallic element of high density, such as antimony, bismuth, cadmium, copper, gold, lead, mercury, nickel, silver, tin, and zinc. These metals, which are toxic even in low concentrations, persist in the environment and can accumulate to levels that stunt plant growth and interfere with animal life. *See* pollutant.

heavy metal tolerance Biochemical and physiological adaptation to concentrations of heavy metals, such as copper, lead, and zinc, that would prevent the growth of most plant species or genotypes. Such adaptation usually involves converting the metal to a harmless form; some species exclude the metal from the plant altogether, whereas others confine it to the roots, thus protecting the more sensitive shoots. Certain species or strains have evolved such tolerance, and some, for example certain strains of bent grass (*Agrostis capillaris*), benefit from the reduced competition in soil containing heavy metal ions. *See also* biomining; bioremediation; hyperaccumulator.

hectare (ha)   A metric unit of area, denoting an area of land 100 meters square (i.e. 10 000 sq m). 1 hectare = 2.47 acres.

heliophyte   A plant typical of sunny habitats.

helophyte   *See* Raunkiaer's life-form classification.

hemicryptophyte   *See* Raunkiaer's life-form classification.

hemiparasite   *See* partial parasite.

Hepatophyta (liverworts)   A phylum of spore-bearing plants containing leafy and prostrate forms, commonly known as liverworts. *See also* alternation of generations. *Compare* Bryophyta.

heptachlor   A chlorinated hydrocarbon pesticide formerly used to treat seeds. Suspected of being a human carcinogen and know to adversely affect the livers of livestock that consume treated foodstuffs, its use has been banned in many countries.

herb   1. A small seed-producing plant that does not produce persistent woody tissue, whose aerial parts die back at the end of the growing season. In some herbs the whole plant dies. *Compare* annual; shrub; tree.
2. A plant or plant part valued for its medicinal, aromatic, or culinary properties.

herbaceous   Lacking perennial stems above ground.

herbaceous perennial   *See* perennial.

herbarium   A collection of dried pressed plants, mounted and arranged systematically together with collection data such as the plant's name, date and place of collection, and collector's name. Other notes of use in taxonomic studies or in future uses of the plant (for example, as a source of pharmaceuticals) may also be kept. Herbaria provide a resource for both anatomical and molecular reference.

herbicide (weedkiller)   A chemical weedkiller that inhibits plant growth or kills the plant. There are two kinds: *contact herbicides* kill or damage the plant on contact with its surface, whereas *systemic herbicides* are taken up by the plant and transported through its tissues. Systemic herbicides are longer lasting than contact ones, and usually more successful in eliminating the roots – important for species such a dandelions, which have deep taproots and can regenerate from fragments of root. Selective systemic herbicides, such as 2,4-D and 2,4,5-T, kill most broad-leaved plants, but leave cereals and grasses unaffected.

herbivore   A plant-eating animal, especially one of the herbivorous mammals, such as cattle, rabbits, etc. There may be various modifications associated with this diet, e.g. to the teeth and digestive system. *Compare* carnivore; omnivore.

herd immunity   A level of immunization (which may be less than 100%) against a particular disease in a particular population at or above which the disease is unable to persist in that population.

heredity   The transmission of genetically determined characteristics from one generation to the next.

heritability   The proportion of phenotypic VARIATION due to genetic factors. It can be estimated from measurements of individuals from different generations, and is used in plant and animal breeding to predict how successful genetic selection will be in improving a particular trait. *See* phenotype; genotype.

hermaphrodite 1. An animal possessing both male and female reproductive organs. The earthworm (*Lumbricus*) is a common example.
2. A plant bearing stamens and carpels in the same flower. In many such plants, self-fertilization and inbreeding are prevented by specific self-incompatibility systems.

heterogeneous Nonuniform: made up of dissimilar components, or varying in structure or composition in different locations.

heteromorphism The existence of more than one form, used especially with reference to life cycles in which the alternating generations are markedly different morphologically, as in ferns and jellyfish. *See* alternation of generations. *Compare* polymorphism.

heterosis *See* hybrid vigor.

heterotroph An organism whose principal source of carbon is organic, i.e. the organism cannot make all its own organic requirements from inorganic starting materials. Most heterotrophic organisms are chemotrophic (*see* chemoheterotroph); these comprise all animals and fungi, most bacteria, and parasitic plants. A few heterotrophic organisms are phototrophic (*see* photoheterotroph). The nonsulfur purple bacteria, for instance, require organic molecules such as ethanol and acetate. *Compare* autotroph. *See also* chemotroph; phototroph.

heterotrophic succession *See* succession.

heterozygous Having two different ALLELES at a given gene locus. Usually only one of these, the dominant allele, is expressed in the phenotype. Such an organism is called a *heterozygote*. On selfing or crossing heterozygotes some individuals with two recessive alleles (double recessives) may appear, giving viable offspring. Selfing heterozygotes halves the heterozygosity, and thus OUTBREEDING maintains heterozygosity and produces a more adaptable population. *Compare* homozygous.

HFC *See* hydrofluorocarbons.

hibernation A state of sleep and greatly reduced metabolic rate that enables certain mammals to survive prolonged periods of low temperature and food scarcity. Stored body fat supplies enough energy for their bodies to work slowly and maintain their temperatures just higher than their surroundings. Some, such as bats, wake and feed on warm days.

Day length or food shortage may stimulate the hibernation mechanism, but in temperate and arctic animals the stimulus is cold. A 'hibernation hormone' has been suggested but not isolated. Temperature regulation is maintained but at a lower level. When hibernation ends, body temperature rises spontaneously, starting at the body core.

hierarchical classification A system of CLASSIFICATION in which individuals are grouped into a series of progressively larger and broader categories; lower groups are always subordinate to and included in groups of higher ranking. The groupings are based on the affinities of the individuals and groups. Hierarchical classification is the basis of CLADISTICS. The INTERNATIONAL CODES OF NOMENCLATURE define the sequence of ranks and their names in particular groups of organisms. For example, in plants the ranks are as follows, in ascending order: form, variety, species, series, section, genus, tribe, family, order, class, phylum or division, and kingdom. Many taxonomists also recognize an even higher rank, the domain.

hierarchy 1. A graded or ranked series, as in an EVOLUTIONARY TREE.
2. (dominance hierarchy) A form of social organization in which certain individuals or groups of individuals in a population

have different status or rank, which affects access to food, mates, or other resources. *See* dominance.

high-level waste *See* radioactive waste.

histogram *See* graph.

Holarctic The circumpolar region that encompasses North America, Europe, and Asia. It is a recognized BIOGEOGRAPHICAL REGION, with significant similarities of flora and fauna, because these continents were once joined in the supercontinent Laurasia. *See also* continental drift.

Holarctic floral province Alternative name for the Boreal floral region, which consists of northern Eurasia and North America. *See* floral province.

holding pond A pond or reservoir that stores polluted RUNOFF.

Holocene (Recent) The present epoch in the geological time scale, being the second epoch of the QUATERNARY period. It dates from the end of the last GLACIATION, about 10 000 years ago, to the present day.

holophytic Describing the type of nutrition in which complex organic molecules are synthesized from inorganic molecules using light energy. It is another term for phototrophic. *See* phototroph.

holozoic (heterotrophic) Describing organisms that feed on other organisms or solid organic matter, i.e. most animals and INSECTIVOROUS PLANTS.

homeostasis The maintenance of a constant internal environment by an organism. It enables cells to function more efficiently. Any deviation from this balance results in reflex activity of the nervous and hormone systems, which tend to negate the effect. The degree to which homeostasis is achieved by a particular group, independent of the environment, is a measure of evolutionary advancement.

homeothermy *See* homoiothermy.

homing The returning of an animal to a site it regularly uses for sleeping or breeding. The term is applied not only to seasonal migrations between feeding and breeding grounds, but also to foraging trips.

hominid A member of the Primates, family Hominidae, which includes modern humans (*Homo sapiens*) as well as extinct forms found in great number and variety of fossils.

homogeneity In statistics, the property of samples or individuals from different populations whereby they are similar with respect to the variable being studied. Such populations – and the resulting data – are said to be homogeneous.

homogeneous Uniform: made up of similar components, or not varying in structure or composition in different locations. *See also* homogeneity.

homoiothermy (homeothermy) The maintenance of body temperature within narrow limits. Organisms that achieve this by behavioral means, such as moving in and out of shade, are termed *ectotherms*, while those that have internal temperature-regulating mechanisms are termed *endotherms*. *Compare* poikilothermy.

homologous chromosomes CHROMOSOMES that pair at MEIOSIS. Each carries the same GENES as the other member of the pair but not necessarily the same ALLELES for a given gene. One member of each pair is of maternal origin, the other of paternal origin. With the exception of the sex chromosomes (X and Y) they are morphologically similar. During the formation of the germ cells only one member of each pair of homologues is passed on to the gametes. At fertilization each parent contributes one homolog of each pair, thus restoring the diploid chromosome number in the ZYGOTE. *See also* meiosis.

homologous structures Structures that, though in different species, are believed to have the same origin in a common ancestor. Thus the forelimbs and hindlimbs of all land vertebrates are said to be homologous, being constructed on the same five-digit (pentadactyl) pattern.

homozygote *See* homozygous.

homozygous Having identical ALLELES for any specified GENE or genes. A *homozygote* breeds true for the character in question if it is selfed or crossed with a similar homozygote. An organism homozygous at every locus produces offspring identical to itself on selfing or when crossed with a genetically identical organism. Homozygosity is obtained by inbreeding, and homozygous populations may be well adapted to a certain environment, but slow to adapt to changing environments. *Compare* heterozygous.

horizon *See* soil.

host An organism used as a source of nourishment by another organism, the parasite, which lives in or on the body of the host. In a *definitive host* the parasite reaches sexual maturity. In an *intermediate host* the resting stage or young of the parasite is supported. *See* parasitism.

host specificity The degree to which a parasite, parasitoid, or symbiont is found on more than one species of host. If it is found on only one species, it is said to be highly specific.

hot desert *See* desert.

hot spring (geothermal spring) A spring heated by the Earth's crust, for example, by volcanic activity or radioactivity in the rocks below.

Human Genome Project An international research project, sponsored by the US Department of Energy, the National Institutes of Health and the Wellcome Trust, started in 1990 to sequence the DNA of the human GENOME and map the location of all the genes. The first working draft was produced in June 2000, and the final draft (with an error rate less than 1 in 100,000 bases, an accuracy of 99.9%) was completed in 2003: some 3.2 billion bases and 31 000 genes. The project used automated sequencing machines involving advanced robotics, together with large amounts of computing power, achieving sequencing rates of 12 000 bases a second. Results were published on the Internet as soon as they were available, thus ensuring free public access to the genome.

humic acid A natural mixture of dark brown acidic organic substances that can be extracted from HUMUS with dilute alkali and precipitated by acidification to pH 1–2. *See* soil.

humidity The moisture content of the atmosphere. The *absolute humidity* is the amount of water vapor present in a unit volume of air (usually 1 cubic meter of air). The *relative humidity* is the ratio of the amount of water vapor actually present in the air to the greatest amount that the air could hold at the same temperature, and is usually expressed as a percentage.

Various instruments are used to measure humidity. In the *wet-and-dry-bulb thermometer*, a pair of thermometers is used, one – the wet-bulb thermometer – being kept moist by a thin muslin bag dipped in a container of distilled water. Provided the air is not saturated, evaporation from the cloth keeps the wet-bulb thermometer at a lower temperature than the dry-bulb thermometer, and the difference between them is a measure of the saturation deficit, and hence the relative humidity of the atmosphere. Such a pair of thermometers may be enclosed in a frame with a handle (a whirling psychrometer or *sling psychrometer*), which can be whirled in the air like a football rattle, ensuring good airflow around the bulbs. A *hydrograph*, which uses specially treated hair that changes its length as the humidity changes, records changes in humidity on a revolving drum. *See also* hygrothermograph.

humification The process by which HUMUS is formed from dead organic material. It involves the action of DECOMPOSERS, SAPROBES that feed on dead organic material, breaking down complex organic molecules by oxidation reactions to form simple inorganic molecules that can be taken up by plant roots. *See* soil.

hummock and hollow cycle A cyclical change in the vegetation of a developing raised peat bog. *Sphagnum* mosses colonize a pool of water on the peat, building up a low hummock. This allows plants such as ling (*Calluna vulgaris*) and cross-leaved heath (*Erica tetralix*) to move in. When they die, the hummock erodes, forming a water-filled hollow, so the cycle starts again. The BOG consists of a mosaic of hummocks and hollows in various stages of development.

humus The nonliving finely divided organic matter in SOIL derived from the decomposition of animal and plant substances by soil bacteria (*see* humification). Humus consists of 60% carbon, 6% nitrogen, and small amounts of phosphorus and sulfur, and is valued by horticulturalists and farmers because it improves the fertility, water-holding capacity, and workability of the soil. Humus has colloidal properties (*see* colloid) that enable it to retain water, so it can improve the moisture content of sandy soils. It aids the formation of crumbs in the soil (*see* crumb structure), and is often added to clay soil to increase the particle size, promoting drainage and aeration. Different types of humus are recognized, according to the types of organisms involved in its decomposition, the vegetation from which it is derived, and the degree of incorporation into the mineral soil. *Mull* humus is found in deciduous and hardwood forests and grasslands in warm humid climates. It is alkaline and bacteria, worms, and larger insects are abundant. Decay is rapid and layers are not distinguishable. *Mor* or raw humus is usually acidic and characteristic of coniferous forest areas. Few microorganisms or animals exist in this type of humus, small arthropods and fungi being the most common.

hunter-gatherers Humans that live by hunting wild animals, fishing, and collecting wild fruits, berries, nuts, mushrooms, and other natural sources of food, but do not cultivate the land or own domestic livestock.

hurricane *See* depression.

Huxley's line *See* Wallacea.

HWRP *See* Hydrology and Water Resources Program.

hybrid An organism derived from crossing genetically dissimilar parents. Thus most individuals in an OUTBREEDING population could be called hybrids. However, the term is usually reserved for the product of a cross between individuals that are markedly different. If two different species are crossed the offspring is often sterile, for example, the mule, which results from a cross between a horse and a donkey. The sterility results from the nonpairing of the chromosomes necessary for gamete formation. In plants this is sometimes overcome by the doubling of the chromosome number (*see* chromosome), giving a polyploid (*see* polyploidy), which may eventually give rise to a new species. By contrast, hybrids derived from different varieties of the same species are often more vigorous than their parents, and are selected and propagated by vegetative means by agriculturists and horticulturists. *See* chimera; hybrid vigor.

hybrid breakdown The situation in which a HYBRID is fertile, but its offspring are not.

hybridization Processes that lead to the formation of a HYBRID. Hybridization is common among angiosperm plant species, though often the offspring are partially or wholly sterile. In some the hybrids may be able to backcross with one or both parental species, leading to HYBRID SWARMS. *Artificial hybridization* of normally self-pollinating species involves the transfer by hand of pollen from one plant to another, often using a paintbrush. Its success depends on the presence or absence of incompatibility systems that prevent successful fertilization between individuals of similar genetic makeup. Such techniques have been important in the development of crop CULTIVARS and ornamental plants. *See* speciation.

hybrid swarm A very variable series of organisms resulting from the continual crossing, recrossing, and backcrossing of the HYBRID generations of two original species or varieties that may occur when the original barrier to reproduction breaks down. *See* apomixis.

hybrid vigor (heterosis) The condition in which the expression of a characteristic is greater in the heterozygous offspring than in either of the homozygous parents. The effect arises from an accumulation of dominant genes in the F_1 (first filial) generation. Thus, if height is controlled by two genes, A and B, and tall and short forms are determined by dominant and recessive alleles respectively, then the crossing AAbb $\times$ aaBB would give an F_1 AaBb, containing both dominant genes for tallness. Usually the more unlike the parents are, the more hybrid vigor is released, but the effect diminishes in subsequent generations as more recessive homozygotes (*see* homozygous) reappear. For crop plants, hybrid vigor is particularly important for characteristics such as growth rate, yield, and disease resistance. Selfing the F_1 generation reduces the amount of heterozygosity (*see* heterozygous), so optimum vigor is best maintained by producing seed by crossing two different parental pure lines.

hybrid zone A region where HYBRIDS between two geographical races occur, usually at the boundary between two separate but closely related populations. *See* hybridization; hybrid swarm.

hydration The taking up of or combining of substances with water.

hydraulic lift The ability of plant roots to take up water from deep soil layers and release it into shallower surface layers. Uptake from the deep moist soil layers is aided by ROOT PRESSURE; water then diffuses out of roots in the drier surface soil. Hydraulic lift tends to happen particularly at night, when the transpiration streams ceases. It is thought to benefit the plant by allowing nutrients in the upper soil layers to dissolve and be available for uptake. Mature trees may move hundreds of liters of water every night in this way, to the benefit of adjacent shrubs and herbs. *See* transpiration.

hydric Describing a habitat that is extremely wet. *Compare* mesic; xeric.

hydrocarbons Organic compounds that contain carbon and hydrogen only, e.g. acetylene, butane. Many of the components of natural gas, petroleum, and coal are hydrocarbons.

hydroelectric power (hydropower) A renewable source of energy derived by falling or flowing water, which powers turbines that generate electricity. Some hydroelectric power plants are powered directly by river water, while others are powered by water stored in a reservoir behind a dam at some height above the natural riverbed. Tidal power, in which the difference in height between high and low tides form the main driving force, is a form of hydroelectric power.

hydrofluorocarbons (HFCs) Compounds that contain only hydrogen, fluorine, and carbon. They are used in manufacturing, and are also by-products of many industrial processes. HFCs were introduced as substitutes for CFCs because they do not significantly affect the OZONE LAYER, but they are powerful GREENHOUSE GASES with GLOBAL WARMING POTENTIALS that range up to 11 700. *See* global warming; ozone layer.

hydrogen An essential element in living tissues. It enters plants, with oxygen, as water and is used in building up complex reduced compounds such as carbohydrates and fats. Water itself is an important medium, making up 70–80% of the weight of organisms, in which chemical reactions of the cell can take place.

hydrogen bond An electrostatic attraction between a hydrogen atom and an electronegative atom in another molecule.

Hydrogen bonding is responsible for the properties of water, including its relatively high boiling point. It is important in many biological systems for holding together the structure of large molecules.

hydrogen economy An idealized economy in which the only fuel is hydrogen, which is burned or oxidized in fuel cells to give electricity directly. The hydrogen could be produced by electrolysis of water, the electricity for this being generated by wind power, water power, or similar renewable sources. Hydrogen fuel would cause no pollution and minimize the emission of GREENHOUSE GASES.

hydrogen ion A positively charged hydrogen atom, H$^+$, i.e. a *proton*. Hydrogen ions are produced by all ACIDS in water. *See* acid; pH.

hydrogen sulfide (H$_2$S) A colorless highly poisonous, highly flammable gas that smells like bad eggs. Some photosynthetic bacteria derive their electrons from hydrogen sulfide instead of water:
$$2H_2S + CO_2 \rightarrow H_2O + CH_2O + 2S$$
See autotroph.

hydrogeological maps Maps that show the distribution of AQUIFERS, AQUICLUDES, and bore holes, the depth to the WATER TABLE, the groundwater quality (ions present), and the yield of groundwater from bore holes.

hydrogeology The study of the chemistry and movement of GROUNDWATER. *See* water cycle.

hydrograph *See* humidity.

hydroid *See* Cnidaria.

hydrological cycle *See* water cycle.

hydrology The study of the WATER CYCLE: the properties, distribution, and circulation of water in the ATMOSPHERE, BIOSPHERE, LITHOSPHERE, and HYDROSPHERE, and how this changes with time. It encompasses geology, meteorology, and oceanography.

Hydrology and Water Resources Program (HWRP) A program of the World Meteorological Association that aims to collect and analyze hydrological data as a basis for assessing and managing freshwater resources, e.g. for human consumption, sanitation, irrigation, hydroelectric power generation, and water transport, and for flood forecasting and the prediction of droughts.

hydrophilic Describing a molecule or surface that has an affinity for water. Such molecules are usually polar, for example proteins. *Compare* hydrophobic.

hydrophily POLLINATION in which water carries the pollen from anther to stigma.

hydrophobic Describing a molecule or surface that has no affinity for water. Such molecules are nonpolar, e.g. benzene. *Compare* hydrophilic.

hydrophyte A plant that grows in water or in extremely wet areas, for example arrowhead (*Sagittaria*) and water lilies. Hydrophytes show certain adaptations to such habitats, notably development of aerenchyma, reduction of cuticle, root system, and mechanical and vascular tissues, and divided leaves. Large intercellular air spaces in leaves, stems, and roots allow oxygen to diffuse through the plant. Hydrophytes with floating leaves can exchange gases with the atmosphere; many submerged hydrophytes, such as spiked water milfoil (*Myriophyllum spicata*), have no stomata, absorbing water and gases over their entire surface. *Compare* mesophyte; xerophyte. *See also* Raunkiaer's life-form classification.

hydroponics (water culture) The growth of plants in liquid culture solutions rather than soil. The solutions contain the correct balance of all the essential mineral requirements. The method is used commercially, especially for glasshouse crops, and also in experimental work in determining the effects of mineral deficiencies.

hydropower *See* hydroelectric power.

hydrosere Any plant community in a SUCCESSION that starts in fresh water. Gradual silting leads to mesophytic conditions in the later seres. *See* sere.

hydrosphere All the water contained in the atmosphere, land, sea, rivers, and lakes.

hydrostatic pressure The pressure exerted by a fluid as a result of its incompressibility. For example, water in a plant cell exerts a pressure on the cell wall; *turgor* is due to the hydrostatic pressure caused by the uptake of water into cells whose expansion is limited by the inelasticity of the cell wall.

hydrothermal activity The production of hot springs and other bodies of water heated by cooling magma or by radioactive decay in rocks. *See also* geothermal power; hydrothermal vent.

hydrothermal vent (smoker) A site on the seabed from which hot springs arise that have been heated by contact with molten rock, usually along a mid-oceanic ridge (*see* plate tectonics). Temperatures may reach 300ºC. These vents are often rich in minerals, especially sulfides, and support communities in which the primary producers are chemotrophic bacteria. Vents rich in copper, iron, and manganese are particularly hot, and the water is black; they are known as *black smokers*. The cooler white smokers flow more slowly and usually contain quantities of arsenic and zinc.

hygrometer An instrument used to measure atmospheric HUMIDITY.

hygroscopic water *See* soil moisture.

hygrothermograph An instrument that measures temperature and relative humidity and records it on a scaled paper chart mounted on a cylinder that turns to give a record over a period of time, usually a week.

hyperaccumulator A plant capable of accumulating HEAVY METALS in its tissues, to the point where it can be 'mined' for the metal. *See also* bioaccumulation; bioremediation.

hyperosmotic *See* hypertonic.

hyperthermophile An EXTREMOPHILE that thrives in a habitat with high temperatures, sometimes requiring a temperature of at least 105°C. Some archaea (e.g. *Pyrolobus*) are able to survive temperatures of 113°C, and may even fail to multiply at temperatures much below 90°C. *See* Archaea.

hypertonic (hyperosmotic) Describing a solution with an osmotic pressure greater than that of a specified other solution, the latter being HYPOTONIC. When separated by a selectively permeable membrane (e.g. a cell membrane) water moves by OSMOSIS into the hypertonic solution from the hypotonic solution. *Compare* isotonic.

hypha (*pl.* hyphae) In fungi, a fine non-photosynthetic tubular filament that spreads to form a loose network termed a *mycelium* or aggregates into fruiting bodies (e.g. mushrooms).

hypolimnion The lower noncirculating water in a thermally stratified lake. If it lies below the COMPENSATION POINT, oxygen becomes depleted, and reoxygenation occurs only when stratification breaks down in the fall. *See* stratification.

hypo-osmotic *See* hypotonic.

hypotonic (hypoosmotic) Describing a solution with an osmotic pressure less than that of a specified other solution, the latter being HYPERTONIC. When separated by a selectively permeable membrane (e.g. a cell membrane) water is lost by osmosis from the hypotonic to the hypertonic solution. *Compare* isotonic.

hypoxic Deficient in oxygen.

IBP *See* International Biological Program.

ICBN *See* International Codes of Nomenclature.

Ice Age A period in the latter part of the PLEISTOCENE characterized by successive coolings and warmings of the earth. In at least four major GLACIATIONS (cold periods), ice caps spread south from the Arctic and north from the Antarctic. Large areas of Britain, Europe, and North America were covered by ice from the North. The cause of the Ice Ages is not known, but it is assumed that others will occur in the future. The term ice age (no capitalization) is used for similar climatic fluctuations that occurred prior to the Pleistocene.

ice core A cylindrical section of ice taken from a bore hole in a glacier or ice sheet in order to study past climates by analyzing the gases in the air trapped between the ice crystals.

ideal free distribution The distribution of individuals of a foraging species that results if the consumers select the most profitable sites at a given time, assuming all are free to move form one patch to another and are equally discriminating. As the most favorable sites fill up, later arrivals select habitats that have lower overall suitability, but which are less crowded.

IGBP *See* International Geosphere–Biosphere Program.

igneous rock Rock formed by crystallization from a molten magma, e.g. basalt and granite.

illuviation The accumulation of dissolved or suspended materials in a soil horizon as a result of LEACHING from another part of the soil profile. *Compare* hardpan.

immigration The flow of genes into a population when individuals from other populations move in and interbreed with members of the population. *See* gene flow. *Compare* isolating mechanisms.

immunity The ability of plants and animals to withstand harmful infective agents and toxins. In most plants this is achieved by physical barriers preventing entry of PATHOGENS and by physiological reactions to isolate the pathogen and its effects. In animals it may be due partly to a number of nonspecific mechanisms, such as inflammation and phagocytosis (cell eating) or an impervious skin (nonspecific immunity). In vertebrates it is largely the result of a specific mechanism, whereby certain substances (antibodies) or lymphocytes present in the body combine with an introduced foreign substance (antigen) – specifically acquired immunity.

impact assessment An evaluation of the effect of a proposed project on the human and natural environment. As well as taking into account its effects on natural or cultivated habitats and landscapes, wildlife, and livestock, the assessment may also consider effects on the quality of life (noise, pollutants, visual impact, and so on), and on the local economy (numbers of jobs created/lost, effects on local businesses, etc.). It may also propose ways of ameliorating potential deleterious effects.

impermeable Describing a substance

that does not allow another substance, especially water, to penetrate or pass through it.

inbreeding Breeding between closely related individuals. The most extreme form of inbreeding is self-fertilization, which occurs in some plants. In animals, mating between siblings or between parents and offspring is generally the closest form of inbreeding. Inbreeding increases homozygosity (*see* homozygous) so that deleterious recessive genes are expressed more often in the PHENOTYPE, and decreases heterozygosity (*see* heterozygous) and hence the potential genetic variability of the population. There is also a general lowering of vigor in inbred stock (INBREEDING DEPRESSION), which is especially pronounced among normally outbreeding populations. In human societies there are usually cultural restraints on marriage between close relatives. *See* inbreeding depression. *Compare* outbreeding.

inbreeding depression The tendency for species to suffer a decline in vigor on inbreeding, probably due to the increased occurrence of individuals HOMOZYGOUS for deleterious or lethal ALLELES. This is most marked in individuals resulting from self-fertilization, especially in plants that are normally OUTBREEDING. It may be manifest at any stage of the life cycle. *See* inbreeding. *Compare* hybrid vigor.

incineration A WASTE treatment process that involves controlled burning at high temperatures. Hazardous components are reduced to carbon dioxide, water, and ash containing very little combustible material, which can then be stored. In addition to reducing toxicity, incineration also reduces the volume of waste. The furnace used for controlled incineration is called an *incinerator*. In *mass burn incineration*, bulk waste volume is reduced by about 90%, and its mass reduced by about 70%. Unsorted waste (except for large items such as refrigerators and hazardous items such as aerosols) is fed in to the furnace, and the hot combustion gases are passed through a

boiler system to extract energy. *See also* combined heat and power plant.

incinerator *See* incineration.

inclusive fitness *See* fitness.

indefinite growth *See* growth.

independent assortment The law, formulated by Austrian geneticist Gregor Mendel, that GENES segregate independently at MEIOSIS so that any one combination of ALLELES is as likely to appear in the offspring as any other combination. The work of T. H. Morgan later showed that genes are linked together on chromosomes and so tend to be inherited in groups. The law of independent assortment therefore applies only to genes on different CHROMOSOMES. The term can also be applied to whole chromosomes. *See also* linkage.

independent variable *See* variable.

indeterminate growth *See* growth.

indicator A substance used to test for acidity or alkalinity of a solution by a color change. Examples are litmus and phenolphthalein. A *universal indicator* shows a range of color changes over a wide range from acid to alkaline, and can be used to estimate the pH.

indicator dyes *See* pH.

indicator species (ecological indicator) An organism that can be used to measure the environmental conditions that exist in a locality. Lichen species are indicators of levels of pollution, because different species are sensitive to different levels and types of pollutants. Certain flowering plants are indicators of acid (*Sphagnum* mosses) or alkaline (*Gypsophila* species) conditions. Indicator species also provide information about past environments; for example, certain species may be typical of ancient woodland or sites where woodlands formerly stood. *Tubifex* worms indicate low levels of oxygen and stagnant water.

indigenous (native) Describing a species, race, or other taxon that occurs naturally in an area and has not been introduced by humans. *Compare* exotic.

indirect competition *See* competition.

industrial ecology The study of the flow of energy and materials from their natural sources through the manufacturing process and its products to their final disposal or recycling. It provides useful information for devising strategies to minimize pollution and reduce the consumption of energy, water, and other resources.

industrial ecosystem An industrial system that mimics a natural ecosystem in that the wastes from one industrial process form the raw materials for another.

industrial melanism An increase in dark forms, for example in the moth *Biston betularia*, in industrial soot-polluted environments. Natural selection against normal pale forms by predators results in dark offspring being at a selective advantage in such environments. This results in an increase in the numbers of the better camouflaged dark forms. *See also* POLYMORPHISM.

industrial smog *See* smog.

infauna Animals that live under the surface of the seabed or other sediments, either by tunneling or by constructing tubes.

infiltration 1. The movement of water through the SOIL surface. The properties of different soils, especially surface structure and texture, give them different capacities for infiltration. *See also* percolation. 2. The leakage of water from the soil into manmade underground structures such as pipes through cracks and damaged surfaces. 3. The application of waste water to land, so that it penetrates the underlying soil.

influent 1. Water, waste water, or another liquid that flows into a stream, reservoir, treatment plant, or industrial process. 2. A factor, such as an immigrating individual, that perturbs the balance of a community.

ingestion 1. A form of heterotrophic nutrition in which prey or pieces of prey or detritus are taken in through the mouth. 2. The taking into the body of toxic substances through the mouth or lungs.

inheritance The receiving of characteristics by transmission of genetic material from parent to offspring.

inhibition Resistance to a colonizing species by an already-established species, especially during a SUCCESSION.

inorganic Describing a substance that is not derived from living organisms or their remains, or which does not contain carbon. However, simple carbon-containing compounds such as oxides and sulfides of carbon are usually considered to be inorganic compounds.

Insecta The largest class of arthropods (*see* Arthropoda) and the largest class in the animal kingdom. Most insects can fly. The body is characteristically divided into a head, thorax, and abdomen. The head bears a pair of antennae, compound eyes, and simple eyes (*ocelli*). The mouthparts are modified according to the diet. The thorax bears three pairs of five-jointed legs and, typically, two pairs of wings. The abdomen is usually limbless. Most insects are terrestrial and respiration is carried out by *tracheae* (branching tubes) with segmentally arranged *spiracles* (tiny holes opening to the outside). Usually the life cycle includes complete METAMORPHOSIS, with a larval and pupal stage, but in some species metamorphosis is incomplete – the larvae (*nymphs*) resemble the adult and there is no pupal stage. Many insects are beneficial, being pollinators of flowers and predators of pests; others are harmful, being pests of crops, disease carriers, and destroyers of clothes, furniture, and buildings. *See also* biological control; pollination.

insecticide A chemical used to kill insects. It may be a natural substance, such as derris or pyrethrums, or a synthetic chemical such as organophosphates (e.g. malathion), carbamates, CHLORINATED HYDROCARBONS, or dinitrophenols. Synthetic insecticides tend to persist in the environment and accumulate in food chains. *See also* bioaccumulation; botanical insecticide.

insectivore Any animal that feeds mainly on insects.

insectivorous plant (carnivorous plant) A plant adapted to supplementing photosynthesis by obtaining nutrients, especially nitrates, from small animals, especially insects. The plants have various means of trapping and killing the insects, which are then digested by enzymes secreted by the plant. Examples include: butterworts (*Pinguicula*) with slippery, inrolling leaves; sundews (*Drosera*), with sticky glandular hairs that close over the trapped insects; Venus's-flytrap (*Dionaea muscipula*), with hinged leaves that snap shut over the victim, triggered by sensitive hairs; and pitcher plants (Nepenthaceae and Sarraceniaceae), with leaves modified to form containers into which insects and other small animals fall and drown.

in situ In its natural or original position or place.

insolation The radiation received from the Sun, measured as the rate of delivery of direct solar radiation per unit of horizontal surface. The radiation received at a given point on the Earth's surface depends on the position of the Earth in its orbit, the inclination of the surface to the Sun's rays, and the thickness and transparency of the atmosphere (again affected by latitude). It varies with the season and with latitude, being about 2.4 times greater at the equator than at the poles.

integrated pest management (IPM) The use of both chemical and biological control and good farming practice to reduce pest numbers, i.e. the use of resistant crop varieties and selective use of pesticides, so that natural predators are not harmed, as well as practices such as rotation of crops, which minimize the chances of pest populations building up.

integrated pollution control (IPC) A system of emissions controls imposed on industrial and other processes, usually imposed by a licensing scheme, that encompass the effects of emissions on all environmental media: air, land, and water. Such controls take account of the potential for emissions to move from one medium to another.

Integrated Risk Information System (IRIS) An electronic database of the human health effects that may result from exposure to various substances found in the environment. Maintained by the United States ENVIRONMENTAL PROTECTION AGENCY, it is a valuable tool for use in risk assessment and other decision-making, and for regulatory activity.

integrated waste management A waste management strategy that encompasses the avoidance or minimization of waste production as well as the maximization of recovery, reuse, and recycling of appropriate waste materials and the minimization of the volume of the final waste for disposal.

intensive farming Farming that aims to get the maximum output from a given area of land by keeping animals in crowded conditions or growing crops in quick succession, often to the detriment of the environment. Growing several crops a year requires large amounts of fertilizer. If the rotation of crops is abandoned, this will necessitate more fertilizer and pesticide applications. Keeping livestock indoors and feeding them on concentrated foodstuffs requires the use of more drugs to control diseases associated with crowding.

interactional food web *See* food web.

interference coefficient A measure of the extent to which interference between

CONSUMERS reduces individual rates of consumption as consumer density increases.

interglacial period A period of warmer climate between two glacial periods, mainly associated with the PLEISTOCENE period. This can be detected by pollen analysis (*see* palynology), which reveals the vegetation succession as the climate changes. *See* Ice Age. *Compare* interstadial period.

Intergovernmental Panel on Climate Change (IPCC) A body of international scientific experts set up in 1988 to correlate and assess information on climate change and propose suitable responses.

intermediate host *See* host.

intermittent growth *See* growth.

internal dose *See* absorbed dose.

International Biological Program (IBP) An international research program (1966–74) instigated by the International Union of Biological Sciences (IUBS), and later taken over by the International Council of Scientific Unions (ICSU), which aimed to study the biological productivity of the Earth's terrestrial, marine, and freshwater systems, human adaptability, and resource use and management. It was important in stimulating worldwide environmental awareness, and was a forerunner of the UNESCO's Man and Biosphere Program. Data were collected on ecosystems (biomes) such as deserts and tropical rainforests, and mathematical models of ecosystem structure and function and the environmental factors affecting them were developed. These provided a basis for developing suitable management systems to improve productivity and environmental quality.

International Codes of Nomenclature A set of rules for the scientific CLASSIFICATION of taxa. The naming of families and lower taxa is based on the first valid name published, the starting-point being taken as the publication of certain authoritative books, for example for plants, Linnaeus's *Species Plantarum*, published in 1753. Different starting dates and authorities apply for different groups of organisms. But if an organism is later assigned to a different taxon or rank, a different name may be given. Thus an organism may legitimately have two names, according to which genus different authors place it in. If several taxa are combined into one, that taxon takes on the name of the former TAXON with the oldest name. There are several separate codes for different groups of organisms: The *International Code of Botanical Nomenclature* (ICBN), which also includes algae, fungi, and slime molds; the *International Code of Nomenclature of Cultivated Plants* (ICNCP); the *International Code of Bacteriological Nomenclature*; and the *International Code of Zoological Nomenclature* (ICZN).

International Geosphere–Biosphere Program (IGBP) A program of research and study established in 1986 by the International Council of Scientific Unions to describe and understand the interactive physical processes that regulate the total earth system, the unique environment that it provides for life, the changes that are occurring in this system, and the manner in which they are influenced by human actions. Resources

International Union for the Conservation of Nature and Natural Resources *See* World Conservation Union.

International Whaling Commission (IWC) An international organization established in 1946, following the signing of the International Whaling Convention by the major whaling nations, to sponsor scientific studies of whale populations and recommend restrictions on the commercial killing of whales to ensure the sustainability of the whale populations and hence of the whaling industry.

interspecific Between species (e.g. COMPETITION between species is interspecific competition). *Compare* intraspecific.

interstadial period A period of warmer climate within a glacial period, usually shorter and cooler than an interglacial. Because of the relatively short timescale, some warmth-loving plants may not migrate back, so will not appear in the pollen record. *See* Ice Age. *Compare* interglacial period.

interstate waters Waters that flow across or form part of state or international boundaries, the Great Lakes, and coastal waters of the United States.

interstitial air The air trapped in the spaces between particles in sediments and soils.

interstitial fauna *See* meiofauna.

intertidal The part of the LITTORAL zone between the lowest and highest tide levels.

interval scale *See* measurement scales.

in the field A term that refers to work carried out *in situ* in a particular habitat.

intraspecific Within a species (e.g. within the gene pool or a population of a species). *Compare* interspecific.

intrinsic rate of increase (*r*) The per capita rate of increase of a population that has reached a stable age structure, in which fecundity and survivorship are stable and which is not crowded and is not experiencing any competitive or other pressures.

introduced species A species taken by humans to a new location in which it did not previously exist. Such species can have a devastating effect on the native flora and fauna: introduced predators have destroyed flightless ground-nesting birds in New Zealand, and introduced diseases for which there are no natural BIOLOGICAL CONTROLS can decimate domestic crops and livestock.

intron *See* exon.

inversion An atmospheric condition in which cold air is trapped by a layer of warm air above it. This may occur, for example, in still conditions when the air at ground level cools rapidly by radiation, becoming cooler than the air above. This can lead to frost pockets occurring in valleys. It also traps pollutants forming SMOG.

invertebrates Animals without backbones, e.g. insects, snails, starfish, and worms.

in vitro Literally 'in glass'; describing experiments or techniques performed in laboratory apparatus rather than in a living organism. Cell tissue cultures and *in vitro* fertilization (to produce test-tube babies) are examples. *Compare in vivo.*

in vivo Literally 'in life'; describing processes that occur or tests made within the living organism. *Compare in vitro.*

iodine A MICRONUTRIENT essential in animal diets mainly as a constituent of the thyroid hormones. Iodine is not essential to plant growth although it is accumulated in large amounts by the brown algae.

ion An atom or molecule that carries an electric charge as a result of the loss or gain of electrons.

ion exchange The reversible exchange of ions of similar charge between an insoluble solid and a solution that is in contact with it. Ion exchange is used to purify chemicals, to soften or demineralize water, to separate substances from mixtures in solution, and to lower potassium levels in patients suffering from kidney failure. *See also* cation exchange capacity.

ionization The formation of ions by the addition or removal of electrons from atoms, molecules, or radicals. It is a major route for the transfer of energy from electromagnetic radiation to matter. *See also* electromagnetic energy; ionizing radiation.

ionizing radiation Radiation that causes atoms or molecules to lose electrons, so forming ions. There are two main kinds: particulate (e.g. alpha and beta radiation, neutrons, positrons) and electromagnetic (e.g. gamma radiation and x-rays). Alpha particles are made up of two neutrons and two protons and carry a 2^+ charge. They do not penetrate tissues well, but produce intense tissue ionization. Beta particles are electrons or positrons, carrying only a single negative charge. They can penetrate several millimeters into tissues, but induce only moderate ionization. Gamma rays and x-rays do not carry a charge, and have no mass, so can penetrate deeper into tissues, creating moderate ionization. Gamma rays have higher energy than x-rays. The deleterious effects of ionizing radiation on living tissues is related to the high water content of tissues. The radiation generates highly reactive H and OH radicals that can damage DNA and proteins. See also radical.

ionosphere The part of the Earth's atmosphere that extends from about 50 km to the outer edge of the atmosphere at around 500 km, in which ionization of atmospheric gases is sufficient to affect the propagation of radio waves. The amount of ionization varies with the time of day, seasons, and solar cycle.

IPC See integrated pollution control.

IPCC See Intergovernmental Panel on Climate Change.

IPM See integrated pest management.

IRIS See Integrated Risk Information System.

iron (Fe) A metallic element that makes up some 5% of the Earth's crust. It is an essential nutrient for animal and plant growth. Iron is a constituent of the red blood pigment hemoglobin and plays an important part in many metabolic activities.

iron pan See hardpan.

irradiation The exposure of an object, organism, or chemical to IONIZING RADIA-TION such as high-energy particles or gamma rays or x-rays.

irrigation The supplying of water or wastewater to land by artificial means to enable plants to grow.

irruption A sudden unpredictable increase in the size of a population, leading to the emigration of large numbers of individuals.

island biogeography theory The premise that the distribution of plant and animal species on islands or areas that are similarly isolated, such as small patches of natural vegetation surrounded by cultivated land, shows a relationship between island area and species numbers, representing the equilibrium between IMMIGRATION and EXTINCTION. See isolating mechanisms.

isocline A line linking points that have the same rates of population increase for a particular species. For example, points that represent certain combinations of resources or species densities.

isoenzyme (isozyme) Any of multiple forms of the same enzyme, each having different kinetic characteristics. Isoenzymes are coded for by different ALLELES, and are usually formed by different combinations of the same subunits. They are readily separated by ELECTROPHORESIS, and can be used to determine the genetic relationships between populations and provide evidence for GENETIC DRIFT. For example, in forestry, they can confirm or refute the identity of seeds purporting to come from different local races of trees, each adapted to specific environmental conditions.

isolating mechanisms Structural, physiological, behavioral, genetic, geographical, or other factors that restrict the interbreeding of one population with another, hence restricting gene flow. The development of isolating mechanisms promotes the formation of new varieties and species. See also ecological isolation; geographical isolation; reproductive isolation.

iso-osmotic   *See* osmosis.

isopods   Members of a large order of crustaceans (*see* Crustacea) whose bodies are flat, elongated, and covered in a series of wide, armorlike plates. Isopods occur in the sea, in fresh water, and on land. Some are internal or external parasites of fish and prawns. Isopods include pill bugs (*Armadillo* and *Armadillidium* spp.) and sow bugs (*Oniscus* spp.), both of which live in leaf litter, and gribbles (*Limnoria* spp.), which burrow into underwater timbers.

isotherm   A line on a map or chart that connects points on the Earth's surface that have the same temperature at a given time, or the same mean temperature for a given period.

isotonic   Describing a solution with an osmotic pressure or concentration equal to that of a specified other solution, usually taken to be within a cell. It therefore neither gains nor loses water by osmosis. *Compare* hypertonic; hypotonic.

isotope   One of two or more atoms of the same element that differ in atomic mass, having different numbers of neutrons. For example ^{16}O and ^{18}O are isotopes of oxygen, both with eight protons, but ^{16}O has eight neutrons and ^{18}O has ten neutrons. A natural sample of most elements consists of a mixture of isotopes. Many isotopes are radioactive and can be used for labeling purposes. The isotopes of an element differ in their physical properties and can therefore be separated by techniques such as fractional distillation, diffusion, and electrolysis. *See also* isotopic dating.

isotopic dating   (radioactive dating; radiometric dating)   A method of determining the age of materials by measuring the concentrations of certain radioactive elements and their decay products. It is based on the principle that radioactive elements decay to other elements (which may or may not be radioactive) at a constant rate (the decay constant). Using the ratio of the amount of the daughter element to the radioactive parent element, the age can be calculated. Such techniques can be used to date ancient rocks, using *potassium–argon dating* (the decay of potassium-40 to argon-40 has a half-life of 11.8×10^9 years) or *rubidium–strontium dating* (half-life of rubidium-87 to strontium-87 is 48.8×10^9 years).

isotropic   Describing a substance whose physical properties (e.g. magnetic susceptibility or ability to refract light) do not vary with direction.

isozyme   *See* isoenzyme.

iteroparous   Describing a plant that reproduces more than once in its lifetime, for example most perennials. Some species, such as the century plant, *Agave americana*, live for many years, then flower and fruit only once and die – such species are described as *monocarpic*.

IUCN   *See* World Conservation Union.

IWC   *See* International Whaling Commission.

J *See* joule.

jet streams Concentrated high-speed high-altitude belts of air flowing broadly east to west. The polar-front jet stream occurs at 10–13 km altitude in middle to high latitudes, associated with the boundary between polar and tropical air and the related depressions. It averages 60 m/sec, and is most persistent and intense in winter, when the temperature gradient between the two air bodies is greatest. The subtropical jet stream flows at 12–15 km altitude, and changes direction from easterly in summer to westerly in winter after the monsoon. It is associated with the sinking air masses formed as warm air above the equator rises and disperses laterally, then cools in the subtropics. This jet stream moves closer to the equator in winter. In the northern hemisphere there is also a tropical easterly jet stream not found in the southern hemisphere, because the lack of large landmasses leads to different airflow patterns. Jet streams are so powerful that they can exert a steering influence on depressions traveling lower in the atmosphere.

joule (J) The SI unit of measurement of work, energy, and heat. 1 joule is the work done when a force of 1 newton moves its point of application 1 m in the direction of the force. It is equal to 0.2392 calorie.

jungle A form of dense tropical RAIN-FOREST or MONSOON FOREST, with tangled lianas and other climbers, palms, and bamboo scrub. Jungle is not used as a scientific term. It is a subclimax forest, found along riverbanks and on the sites of former clearings. The dense undergrowth is a result of greater light penetration than in climax rainforest. *See* climax community.

Jurassic The middle period of the Mesozoic era, 215–145 million years ago. During the Jurassic dinosaurs were becoming large and abundant and bony fishes (teleosts) were also evolving rapidly. Fossils of one of the earliest known birds, *Archaeopteryx*, and also of the first mammals are found in the late Jurassic. The climate was warm and humid, and there was widespread expansion of forests, which gave rise to thick deposits of COAL. The Jurassic has been called the 'Age of Cycads'. The seed-ferns declined, and there was a great increase in the number and diversity of cycads, palmlike GYMNOSPERMS, and conifers. Fossils of monocotyledonous pollen and stems of palms suggest that the angiosperms (*see* Angiospermophyta) evolved during this period.

K

Kampfzone *See* elfin forest.

karst Any region underlain by LIME-STONE having a characteristic set of barren landforms with bare pavements, sink-holes, smaller holes, and deep fissures due to solution weathering of well-jointed limestone, and underground caves and streams. The term was originally applied only to the landscape of the Karst in the Dinaric Alps bordering the Adriatic Sea.

kelp Large brown seaweeds that grow below the low-tide mark in relatively shallow water, anchored to the seabed or to rocks. Kelp are harvested for fertilizer, either pulped into a liquid manure or burned to provide ash. The ash is a rich source of iodine, potash, and soda.

Kelvin scale *See* temperature.

keranga *See* heath forest.

kerosene *See* petroleum.

key A hierarchical list of characteristics organized so as to enable rapid identification of a specimen. Most keys consist of a succession of contrasting choices, each leading to further selections until by process of elimination the specimen is finally identified. A key is not a natural classification – it is not based on phylogenetic relationships – but simply an aid to identification. An error made at any stage, but particularly near the top of the hierarchy, can lead to serious misidentification.
   A simple key, which at each stage presents the user with a pair of choices, is called a *bracketed key*. In an *indented key*, successive choices are indented below the immediate parent feature. This shows more clearly the characteristics of different taxa, making it easier to retrace your steps if you have made an error. Alternatively the choices may be organized as a flow-chart.
   In a *multi-access key*, different alternatives for each character of the organism are offered, and each is given a letter or number code. Having made your selections for all the specimen's characteristics, you then look up the combination of letters or numbers to find the species.

key-factor analysis A statistical treatment of population data that includes information on births, deaths, immigration, and emigration, and sometimes also various stages of the life cycle such as pupation or hibernation, that aims to identify the factors most likely to cause changes in population size.

keystone species An organism that plays a pivotal role in maintaining the status quo of its ECOSYSTEM, and especially the composition of its local COMMUNITY. For example, the removal of a keystone predator from an ecosystem causes a reduction in SPECIES DIVERSITY among its former prey. Monitoring a keystone species can give valuable information about the health of the ecosystem and the degree to which its resources are being exploited or overexploited. *See also* species diversity.

kieselguhr *See* diatoms.

kinetic rate coefficient A number that describes the rate at which a water attribute such as BIOLOGICAL OXYGEN DEMAND rises or falls.

kingdom Formerly, the highest ranking

category in most classification systems. However, most taxonomists today recognize the rank of DOMAIN as higher than kingdom. The most widely accepted classification system, the FIVE KINGDOMS CLASSIFICATION, recognizes five kingdoms, of which four (BACTERIA, FUNGI, PLANTAE, and ANIMALIA) are natural on phylogenetic grounds and the fifth (the PROTOCTISTA) contains unrelated organisms that do not fit into the other four kingdoms. Two superkingdoms are often recognized – the Prokarya (*see* prokaryote) and Eukarya (*see* eukaryote), based on fundamentally different cell structure, but increasingly these groups are being replaced by the three domains of Bacteria, ARCHAEA, and Eukarya, based on more recent molecular evidence. *See* phylogeny. *See also* cladistics.

kin selection   The concept that differences in reproductive success among lineages of related individuals can be explained in terms of the propagation of genes identical to their own – a kind of self-interest that is served by promoting the reproduction of relations that share all or a large proportion of their genes. For example, worker bees are all daughters of the same queen, developed from unfertilized eggs. They thus share all their genes. By caring for each other, and for their egg-laying mother, they promote the production of more workers that also carry identical genes to their own. Were they themselves to mate, their offspring would carry only half their genes. The same theory has been used to explain the sharing of care of the young between related animals that are not the parents, and other communal behavior.

Köppen climate classification system   *See* climate.

Kranze anatomy   The specific arrangement of the photosynthetic tissues in the leaves of C_4 PLANTS. *See* photosynthesis.

Krebs cycle   *See* respiration.

krill   Shrimplike marine crustaceans (*see* Crustacea).

Krummholz   *See* elfin forest.

***K* selection**   *See* life-history strategy.

***K* species**   A species that responds to *K* selection (*see* life-history strategy) and has low birth rates, a relatively long period of development, and high survival rates. *See also* density dependence.

kurtosis   The peakedness of a statistical distribution, as measured against a normal distribution. Kurtosis may be positive or negative. The larger the kurtosis, the more peaked the distribution, the smaller the kurtosis, the flatter the distribution.

***k*-value**   The effectiveness of a factor causing mortality, given by the difference in numbers in the population before and after the factor acts, expressed as $\log_{10}$ (before) – $\log_{10}$ (after). This may be given for the population as a whole or for a particular stage in the life cycle.

Kyoto Protocol   An international treaty signed in 1997 by which consenting governments agreed to reduce their emissions of greenhouse gases to less than their 1990 levels by the year 2010. *Compare* Montreal Protocol.

L

lagoon 1. A body of shallow coastal water with restricted access to the sea, from which it is usually separated by a reef, spit, or sand bar, or by a series of islands. 2. A shallow pond used as temporary storage for WASTEWATER, SEWAGE SLUDGE, liquid waste, or spent nuclear fuel rods. 3. An artificial treatment pond where the natural decomposition of biological waste is accelerated by the action of sunlight, bacteria, and oxygen.

lag phase The period of relatively slow growth shown by microbial populations when they are first introduced to new media.

Lagrangian measurements *See* waterflow measurements.

Lagrangian models A system of modeling based on a moving reference point. Such models may be use to track moving bodies of air or gaseous POLLUTANTS. Statistical Lagrangian models use long-term averages of concentration and deposition fields of pollutants and of precipitation, as well as climate data and annual frequencies of wind direction and speed, and also take into account the different species of products resulting from breakdown of the pollutants. Alternatively the model may be based around a comparative study of a succession of air mass trajectories arriving at a receptor at relatively short intervals (e.g. every 6 h). Lagrangian models are in stark contrast to the more complex *Eulerian models*, in which the governing equations are solved for each grid point on the trajectory at each time interval.

LAI *See* leaf area index.

lake A large body of water surrounded by land. It is an aquatic BIOME.

Lamarckism A largely discredited theory of evolution postulated in 1809 by French biologist Jean-Baptiste de Lamarck (1744–1829) that acquired characteristics can be inherited, so resulting in permanent changes in populations.

land bridge (isthmus) A piece of land connecting two large land masses, which permits the migration of plants and animals between the land masses. For example, the Isthmus of Panama, which was submerged by the sea for most of the TERTIARY, allowing North and South American species to evolve separately, was raised up to become land again during the PLIOCENE, allowing mixing of the species. It is thought that humans may have first moved into North America from Siberia across a land bridge where the Bering Sea now lies. *See also* geographical isolation.

land farming *See* land spreading.

landfill A land-based disposal facility for solid waste. The waste is deposited in an excavated pit sealed to prevent the leaching of contaminants into the surrounding soil, rocks, or water, compacted or leveled, then buried under soil. The are two main types: *sanitary landfills*, which take municipal waste, and *hazardous waste landfills* to contain hazardous waste.

landfill gas Gas generated by the anaerobic decomposition of solid waste in landfill sites. It is a mixture of gases, including odorous organic sulfur compounds, and contains significant amounts of methane. If the METHANE concentration exceeds a concentration in air of 5%, it poses a risk of explosion, so landfill gas is usually actively or passively vented, or siphoned off for use in COMBINED HEAT AND POWER PLANTS. *See* decomposition.

Landsat satellite *See* remote sensing.

landscape architecture The creation of new outdoor landscapes, such as gardens and parks, and other green open spaces, in order to create a natural setting for urban life and to enhance the environment for human use and enjoyment.

landscape ecology The study of landscapes taking into account the distribution patterns of communities and ecosystems and the ecological processes that affect them over time. *See* ecology.

land spreading 1. The spreading of manure, mulch, or fertilizer over the land. 2. (land farming) The spreading of HAZARDOUS WASTE onto the soil (sometimes covering it with a further layer of soil) to allow it to be naturally decomposed by microbes.

land-use capability (agricultural potential) The suitability of land for agriculture. This takes into account soil type, climate, and topography, and the presence of problems such as industrial waste sites. Land is ranked into classes according to the anticipated sustainability of crop production on the land, the degree to which is can be cultivated, and the effort needed to bring it into cultivation. Maps showing land-use capability rankings are used by planners. *Environmental potential* maps are similar, but are intended for assessing multiple-use locations, so other factors such as potential subsidence, rock foundations, mineral resources, and susceptibility to flooding are also taken into account.

Langelier index (LI; saturation index; water balance index) A formula used to determine the overall tendency of a sample of water to be either scale-forming or corrosive. It takes into account pH, water temperature, calcium hardness, and total alkalinity. If all these factors are in balance, the water will be neither scale-forming nor corrosive.

La Niña An irregularly occurring climatic perturbation characterized by below-normal sea surface temperatures in the eastern and central Pacific, and higher-than-normal temperatures in the western Pacific. It causes upwelling of cold water along the western coast of South America that disrupts local and global weather patterns in the opposite fashion to EL NIÑO. La Niña often follows an El Niño event.

lapse rate The rate at which temperature decreases with altitude. In the TROPOSPHERE the average lapse rate is about 6.5°C per 1000 meters.

larva *See* Insecta.

latency The interval between first exposure to a toxic chemical and the appearance of an effect.

lateral Situated at or on a side, or relating to a side.

laterite A product of rock decay that is red in color due to its high content of oxides of iron and aluminum. It may form a crust on the surface of the SOIL in tropical regions with alternating wet and dry seasons. During the wet season soluble mineral salts are washed down into the lower soil horizons, but in the dry season evaporation from the surface draws the solution back up by capillarity, and iron and aluminum oxides accumulate, forming the crust, a process called *laterization*. Such a soil is called a *latosol*. *See* red-and-yellow forest soil. *Compare* hardpan.

laterization *See* laterite.

Latin square An experimental design grid in which the number of treatments equals the number of replications, so each treatment occurs once in every column and row. In field trials this design eliminates the effects of differences in environmental factors such as soil nutrients and water content, which may vary up, down, and across the experimental area. However, a Latin square is not suitable for large numbers of treatments, because it would require impracticably large numbers of replications.

latitude The angular distance north or south of the EQUATOR (latitude 0°) of a point on the Earth's circumference, which is equal to the angle subtended at the center of the globe by the meridian between the Equator and the point in question. Together with the LONGITUDE this gives the point's exact location.

latosol *See* laterite.

Laurasia The northern continent produced by the splitting of the supercontinent PANGAEA along the line of the northern Atlantic Ocean and the Tethys Sea. It included the land masses that eventually separated to become North America, Greenland, Europe, Asia, and the Pacific islands east of Sulawesi.

leachate *See* leaching.

leaching The removal of HUMUS and SOIL nutrients in solution by water moving down the soil profile. It makes soils more acid because cations (e.g. potassium and magnesium ions) are replaced by hydrogen ions. Leaching leads to the formation of podsolized and lateritic soils. Leaching of nutrients, especially nitrates, from agricultural land can lead to serious pollution of water supplies. The term is also applied to water percolating through solid waste and extracting dissolved or suspended materials from it. In both cases the water containing the leached materials is called the *leachate*. *See also* laterite; podzol.

lead (Pb) A dense gray soft metallic element that occurs in small quantities in a wide variety of minerals, especially galena. It oxidizes readily when cut, forming a surface layer of lead monoxide, which protects the metal from further corrosion. Lead is hazardous to health if breathed or swallowed, and most countries have legislation to restrict its use in paints, gasoline,

and pipework, and its use in pesticide sprays is being reduced. Emissions of *lead tetraethyl* from vehicle fuel is thought to be toxic, especially to children. Lead poisoning affects particularly the nervous system, gastrointestinal tract, and blood-forming tissues, and it is believed to affect brain development in children.

lead tetraethyl *See* lead.

leaf A flattened appendage of a plant stem that arises as a superficial outgrowth from the apical meristem. Leaves are arranged in a definite pattern, have buds in the angle between leaf and stem, and show limited growth. It is the main photosynthesizing organ of a plant.

leaf area index (LAI) The ratio of the total surface area of a plant's leaves to the area of ground available to that plant:
LAI = leaf area/ground area.
The LAI is a useful guide when assessing the density of plants that can be successfully cultivated on a given area of land.

leaf fall The shedding of leaves as a result of the formation of an abscission zone at the base of the leaf stalk (*see* abscission). In both evergreen and deciduous trees old leaves are continually shed throughout the growing season, but in deciduous species the remaining leaves are also shed at the onset of winter or a dry season.

Leakey, Richard Erskine (1944–) Kenyan anthropologist and conservationist. Richard Leakey was born in Nairobi, the son of the famous anthropologists Louis and Mary Leakey. Having left school at sixteen, he first worked as a hunter and animal collector before turning in 1964 to the search for fossil man. In the 1960s and 1970s, Leakey did significant anthropological research at Lake Turkana in northern Kenya. In the 1980s, following a period of ill-health, he changed his career and devoted much of his time to conservation. In particular, Leakey has been a leading figure in the fight to preserve the African elephant by banning the trade in ivory.

lean burn A combustion engine technology that reduces unwanted exhaust emissions by burning unusually weak mixtures of gasoline and air. It achieves this by having a pool of enriched mixture close to the sparking plugs, which enables the plugs to ignite the fuel; ignition then spreads to the weaker remainder.

least significant difference In analysis of variance, a statistical test of the degree to which the means of two samples or populations differ, used once the analysis of variance has already shown that they do differ. A level of significance is selected, for example 5%. In this case, the least significant difference at the 5% level is the difference between two samples or means that would be exceeded only 1 in 20 times under random sampling conditions. *See also t*-test.

leghemoglobin A protein found in the ROOT NODULES of leguminous plants (Fabaceae) that contain the nitrogen-fixing bacterium *Rhizobium*. It is coded for by a plant gene, but is synthesized only in the presence of the bacteria. It is thought to transport oxygen to the bacteria in a way that avoids destruction of the nitrogenase (nitrogen-fixing) enzyme, which is sensitive to oxygen. *See* nitrogen fixation.

Legionella A genus of rod-shaped bacteria (bacilli) that include *L. pneumophila*, the causative organism of Legionnaires' disease. This is a pneumonia-like infection that may prove fatal to vulnerable people such as the elderly. It appears to spread by inhaling contaminated droplets of air from air conditioning systems, the bacterium multiplying in the condensing towers of the cooling system.

legume Any plant of the family Fabaceae, e.g. peas, beans, etc., also known as *pulses*.

lek A territory held by males of certain animal species during the breeding season for the purpose of displaying to the females and attracting them to mate. It is defended against rival males. In some species the lo-cation of the territory may be almost as important as the display itself in attracting the attention of a potential mate, dominant males holding leks in the center of the display ground. Examples of such species include sage grouse (*Centrocercus urophasianus*), dugongs (*Dugong dugon*), and dung flies (family Scatophagidae). *See* territoriality.

leptokurtic A distribution of values that is symmetrical, similar in shape to a NORMAL DISTRIBUTION, but which has a much higher central peak and consequently flatter tails on either side.

lethal dose The dose of a toxic substance that will kill 50% of test organisms within a designated period of time.

leukemia A cancer of the blood-forming organs in humans and other warm-blooded animals, characterized by a large rise in the number of white blood cells in the tissues, and often also in the blood. The disease is due to damage to the genetic material. Some incidents can be related to high doses of radiation.

Lewis–Leslie matrix A simple matrix model used to predict the structure of a POPULATION through time. It consists of a matrix of values of fecundity and survivorship for individuals of each age.

LI *See* Langelier index.

liana A long-stemmed woody climbing plant that grows from ground level to the CANOPY of trees, and often hangs down freely from branches, e.g. curare (*Strychnos toxifera*). Lianas may have stems up to 100 m long, and may climb over the canopy to gain maximum sunlight. Many flower only when they reach the top of the canopy. Up to a quarter of all the woody growth in some RAINFORESTS may be lianas. They often bind the forest trees together, and if one tree falls or is felled, it is held up by the others until it rots away. A temperate example is old man's beard (*Clematis vitalba*).

lichens Symbiotic associations between an alga or cyanobacterium (the *photobiont*) and a fungus (the *mycobiont*). They are slow-growing but can colonize areas too inhospitable for other plants.

There are three main forms of lichens: *crustose* lichens have relatively simple, usually unbranched thalli and grow closely attached to the substrate; *foliose* lichens are often only loosely attached, and have lobed 'leafy' thalli; *fruticose* lichens are attached at only one point, and may be upright and bushy or dangling and tassel-shaped. Some lichens have an intermediate form in which a crustlike thallus gives rise to upright branched structures.

Certain lichens are important pioneers of newly exposed areas, and can tolerate extremes of desiccation. Their metabolic products contribute to rock disintegration. Many lichens grow extremely slowly, maybe only 1 mm a year, so large lichens may be hundreds or even thousands of years old. Some species are tolerant or intolerant of POLLUTION, and are valuable INDICATOR SPECIES.

Liebig's law of the minimum The concept, first proposed in 1840 by Justus von Liebig, that the health and growth rate of a plant and the size it reaches depend on the availability of the scarcest of its essential nutrients. The modern version of this is the concept of LIMITING FACTORS.

life cycle The sequence of developmental changes making up the span of an organism's life from the fertilization of gametes to the same stage in the subsequent generation. The cycle may involve only one form of the organism, as in higher animals and plants, involving a simple progression from the formation of the ZYGOTE to the release of GAMETES by the same individual (*haplobiontic*). In most plants, many algae, and some animals there is a succession of individuals in the life cycle, linked by sexual or asexual reproduction, usually an alternation between a HAPLOID and a DIPLOID generation, as in ferns. This type of life cycle is called *diplobiontic*. In different species the haploid or the diploid stage may the dominant form. Haploid and diploid generations may or may not differ morphologically. *See* alternation of generations. *See also* polymorphism.

life form The overall morphology, habits, and lifestyle of an organism. A description of the vegetation of an area usually includes reference to life forms – trees, shrubs, succulents, annuals, etc. Different life forms are associated with different types of environment. *See* Raunkiaer's life-form classification.

life history All the changes an organism goes through from fertilization until death, including its pattern of growth and reproduction.

life-history strategy The idea that SELECTION PRESSURE acts on an organism's life history – in particular the number of young it produces, their rate of development, amount of parental care, and survival rates – to adapt it for survival in either a stable population near the carrying capacity of its environment or an unstable population in an unpredictable or short-lived environment. There are two contrasting strategies – in *K*-selection, individuals produce small numbers of young that require considerable parental care or investment and take a relatively long time to develop but have a high survival rate, while in *r*-selection large numbers of young are produced, which are smaller at birth/hatching, and which develop rapidly to sexual maturity but have low survival rates. *r*-selection provides for rapid expansion of the population under favorable conditions, while *K*-selection makes it less likely that a population will outstrip its resources under relatively stable conditions.

life table A summary of the survival rates of individuals in a population at different ages or at different stages in the life cycle. In a *cohort life table* or *dynamic life table* a group of individuals born within the same short time interval (i.e. a COHORT) is followed from birth to death of the last survivor. This type of life table is suitable for ANNUAL species in which there is little or no overlap of generations. It is also suitable for sessile animals and for organisms in a restricted area such as an island, where dispersal is limited and it is possible to sample marked individuals throughout their lifetimes.

For more mobile animals and populations containing several generations, a life table can be constructed from knowledge of the survival of individuals of known age over a single time interval. Such a *time-specific, vertical*, or *static life table* requires a knowledge of the ages of individuals, and also makes the assumption that the survivorship rates and the numbers in each cohort remain the same from year to year, and that birth rates, death rates, and population size are static. In a *dynamic-composite life table*, instead of a true cohort, a composite number of the animals marked over a period of years is used and the data are pooled. This method again makes the same assumption as the static life table.

life zone   A zone in which the flora and fauna are typical of a particular latitude or range of altitude. These zones reflect environmental gradients. The term is most commonly used in relation to relatively local changes such as the zonation of communities on a mountainside. *Compare* biome.

light   Electromagnetic radiation in the visible spectrum, i.e. with wavelengths ranging from 400 nm (extreme violet) to 770 nm (extreme red). The *wavelength* of light is the distance from one peak of the electromagnetic wave to the next. The *spectrum* of light is the span of wavelengths visible to the human eye. White light contains light of all wavelengths. Objects appear colored because they reflect certain wavelengths of light, but not others. While light may be considered as waves of electromagnetic energy, it may also be viewed as discrete packets of energy that carry the electromagnetic field, called *photons*. A photon has no charge or mass. Its energy depends on the wavelength of the light, shorter wavelengths having higher energy. All photons travel at the speed of light, about 299,492 kps per second. When light is reflected or refracted by

certain substances, it is separated into two or more components, such that the wave action of each is concentrated in a different plane. This is called *polarized light. See* electromagnetic energy.

light attenuation instrument (transmissometer)   A device that measures the attenuation of light in water due to suspended particles. The degree of attenuation of light is related to the concentration of suspended materials, but not to the number of suspended particles: a single large particle might produce the same attentuation as hundreds of smaller ones.

light bottle   *See* dark bottle.

light compensation point   *See* compensation point.

light meter   An instrument for measuring the intensity of incident or reflected light, consisting of a photosensitive cell and a milliammeter or millivoltmeter calibrated in light levels.

lightning   A discharge of electricity between clouds and the earth, which produces a brilliant flash and the sound of thunder. It occurs when electrostatic charges build up on the undersurface of clouds. *See* nitrogen cycle.

light pollution   The emission of light into the night sky by street lights and other artificial lighting, which produces a glow in the sky and reduces the visibility of the stars.

light reaction   *See* photosynthesis.

light saturation point   *See* compensation point.

lignin   *See* wood.

limestone   A SEDIMENTARY ROCK containing at least 50% by weight of calcium carbonate, and sometimes also magnesium carbonate. It may be organic, having been formed from the calcareous skeletal remains of living organisms, such as corals or algae; formed from the accumulated fragments of earlier limestones; or derived by solution and precipitation of preexisting limestones.

limestone pavement A large area of limestone that has been weathered into blocks (*clints*) by solution processes that have carved out a series of deep, smooth-sided clefts (*grikes*). The grikes have a characteristic flora that often includes rare species.

limestone scrubbing The use of LIMESTONE to remove sulfur and other pollutants from waste gases.

lime-sulfur spray An orange liquid made from elemental sulfur combined with calcium hydroxide and water, which is used to kill various mites and to control peach leaf curl, scab, and powdery mildews.

liming *See* flocculation.

limiting factor Any factor in the environment that governs the behavior or metabolic activity of an organism or system by being above or below a certain level. In general, a number of different factors (e.g. temperature, light intensity, carbon dioxide concentration, availability of prey) may be limiting, but at any one time the factor that is closest to its critical minimum is the limiting factor. For instance, the rate of photosynthesis rises with increasing light intensity so long as there is sufficient carbon dioxide available, but at high light intensities carbon dioxide may become the limiting factor instead.

limiting resource A resource present in insufficient quantity to the extent that growth or reproduction of individuals in a population is restricted. The term is usually given to the RESOURCE that is having the greatest restrictive effect at any one time. *Compare* limiting factor.

limnetic Relating to deep fresh water.

limnology The scientific study of fresh-

water and its flora and fauna, and the chemical and physical properties of inland waters.

lindane An organochlorine PESTICIDE used as an INSECTICIDE and wood preservative. Its use is banned in many countries, because it is harmful to bees and to freshwater and marine aquatic life, and is also toxic to humans. *See* organochlorines.

line of best fit *See* graph.

link *See* food web.

linkage The occurrence of GENES together on the same CHROMOSOME so that they tend to be inherited together and not independently. Groups of linked genes are termed *linkage groups* and the number of linkage groups of a particular organism is equal to its HAPLOID chromosome number. Linkage groups can be broken up by crossing over at MEIOSIS to give new combinations of genes. Two genes close together on a chromosome are more strongly linked, i.e. there is less chance of a crossover between them, than two genes farther apart on the chromosome. Linkage is indicated when the associated inheritance of two or more non-allelic genes is greater than would be expected from INDEPENDENT ASSORTMENT. The genes on a single chromosome form one linkage map. *See* allele; gene; independent assortment.

linkage density *See* food web.

lipid Any of various fatty or oily substances found in plants and animals. Lipids are classified into simple lipids and compound lipids. The simple lipids include *glycerides (acylglycerols)*, and the *waxes*. Compound lipids include the *phospholipids (phosphoglycerides* or *phosphatides)*.

Simple lipids include plant pigments such as carotene and xanthophylls, and also more complex substances such as natural rubber. Glycerides are important storage lipids in seeds, especially in such species as the castor oil bean (*Ricinis communis*). Lipids are also important energy storage compounds in animals, forming 'fat', which may also help insulate against the cold. Waxes coat plant surfaces, helping to provide a waterproof layer and mechanical barrier.

liter A unit of volume equal to 10 cubic decimeters or 0.264172 US gallon.

lithology The characteristics of a rock formation: its mineral composition, grain size and distribution, and bedding. The term is usually applied to visible features.

lithosere A succession of plant communities that start on a bare rock surface. *See* sere.

lithosphere The solid outer part of the Earth, including its crustal plates (*see* crust; plate tectonics) and the brittle uppermost part of the mantle. It thickness varies from about 1–2 km at the crests of the MID-OCEANIC RIDGES to 140 km under older parts of the oceanic crust, and up to 300 km in parts of the continental crust.

lithotroph An organism that derives its energy from the oxidation of inorganic compounds or elements.

litmus A dye obtained from certain lichens (*Lecanora tartarea* and *Roccella tinctorum*). In acids blue litmus turns red and in alkalis red litmus turns blue. *Litmus paper* is prepared by soaking absorbent paper in litmus solution and then drying it. Litmus solution changes over a pH range 4.5–8.3.

litter Dead plant remains on the soil surface.

Little Ice Age The period between about 1550 and 1860 during which there was a worldwide expansion of glaciers and the climate in mid-latitudes became temporarily colder.

littoral 1. The zone of the seashore between the high and low tide mark. The term is also applied to organisms living in this zone. Since tidal ranges vary continu-

ally, the zone is often defined in terms of the upper and lower limits of certain species of organism. *Compare* benthic zone; sublittoral. 2. The zone between the water's edge and a depth of about 6 m in a pond or lake, where light reaches the bottom sediments. Rooted HYDROPHYTES, both emergent and submergent, are found in this zone. *Compare* profundal; sublittoral.

liverworts *See* Hepatophyta.

livestock manure *See* manure.

living fossil A modern organism with anatomical or physiological features found elsewhere only in extinct species. Examples include the deepsea fish the coelacanth (*Latimeria chalumnae*), and the maidenhair tree (*Ginkgo biloba*), discovered in Japan in the 17th century, and later in China, but found only in cultivation.

llanos SAVANNA GRASSLANDS of the Orinoco river basin in Venezuela and northeast Colombia. It is a mainly treeless area covered in swamp grasses and sedges with just a few scrub oaks and dwarf palms, trees being concentrated mainly along the rivers and in areas of higher altitude. Cattle ranching is an important local industry, together with oil extraction.

loam A medium-textured SOIL containing a mixture of large and small mineral particles. Loams are easy soils to work and combine the good properties of sandy and clay soils.

local diversity *See* species diversity.

local guild *See* guild.

local mate competition Competition between males for mates that takes place close to their place of birth, resulting in frequent mating with close relatives.

local population *See* subpopulation.

local stability The tendency for a COM-

MUNITY to return to its original state when subjected to a small perturbation.

locus (*pl.* loci) The position of a GENE on a CHROMOSOME. ALLELES of the same gene occupy the equivalent locus on HOMOLOGOUS CHROMOSOMES. *See* linkage.

loess A fine-textured uniform SOIL comprising mainly quartz particles about 0.015–0.005 mm in diameter. Widespread in central Europe, southern Russia, China, the central USA, and Argentina, it is derived from wind-blown clay and silt particles originally deposited at the edge of the ice sheets at the end of the last Ice Age. Loess is a fertile soil, often calcareous, and is widely used for growing wheat and other crops.

logging The felling of trees, usually for commercial purposes.

logistic equation (logistic model) A mathematical expression of the S-shaped growth curve of a single-species population in which there are no discrete generations, which shows the response of population growth to competition in a confined space with limited resources:
$$dN/dt = rN(N - K)/K$$
where N is the number of individuals in the population, t is time, r is the BIOTIC POTENTIAL of the organism concerned, and K is the carrying capacity (saturation value) for that organism in the environment.

logistic model *See* logistic equation.

lognormal distribution *See* distribution.

LOI *See* loss on ignition.

long-day plant *See* photoperiodism.

longitude The angular distance of a point east or west of the Greenwich meridian, an imaginary line running from the North Pole to the South Pole at right angles to the Equator that passes through the London suburb of Greenwich in the UK, which represents 0° of longitude. It is ex-

pressed as the number of degrees east or west of this meridian, which is equal to the angle subtended at the center of the globe between the Greenwich Meridian and the point in question. Together with the LATITUDE this gives the point's exact location.

longshore drift The lateral movement of sand and shingle along a shoreline due to wave action.

Longworth trap A trap used to catch small mammals without injuring them. It consists of a metal box in which food and/or nesting material are placed, approached by a narrower tunnel. On entering, the mammal trips a door that drops behind it.

loss on ignition (LOI) A technique for measuring the organic content of soil and sediment samples based on the loss in weight of the sample as the organic matter is completely burned off.

Lotka–Volterra model A model of the dynamics of predator–prey or herbivore–plant systems. Fluctuations in the prey population are described by
$$dN/dt = rN - aPN$$
where N is the number of prey, r a constant, P the number of predators, t is time, and a the searching efficiency or attack rate. Fluctuations in the predator population are described by
$$dP/dt = fa\,PN - qP$$
f being the predator's efficiency at turning food into predator offspring and q another constant. The model shows a tendency for predator–prey populations to undergo oscillations, in which increasing predator population leads to a decrease in prey population. This leads to a decrease in predator numbers, allowing prey to build up again, and so on.

Lovelock, James Ephraim (1919–) British scientist. Lovelock had developed a sensitive electron-capture detector and, in the summer of 1966, working as an independent scientist, he used it to monitor the supposedly clean Atlantic air blowing onto the west coast of Ireland. Here he detected the presence of chlorofluorocarbons (CFCS). Although unable to pursue the matter further at the time through lack of funding, Lovelock did further work in the Antarctic in 1971, where, again, he found atmospheric CFCs. It was partly as a consequence of this work that ROWLAND began to ponder their role in the atmosphere. It is, however, as the author of the GAIA HYPOTHESIS, first presented in his *Gaia* (London, 1979), and developed further in several sequels, that Lovelock is best known.

low-emissivity windows Windows constructed to inhibit transmission of radiant heat while allowing sunlight to pass through, thus reducing energy loss from buildings.

lowest achievable emission rate Under the CLEAN AIR ACT, an emission rate that reflects either the most stringent emission limitation contained in the implementation plan of any state for such source (unless the owner or operator of the proposed source demonstrates such limitations are not achievable) or the most stringent emissions limitation achieved in practice, whichever is the more stringent. In any case, application of this term does not permit a proposed new or modified source to emit pollutants in excess of existing new source standards.

low-level radioactive waste *See* radioactive waste.

low-NOx burner A combustion technology that reduces emissions of nitrogen oxides.

Luca *See* origin of life.

luxury consumption A means of exploiting short-term abundance of a nutrient by consuming and storing more than is needed for growth.

Lydekker's line *See* Wallacea.

M

macchia *See* maquis.

MacArthur, Robert Helmer (1930–72) American ecologist. MacArthur made a number of contributions to ecology. His paper *Population Ecology of some Warblers of Northeastern Coniferous Forests* (1958) described his investigations into how five closely related species of warbler could spend the spring in the spruce forests of New England apparently feeding on the same diet and occupying the same niche. MacArthur managed to establish that the warblers tended to occupy and hunt in different parts of the tree, so preserving the current dogma – that in equilibrium communities no two species occupy the same niche. At a more theoretical level MacArthur produced the broken stick model (*see* random niche model), which he used to predict the relative abundance of species in a particular ecosystem. With Edward Wilson he published *Theory of Island Biogeography* (1967). MacArthur and Wilson were struck by the apparent stability of the number of species found on islands. They assumed immigration and extinction to be the major forces operating and further assumed that these two processes are in equilibrium. This allowed them to predict that smaller islands should have fewer species than large ones, as too would distant islands over those nearer to the mainland. In the same 1967 volume MacArthur and Wilson clearly formulated a distinction that was emerging in a variety of forms in the writings of ecologists, that of *r*- and *K*-strategists. *r*-strategists are opportunistic species living in a variable environment; they typically have high reproductive rates, heavy mortality, short lives, and rapid development. *K*-strategists on the other hand are larger, develop more slowly, and are in general more stable. In 1962 MacArthur succeeded in showing that the basic theorems of natural selection apply to both *r*- and *K*-species.

macrobiota The larger soil organisms, especially tree roots and vertebrates such as moles. *Compare* macrofauna; meiofauna; mesobiota; mesofauna; microfauna.

macroevolution EVOLUTION on a scale above the species level. It includes the evolution of new genera, families, orders, classes, and phyla. *Compare* microevolution.

macrofauna The larger animals of a soil community, usually invertebrates of body width approximately 2–20 mm, but excluding larger animals such as earthworms. *Compare* macrobiota; meiofauna; mesobiota; mesofauna; microfauna.

macroflora The larger plant structures in the soil, such as tree roots. *Compare* microflora.

macromolecule A very large molecule, usually a polymer, having a very high molecular weight (with 10,000 or more atoms). Proteins and nucleic acids are naturally occurring examples.

macronutrient A nutrient required in more than trace amounts by an organism. It may be an organic or inorganic compound. *Compare* micronutrient.

macroparasite *See* parasitism.

MACT *See* maximum available control technology.

146

magma   *See* plate tectonics.

magnesium   An element essential for plant and animal growth. It is contained in the CHLOROPHYLL molecule and is thus essential for photosynthesis. Deficiency of magnesium may result in CHLOROSIS, stunted growth, and in some species whitening and puckering of the leaf edges. It may be remedied by adding magnesium sulfate or magnesium oxide to the soils. In animals magnesium is found in bones and teeth. As magnesium carbonate it is found in large quantities in the skeletons of certain marine organisms, and is present in smaller quantities in the muscles and nerves of higher animals.

magnesium sulfate   A colorless crystalline solid that occurs naturally in many salt deposits in combination with other minerals. It is used as a source of magnesium in fertilizers.

magnetic north   The direction in which a compass needle points, as opposed to true north (the North Pole). It is the northernmost pole of the Earth's magnetic field, the location of which may vary over geological time. *Compare* true north.

mainland-island metapopulation   *See* metapopulation.

maintenance threshold   (maintenance rate)   The level or rate of consumption below which a CONSUMER cannot survive.

malathion   An ORGANOPHOSPHATE insecticide used especially to kill aphids, mites, flies, and mosquitoes.

male sterility   1. A condition in plants in which pollen production is prevented by the mutation of one or more of the genes involved. Male sterility is used by plant breeders to ensure cross-pollination takes place. It is also used in GENETIC ENGINEERING to ensure that genetically modified varieties do not release pollen containing modified genes.
2. The release of artificially sterilized male animals into a wild population with the aim of reducing the breeding rate of the population as a form of BIOLOGICAL CONTROL, for example in the control of mosquitoes.

mallee   *See* maquis.

Malthusian theory   The theory, proposed by Thomas Malthus (1766–1834), that population growth will always continue to the limit of subsistence, where resources become exhausted, and that the only way to avoid famine, war, and ill health is to impose strict reproductive controls.

Mammalia   The class of vertebrates that contains the most successful tetrapods. They are homoiothermic, with an insulating body covering of hair and usually with sweat and sebaceous glands in the skin. The socketed teeth are differentiated into incisors, canines, and grinding premolars and molars. Mammals have a relatively large brain and an external ear (pinna), and three auditory ossicles in the middle ear. Oxygenated and deoxygenated blood are separated in the four-chambered heart and a diaphragm assists in respiratory movements. Typically, the young are born alive and are suckled on milk secreted by the mammary glands. A bony secondary palate allows the retention of food in the mouth while breathing. Mammals evolved from active carnivorous reptiles in the Triassic. There are two subclasses: Prototheria, which comprises a single order (Monotremata) containing all the egg-laying mammals; and Theria, which contains all the mammals that bear live young. Subclass Theria is divided into two infraclasses: Metatheria, which comprises the marsupials (pouched mammals); and Eutheria, which contains the placental mammals.

managed ecosystem   An ecosystem in which regular human intervention is used to achieve sustainable or high yields of crops or game.

Mandibulata   *See* Arthropoda.

manganese A metallic trace element required by plants and animals. Manganese is involved in certain metabolic processes. In plants, deficiency of manganese may cause dwarfing, mottling of the upper leaves, CHLOROSIS between the leaf veins, blight of sugar cane, and gray speck of oats. In animals, it is needed for bone formation.

mangrove Any of a range of trees and shrubs that form forests and dense thickets along muddy coast and estuaries and in salt marshes, some of which have prop roots (*see* aerial root). Adventitious roots arch downward to the mud and send up new trunks. The prop roots trap sediment, building up the land and aiding SUCCESSION. Many mangrove species have *pneumatophores*, or breathing knees, that rise up out of the mud. They have abundant *lenticels*, through which air diffuses to supply the waterlogged roots below the mud. Mangrove seeds often germinate while still in the fruit attached to the parent plant, putting down a long root toward the mud. When the seed drops, it spears its way into the mud. Mangroves thus act as pioneer species, colonizing bare mud. The wood of some species is exploited locally for its bark, which can be used for tanning, and as a source of wood chips. In some parts of the world, valuable mangrove forest barriers that protect the coast are being felled for timber. *See also* halophyte; succession.

man-made mineral fibers (MMMF) Manufactured inorganic fibers such as asbestos, glass wool, or fused rock products used in thermal and acoustic insulation, and certain ceramic fibers. Some of these fibers, such as asbestos, pose human health risks because they are fine enough to lodge and persist in the lungs. *See* asbestos.

mantle 1. The part of the Earth that lies between the CRUST and the central CORE. It occupies some 84% of the planet's volume and accounts for 68% of its mass. It ranges from 40 to 2900 km in thickness, with an average of about 2300 km. The outer part of the mantle consists of solid rock, but the rock becomes more plastic and mobile as pressure and temperature increase with depth.
2. *See* Mollusca.

manure An organic FERTILIZER based on animal excreta, often mixed with straw. The term is also used to mean fertilizers in general. The term *livestock manure* is sometimes used to refer to manure derived from domestic animals. *Compare* compost.

map A graphical representation of the Earth's surface on a plane surface. It may include location and distance guides such as projections, grids, and scales. Information is often displayed in the form of symbols. The commonest maps show the surface TOPOGRAPHY, elevations being shown as bands or shades of color, CONTOUR LINES, or *spot heights*.

Maps that depict the geology of an area come in two forms: *solid editions* show the underlying rock types supposing the layer of superficial drift, such as glacial deposits, alluvium or peat, have been stripped away. *Drift editions* show predominantly the superficial deposits, the bedrock being shown only where it lies uncovered at the surface. *Soil survey maps* show the distribution of soil types according to various classifications, and are usually on a similar scale to maps of the local geology. *Land-use capability maps* (*agricultural potential maps*) show the suitability of the soils for agriculture (*see* land use capability). For multi-use purposes *environmental potential maps* may be useful.

map measurer A device for measuring distances to scale on a map, which involves running a knurled wheel along the proposed route or unknown distance.

maquis A broad-leaved evergreen forest characteristic of a Mediterranean-type climate consisting of scrub and found on poor soils in regions with a pronounced dry season. It is made up of thickets of evergreen shrubs with leathery leaves or spiny bushes and shrubs up to about 3 m tall, e.g. gorse (*Ulex*), broom (*Cytisus*, *Genista*), laurels (*Laurus*), and members of the Ericaceae (heath family). These plants are interspersed with aromatic herbs and smaller shrubs, such as thyme (*Thymus*), *Cistus*, and myrtle (*Myrtus*), and scattered small trees such as olive trees (*Olea europaea*) and fig trees (*Ficus*). In many parts of the Mediterranean it is a plagioclimax, the result of burning and grazing of the original evergreen Mediterranean forest. Similar vegetation in California is called *chaparral*. In Italy, it is called *maccia*, in Spain, *mattoral*, in Australia, *mallee*, and in the Cape Province of South Africa, *fynbos*. *See also* garrigue.

mariculture　The cultivation of marine organisms, such as prawns and oysters, in their natural habitats.

marine　Relating to or living in the sea.

maritime climate　A CLIMATE modified by proximity to the sea. Because water warms and cools slower than land, proximity to the sea or prevailing onshore winds has a moderating influence on climate, bringing warmer winters and cooler summers, as well as increased precipitation owing to the moisture picked up as winds travel over the oceans.

mark-recapture method　A technique for estimating the population size of mobile or elusive animals based on the recapture rate of marked individuals. Samples of the population are trapped and marked at intervals, then released. From the assumption the marked individuals become randomly distributed among the wild population, and that subsequent trapping is random, subsequent samples represent a specific proportion of the total marked to unmarked individuals, and from this the total size of the population can be estimated:
$$S_1 \times S_2/S_{2M}$$
where S_1 and S_2 are the numbers of individuals in the first and second samples respectively, and S_{2M} is the number of marked individuals in the second sample.

marl　A soil containing a high proportion of calcium carbonate, as in limestone areas. It is usually an alluvial deposit.

marram grass　A grass (*Ammophila arenaria*) planted on sand dunes to stabilize them. It is capable of rapid growth to compensate for burial by sand, and its rhizomes help bind the loose sand.

marsh　An area of more or less permanently wet ground with no surface accumulation of peat, usually dominated by herbaceous vegetation, especially rushes, reeds, and sedges. It may be freshwater, saltwater, tidal, or nontidal.

marsh gas　The gas METHANE produced by the anaerobic bacterial decomposition of organic matter under water.

marsupial　*See* Mammalia.

mass burn incineration　*See* incineration.

mass extinction　*See* extinction.

masting　The production by trees and shrubs of especially large quantities of seeds only in certain years, which are called *mast years*.

mast years　*See* masting.

mathematical model　A formulation of a quantitative hypothesis that enables prediction of the behavior of a system and can be tested by experiment.

mating strain　*See* mating type.

mating system　*See* breeding system.

mating type (mating strain)　The equivalent of gender in lower organisms, especially microorganisms and fungi, where there are no obvious morphological differences between the sexes. Mating types are identified by their behavior (one donates material to the other) or physiology, and are often identified as plus (+) and minus (−) strains. Only individuals of different mating type will mate together.

matric potential　(y_m) The component of water potential (*see* osmosis) that is due to capillary and imbibitional forces. Matric potential can be significant in cell walls and intercellular spaces, in dry seeds or in dry clay soils, and decreases the movement of water.

matrix (*pl.* matrices) Any medium in which something is embedded, e.g. the stroma in a CHLOROPLAST, and the pectic substance in which cellulose microfibrils are embedded in the plant cell wall.

mattoral *See* maquis.

maturation The process of becoming mature or fully developed. For example, an animal developing full adult characteristics or a fruit becoming ready to be dispersed.

maximum and minimum thermometer A thermometer that records the maximum and minimum temperatures obtained in a given time period. It consists of a U-shaped tube, closed at one end, containing an indicator liquid, with a temperature scale behind each arm. As temperature falls, the air in the closed side of the tube contracts, and liquid rises up that arm, pushing up a small metal index bar. Expansion of the air when the temperature rises forces the liquid down on that side and up on the free side, pushing up another index bar. The two index bars indicate the extremes of temperature. The thermometer is reset by drawing the index bars down to the liquid using a small magnet.

maximum available control technology (MACT) The best available control technology for reducing polluting emissions, taking into account both technical feasibility and cost. Under the United States CLEAN AIR ACT, this must achieve not less than the average emission level achieved by controls on the best performing 12% of existing sources of similar pollution.

maximum sustainable yield (MSY) The maximum yield (number of individuals that can be harvested) that can be removed repeatedly from a population without reducing its numbers and driving it toward extinction, i.e. the maximum yield at which removal can be balanced by recruitment.

May, Robert McCredie (1936–) Australian–American theoretical ecologist. Whereas an earlier generation of theoretical physicists had become interested in biology through their desire to understand the molecules of life, May and his generation were more attracted to the problem of understanding the abundance and distribution of species. Greatly influenced by the pioneering work of Robert MACARTHUR, they could see tempting analogies between the flow of energy in a physical system and the structure and growth of an ecosystem. May has worked on a number of detailed problems on the population dynamics of various species and is noted for his influential *Theoretical Ecology* (1976).

meadow An area of moist GRASSLAND representing a subclimax maintained by mowing, traditionally for hay in midsummer. *See* climax community.

mean A measure of the average value of a set of numbers. The *arithmetic mean* is the sum of a set of values divided by the number of values. For example, the arithmetic mean of the three values 3, 4, and 5 is $(3 + 4 + 5)/3 = 4$. The *geometric mean* is the nth root of the product of a set of n numbers. For example, the geometric mean of 16 and 4 is the square root of $16 \times 4 = 8$. The geometric mean of 2, 4, and 27 is the cube root of $2 \times 4 \times 27 = 6$.

mean sea level *See* sea level.

measurement scales There are many different scales for measuring variables. For qualitative data, a *nominal scale* classifies objects into categories based solely on descriptive features, for example, animal, vegetable, and mineral. There is no logical order. An *ordinal scale* classifies by rank, assigning a numerical value to a descriptive feature, for example, cold, cool, warm, or hot might be marked 1 to 4 on an ordinal temperature scale. For quantitative variables, an *interval scale* is used, the numbers relating to values that differ by a fixed, equal value, for example the degrees on a scale of temperature. Such scales may relate values to an arbitrary zero point, e.g. the Celsius or Fahrenheit scales. Where the zero point represents an absence of the feature being measured (i.e. it is an absolute zero), such a scale is called a *ratio scale*, e.g. the Kelvin scale of temperature.

Mediterranean climate A CLIMATE of middle latitudes characterized by warm, dry summers and cool, wet winters.

megafauna Animals large enough to be seen with the naked eye. *Compare* microfauna.

meiofauna (interstitial fauna) The intermediate-sized animals in a SOIL or sediment community that live within the soil, sediment, or rock fissures. They are usually greater than 40 μm and less than 2 mm in length, e.g. certain mites and springtails (Collembola). Most are detritus feeders.

meiosis The process of cell division leading to the production of daughter nuclei with half the genetic complement of the parent cell (i.e. HAPLOID nuclei). Cells formed by meiosis give rise to GAMETES, and FERTILIZATION restores the correct chromosome complement.

Meiosis consists of two divisions during which the chromosomes replicate only once. As in mitosis, the stages prophase, metaphase, and anaphase occur, but during prophase HOMOLOGOUS CHROMOSOMES attract each other and become paired forming *bivalents*. At the end of prophase genetic material may be exchanged between the CHROMATIDS of homologous chromosomes, an important source of VARIATION between daughter cells. Meiosis also differs from mitosis in that after anaphase, instead of nuclear membranes forming, there is a second division, which may be divided into metaphase II and anaphase II. The second division ends with the formation of four haploid nuclei, which develop into gametes. *See* homolo-

gous chromosomes. *See also* alternation of generations. *Compare* mitosis.

melanin *See* melanism.

melanism 1. The possession of a dark appearance due to the presence of a dark brown pigment, *melanin*. 2. The occurrence in a population of individuals that have a more than usually dark skin, feathers, or fur.

melatonin *See* photoperiodism.

Mendelism The theory of inheritance founded on the work of the Austrian monk Gregor Mendel (1822–84), according to which characteristics are determined by particulate 'factors' that are transmitted by the germ cells. It is the basis of classical genetics. Mendel formulated two laws to explain the pattern of inheritance he observed in plant crosses. The first law, *Law of Segregation*, states that any character exists as two factors, both of which are found in the somatic (body) cells but only one of which is passed on to any one gamete. The second law, the *Law of Independent Assortment*, states that the distribution of such factors to the gametes is random; if a number of pairs of factors is considered, each pair segregates independently.

Today Mendel's 'characters' are termed GENES and their different forms (factors) are called ALLELES. It is known that a DIPLOID cell contains two alleles of a given gene, each of which is located on one of a pair of HOMOLOGOUS CHROMOSOMES. Only one homology of each pair is passed on to a gamete. Thus the Law of Segregation still holds true. Mendel envisaged his factors as discrete particles but it is now known that they are grouped together on chromosomes. The Law of Independent Assortment therefore applies only to unlinked genes. *See* independent assortment; linkage.

mercury (Hg) A silvery metallic element that is liquid at room temperature. It occurs naturally as cinnabar (mercurous sulfide) and as elemental mercury in cinnabar and some volcanic rocks. The vapor is highly poisonous. Mercury is used in thermometers and other scientific apparatus and mercury compounds are used in fungicides and timber preservatives.

meristem A distinct region of actively or potentially actively dividing cells in a plant, primarily concerned with growth.

meromictic Describing a lake in which vertical mixing of the water never occurs.

mesic Describing a habitat that is neither wet (hydric) nor dry (xeric), but has intermediate moisture content.

mesobiota Soil or sediment organisms of intermediate size, e.g. larger insects, earthworms, and nonmicroscopic fungi.

mesoclimate The climate of a relatively small area such as a valley or town.

mesofauna Soil or sediment animals of intermediate size, such as nematodes, earthworms, arthropods, and mollusks.

mesophilic Describing microorganisms (*mesophiles*) that grow best at moderate temperatures, usually taken to be 20–45°C. *Compare* psychrophilic; thermophilic.

mesophilic digestion *See* digestion (def. 2).

mesophyte A plant that is adapted to grow under adequate conditions of water supply and has no particular adaptations to withstand environmental extremes. In drought conditions wilting is soon apparent because the plants have no special mechanisms to conserve water. Most angiosperms are mesophytes. *Compare* hydrophyte; xerophyte.

mesoplankton *See* plankton.

mesothelioma *See* asbestos.

mesotrophic Describing water with levels of plant nutrients intermediate between those of OLIGOTROPHIC and EUTROPHIC water.

Mesozoic The middle era in the most recent (Phanerozoic) eon of the GEOLOGICAL TIME SCALE, dating from about 230–66 million years ago, during which time the supercontinent of PANGAEA broke up into separate continents that began to drift across the globe. Known as the Age of Reptiles, it is divided into three main periods: the TRIASSIC, JURASSIC, and CRETACEOUS. Vertebrate life diversified during this era, and the first mammals, lizards, turtles, crocodiles, and birds appeared. At the end of the Triassic period a mass EXTINCTION wiped out some 35% of existing animal groups. Later diversification culminated in the domination of land, air, and sea by dinosaurs, pterosaurs, and plesiosaurs. As the climate warmed during the Triassic period the dominant vegetation of seed-ferns gave way to cycads, and the Mesozoic is sometimes also called The Age of Cycads. In the Jurassic period, tropical and temperate regions were dominated by forests of palmlike cycads and conifers, and ginkgoes were also widespread. The first angiosperms arose in the Cretaceous period and rapidly diversified, becoming the dominant vegetation by the end of the Mesozoic as the climate cooled and dried. The era ended with another mass extinction, and the demise of the dinosaurs.

metabolic pathway *See* metabolism.

metabolic rate The volume of oxygen consumed by an organism per unit time per unit body mass. This is the *basal or resting metabolic rate* (BMR or RMR), the lower level of energy use under normal conditions in which the organism is resting and is not consuming or digesting food. An kind of activity ('work') or stress, or an increased need to maintain body heat (perhaps due to temperature fall), will lead to an increase metabolic rate.

metabolism The chemical reactions that take place in cells. The molecules taking part in these reactions are termed *metabolites*. Some metabolites are synthesized within the organism, whereas others have to be taken in as food. It is metabolic reactions, particularly those producing energy, that keep cells alive. Metabolic reactions characteristically occur in small steps, comprising a *metabolic pathway*, in which the products of one reaction form the starting materials for the next, and are regulated by various FEEDBACK MECHANISMS. Metabolic reactions involve the breaking down of molecules to provide energy (CATABOLISM) and the building up of more complex molecules and structures from simpler molecules (ANABOLISM).

metabolite See metabolism.

metalimnion (thermocline) The middle layer of water in a thermally stratified lake, between the EPILIMNION and the HYPOLIMNION, which shows a rapid decrease in temperature with depth.

metal tolerance The ability of certain plants to tolerate high concentrations of metals, especially heavy metals (e.g. lead, copper, and zinc) in the soil. The plants may contain modified enzymes capable of working in the presence of the metal; they may have specific ion transport mechanisms that prevent metal uptake by the roots; or they may be able to transfer metals to vacuoles, where they will not interfere with metabolism. *See also* bioaccumulation; biomining; bioremediation; hyperaccumulator.

metamerically segmented Having a body divided along its longitudinal axis into a series of similar units (called segments or metameres), at least in its embryonic stage. Metamerically segmented animals include annelid worms (*see* Annelida) and chordates (*see* Chordata).

metamorphic rock Rock that has been changed as a result of the pressure of other rocks, high temperatures, percolating solutions, etc., for example, marble.

metamorphosis A phase in the life history of many animals during which there is a rapid transformation from the larval to the adult form. Metamorphosis is widespread among invertebrates, especially marine organisms and arthropods, and is typical of the amphibians.

metapopulation A population divided into a series of interacting subpopulations linked by migration, for example where small patches of suitable habitat are separated by unsuitable habitat. For example, in a *mainland-island metapopulation* different local populations are in different habitat patches or islands near a larger patch, the mainland, from which individuals can migrate to the other patches.

Each metapopulation will have its own characteristic spatial distribution and demographics, which together make up the *metapopulation structure*. The degree of interaction between the subpopulations may vary. A *spatially explicit metapopulation model* relates the extent of migration between local populations to the distance between them. A *spatially implicit metapopulation model* simply assumes equal opportunities for migration between local populations. A metapopulation in which some local populations (*sources*) show a positive growth rate at low densities, while others (*sinks*) show a negative growth rate when there is no immigration, is called a *source-sink metapopulation*.

metazoans In some older classifications, a subkingdom (the Metazoa) of multicellular animals whose bodies are composed of specialized cells grouped together to form tissues and that possess a coordinating nervous system. This subkingdom included all animals except single-celled animals and sponges. *Compare* Animalia.

meteorology The study of the Earth's atmosphere and its influence on weather and climate.

meter 1. The SI unit of length, specifically the distance traveled by light in a vacuum in 1/299,792,458 of a second. 1 m is equal to 39.37 inches.

2. An instrument used for counting or measuring, for example an electricity meter or pH meter.

methane A colorless gas produced naturally by the anaerobic bacterial decomposition of organic matter in marshes, coalmines, and the intestines of termites and RUMINANTS such as cattle; it also makes up some 99% of NATURAL GAS. Methane is an important GREENHOUSE GAS. It can be burned for use in heating or power generation.

methanogen An EXTREMOPHILE bacterium of the domain ARCHAEA that produces methane gas as a by-product of its metabolism. *See also* halophile; thermophile.

methanol *See* ethanol.

methoxychlor A CHLORINATED HYDROCARBON used as a contact insecticide. It is highly poisonous, affecting nerve transmission, and accumulates in the fatty tissues of animals and is also toxic to freshwater and marine aquatic life. *See* biomagnification.

methyl bromide A colorless volatile liquid used as a pesticide, particularly to fumigate soil and agricultural products. When released into the atmosphere, it contributes to depletion of the OZONE LAYER. *See* ozone hole.

methylmercury compounds A class of organic compounds in which mercury is combined with a methyl group, which are used as fungicides in seed dressings and timber preservatives. Birds and other animals that eat the seeds may be poisoned by the mercury.

micelle A complex between clay minerals and humus that bears a net electronegative charge on its surface, thus attracting and retaining cations and contributing to the long-term fertility of the SOIL.

microalgae Algae that are too small to see with the naked eye.

microbial film A thin layer of microbes.

microbial loop A nutrient cycle involving only microbes, such as the nutrient cycling between phytoplankton and their consumers in seas and lakes, and around some mica minerals (silicates) in the soil. Such cycles are almost independent of the nutrient cycles involving larger producers and consumers.

microbiology The study of microscopic organisms (e.g. bacteria and viruses), including their interactions with other organisms and with the environment. Microbial biochemistry and genetics are important branches of science because of the use of microorganisms in BIOTECHNOLOGY and GENETIC ENGINEERING.

microclimate The climate of a small area, which may range in size from an urban area to the immediate area around a shrub or under a log.

microconsumer An organism, such as a bacterium or fungus, that feeds by breaking down dead organic matter, releasing simpler organic and inorganic substances to the environment – in other words, a DECOMPOSER. *See* saprobe.

microcosm A small-scale, simplified ecosystem designed for experimental purposes, which is intended to represent the processes that go on in a larger, analogous ecosystem. For example a sample of pond water in a jar, maintained by artificial light and bubbled air.

microenvironment *See* habitat.

microevolution EVOLUTION on a scale within a species, which results from the action of natural selection on the genetic VARIATION between individuals of a POPULATION. *Compare* macroevolution.

microfauna Soil animals so small that they can be seen only with a microscope. *Compare* macrofauna; megafauna; meiofauna.

microflora Plantlike soil organisms so small that they can be seen only with a microscope. The term is often used to include bacteria, fungi, and algae.

microfossil *See* fossil.

microhabitat *See* habitat.

micrometer (μm) A unit of length equal to 10^{-6} m (one-millionth of a meter), formerly called a *micron*.

micron *See* micrometer.

micronutrient (trace element) A nutrient required in trace amounts by an organism. For example, a plant can obtain sufficient of the essential trace element manganese from a solution containing 0.5 parts per million of manganese. Many act as enzyme cofactors or are components of pigments. They include boron, cobalt, copper, manganese, molybdenum, and zinc. *See* deficiency disease. *Compare* macronutrient.

microorganism A microscopic organism: one so small that it can be seen only with the aid of a microscope. The term is usually restricted to bacteria, viruses, protozoans, and microscopic fungi and algae, excluding such organisms as rotifers.

microparasite *See* parasitism.

microplankton *See* plankton.

microsere *See* sere.

mid-oceanic ridge *See* plate tectonics.

migration An instinctive regular two-way movement of part or all of an animal population to and from a given area, usually along well-defined routes. It is closely linked to the cycle of the seasons and is triggered off by seasonal factors such as increasing and decreasing daylengths in spring and autumn. Many birds, hoofed mammals, bats, whales, fish, and insects migrate, often covering immense distances. For example, the Arctic tern breeds on the northernmost coasts of Eurasia and America and winters around the Antarctic pack-ice 11 000 miles to the south. Migratory mammals such as the wildebeest live in habitats with fluctuating climatic conditions and migrate in order to find an adequate food supply.

mildew A fungal disease of plants in which the mycelium is visible on the surface of the plants as white or grayish powdery patches.

millibar (mb) A unit of pressure equal to one-thousandth of a bar. One bar is 10^5 pascals (10^5 newtons per m^2) or 14 lb/in^2, or approximately one atmosphere. The standard pressure of the atmosphere at sea level is 1013 millibars.

millimolar Describing the concentration of a solution expressed as the number of millimoles (one-thousandths of a MOLE) of dissolved substance per dm^3 (liter) of solution, e.g. a 10-millimolar solution contains 10 millimoles of dissolved substance per dm^3. *See* mole.

mimic An animal (the mimic) that resembles another animal (the model), and by so doing gains an advantage from its resemblance to the model. *See also* mimicry.

mimicry The resemblance of one animal to another by which the mimic gains advantage from its resemblance to the model. For example, in *Batesian mimicry* certain edible insects mimic the warning coloration of noxious insects and so are avoided by their predators. Natural selection produces more accurate mimicry as only those individuals closely resembling the model will be mistaken for it and left alone. In *Müllerian mimicry* a group of poisonous animals resemble each other, for example, bees, wasps, and hornets, increasing the likelihood that potential predators will learn to avoid them.

mineralization 1. The process by which the organic constituents of an organism are replaced by inorganic materials for form a fossil.

2. The release of inorganic material from organic matter during aerobic or anaerobic decomposition.
3. Any process by which minerals are introduced into a rock.

minimum viable population   *See* population.

Miocene   An epoch of the Tertiary, 25–7 million years ago. The climate became drier and the grasses evolved and spread rapidly, this perhaps explaining the replacement of early mammals by more modern forms. About half to three-quarters of existing mammalian families are represented in rocks of the Miocene.

mist   A suspension of water droplets in air that reduces the visibility to not less than 1 kilometer. *Compare* fog.

mitigation   Action taken to minimize the effects of environmental damage or POLLUTION. These include avoidance of the harmful agent, means of reducing or compensating for its effects, and the restoration or replacement of damaged ecosystems.

mitochondrion (*pl.* mitochondria)   An organelle of all plant and animal cells chiefly associated with aerobic RESPIRATION. It is surrounded by two membranes separated by an intermembrane space; the inner membrane forms finger-like processes called *cristae*, which project into the gel-like *matrix*. The diameter is always about 0.5–1.0 μm, and the length averages 2 μm. Mitochondria are most numerous in active cells (up to several thousand per cell). The matrix is also involved in amino acid metabolism and in fatty acid oxidation.

It is believed that mitochondria and CHLOROPLASTS may be the descendants of once independent organisms that early in evolution invaded EUKARYOTE cells, leading to an extreme form of SYMBIOSIS. *See* endosymbiont theory.

mitosis   The ordered process by which the cell nucleus divides in two during the division of body (i.e. somatic or nongermline) cells. The CHROMOSOMES replicate prior to mitosis to form two sister CHROMATIDS, which are then separated during mitosis in such a way that each daughter cell inherits a genetic complement identical to that of the parent cell. Mitosis may be followed by cytokinesis (cell division). *Compare* meiosis.

mixed deciduous forest   *See* temperate deciduous forest.

mixed feeder   An organism that feeds on both plants and animals.

mixing height   Where there is an inversion in the atmosphere, in which a mass of stable air caps an unstable layer below, the height to which pollutants can be mixed is called the mixing height. This extends to the level of the INVERSION.

MMMF   *See* man-made mineral fibers.

model   A mathematical expression of the way a system (for example, an ecosystem or the atmospheric circulation) works that can be subjected to testing by experimentation, observation, or analysis of existing data. A successful model allows for prediction of the behavior of the system under various conditions. An *analytical model* attempts to explain existing data, while a *simulation model* attempts to predict the behavior of the system if one or more variables are perturbed. For example a simulation of the effects of increasing habitat loss on a population may give an idea of the timescale in which conservation efforts must be concentrated.

modular organism   An organism that grows by repeatedly producing similar parts, e.g. the leaves, shoots, and branches of plants, or the polyps of corals. Nonmodular organisms are called *unitary organisms*.

molality   The concentration of a substance expressed in moles per unit weight. The *molal concentration* is the number of moles of solute in one kilogram of solution. *Compare* molarity.

molarity A measure of the concentration of a solution based upon the number of molecules or ions present, rather than on the mass of solute, in any particular volume of solution. The molarity (M), or *molar concentration*, is the number of moles of solute in one cubic decimeter (liter). Thus a 0.5M solution of hydrochloric acid contains 0.5 x (1 + 35.5)g HCl per dm^3 of solution. *Compare* molality.

mole (mol; M) The SI base unit of amount of substance, defined as the amount of substance that contains as many elementary entities as there are atoms in 0.012 kilogram of ^{12}C. The elementary entities may be atoms, molecules, ions, electrons, photons, etc., and they must be specified. The amount of substance is proportional to the number of entities. One mole contains 6.022 045 × 10^{23} entities.

molecule A particle formed by the combination of atoms in a whole-number ratio. A molecule of an element (combining atoms are the same, e.g. O2) or of a compound (different combining atoms, e.g. HCl) retains the properties of that element or compound. Thus, any quantity of a compound is a collection of many identical molecules. Molecular sizes are characteristically 10^{-10} to 10^{-9} m.

Molina, Mario José (1943–) Mexican physical chemist who worked with ROWLAND on the chlorofluorocarbons (CFCs) used as the propellant in most aerosol cans and on their affect on the ozone layer.

Mollusca (mollusks) A phylum of bilaterally symmetrical unsegmented invertebrates, including the aquatic bivalves, mussels, octopuses, squids, etc., and the terrestrial slugs and snails. The body is divided into a head, a ventral muscular locomotory organ (foot), and a dorsal visceral hump that houses most of the body organs and is covered by a tissue layer (*mantle*), which typically secretes a calcareous shell

into which the head and foot can retract. The mantle extends into folds forming a cavity containing the gills. The rasping radula is used for feeding. In aquatic forms development usually occurs via a swimming larva.

mollusk *See* Mollusca.

molybdenum A micronutrient (trace element) needed for plant growth. Deficiency of molybdenum may lead to CHLOROSIS between the leaf veins.

monitoring Periodic or continuous observation or measurement of variables or organisms to evaluate performance or progress of a reaction or procedure; for example the testing of water to ensure its safety for human consumption, or the recording of the growth of plants following a specific treatment.

monocarpic *See* iteroparous.

monoclimax *See* climax community.

monocotyledons *See* Angiospermophyta.

monoculture The growing of a single species or variety of crop over a large area. The economic reasons for this are the ability to grow crops of uniform height for harvesting, and economies of scale in the harvesting process. The major disadvantage is increased susceptibility to pests and pathogens, especially if a single variety is grown. *Compare* center of diversity.

monogamy The exclusive pairing of a male and a female for the purpose of mating, in which each partner remains completely faithful to the other. Such a pair bond may last for a single breeding season, or for life, depending on the species.

monomictic Describing a lake in which vertical mixing of the water takes place only once a year. *Compare* dimictic; meromictic.

monosaccharide *See* carbohydrates.

monotreme *See* Mammalia.

monsoon A weather pattern in the subtropics in which the prevailing winds blow in opposite directions at different times of the year. It is due to the different heating of land and sea, which causes seasonal changes in atmospheric pressure systems that are further influenced by shifts in upper wind patterns and jet streams. It is especially well developed over southern and eastern Asia, where the wet summer monsoon from the southwest dominates the climate.

monsoon forest A TROPICAL FOREST that occurs in regions that experience a few months of drought, such as India, Myanmar (Burma), and Indo-China. Species diversity is lower than in RAINFOREST, and monsoon forests tend to have more open canopies and very dense undergrowth. Many of the trees are deciduous, shedding their leaves in the dry season.

Montreal Protocol An international agreement between some 30 countries that came into force on January 1, 1989 with the aim of protecting the ozone layer by controlling emissions of CFCS, HALONS, and other ozone-depleting substances. The goal was to completely phase out the production and use of CFCs by 1996, and to restrict halon consumption to 1988 levels. More than 265 countries have now ratified the agreement. *See also* Clean Air Act.

monuron A persistent herbicide used especially to control broad-leaved weeds.

moorland An area of acidic land dominated by low-growing ericaceous shrubs such as heather (*Calluna vulgaris*) and bilberry (*Vaccinium myrtillus*), mixed with grasses, sedges, and rushes. The term is most commonly used for upland areas, but in some parts of the UK it is also used for low-lying land on deep peat.

mor *See* humus.

morbidity The state of being diseased. The morbidity rate is the number of individuals suffering from a particular disease per 100 thousand of the population.

morph A local population of a species that is phenotypically distinct from other populations, yet remains capable of interbreeding with them.

morphogenesis The development of form and structure.

morphology The study of the form of organisms. The term may be used synonymously with *anatomy* although generally the study of external form is termed morphology whereas the study of internal structures is termed anatomy.

morphometry The measurement of external form. The term may be applied to landscape features such as drainage systems as well as to organisms.

mortality rate (*m*) The number of deaths per capita of the population, or of a specific age group, calculated for a given period of time, often one year.

mosquitofish A small live-bearing fish native to rivers in the southeastern United States. It feeds on mosquito larvae, and has been introduced to many parts of the world to help control mosquitoes.

mosses *See* Bryophyta.

motile Capable of independent locomotion.

mountain coniferous forest *See* boreal forest.

MSY *See* maximum sustainable yield.

mucilage Any substance that swells in water to form a slimy solution. Mucilages are often complex substances, and may contain proteins, cellulose, and derivatives of uronic acids. In plants, mucilages are often used to retain water, e.g. mucilages in the center of succulent desert plants and on the surface of many seeds, helping the seed to take up water during germination. They may also form part of the cell wall matrix. Most bacteria have an outer layer of mucilage, and many seaweeds contain mucilage in their tissues, which helps retain moisture when exposed at low tide, and also on their surfaces, helping reduce resistance to wave action and thus minimize damage.

mud flat   A large flat area of mud, usually in an estuary or around a coast, that is periodically inundated by the sea.

mulch   1. A protective covering of compost, paper, bark chippings, cocoa shells, etc. spread over the ground to reduce evaporation, control weeds, enrich the soil, and protect against temperature fluctuations. Straw is used to keep soft fruits such as strawberries (*Fragaria ananassa*) clean and dry.
2. In agriculture, the process of leaving crop residues on the soil surface or burying them to add to the organic matter in the soil and reduce evaporation. Crops are planted by machinery that penetrates the mulch. In humid areas mulch does not break up readily, and may deprive the crop of nitrogen, so fertilizer is often added below the mulch. Intercropping (the planting of a second crop between the rows of the main crop) may be used to provide plants for mulching after the main crop has been harvested.

mull   *See* humus.

multicropping   The practice of growing more than one crop on the same piece of land in the same year.

multifactorial experiment   An experiment in which more than one factor is varied at time or more than one treatment given to the subject or system under investigation.

multifactorial inheritance   *See* quantitative variation.

multiple regression techniques   *See* regression.

Munsell soil color chart system   A standard scheme for naming the color of a SOIL, a feature important in the identification and differentiation of soil horizons.

muskeg   A type of peat bog characteristic of the BOREAL FOREST of North America. Muskegs are usually dominated by black spruce (*Picea mariana*), and INSECTIVOROUS PLANTS such as sundews (*Drosera*) and pitcher plants (*Sarracenia*) are common. *See* peat.

mutagen   Any physical or chemical agent that induces mutation or increases the rate of spontaneous mutation. Physical mutagens include ultraviolet light, x-rays, and gamma rays.

mutant   *See* mutation.

mutation   A change in one or more of the bases in DNA, which results in the formation of an abnormal protein. An organism that has undergone a mutation is called a *mutant*. Mutations are the ultimate source of all genetic VARIATION. They are inherited only if they occur in the cells that give rise to the gametes; mutations in other (somatic) cells may give rise to CHIMERAS and cancers. *Gene mutations* alter only a single gene, resulting in new ALLELES of the gene and hence new variations upon which natural selection can act. *Chromosome mutations* result in the gain, loss, or rearrangement of whole segments of chromosomes and are usually, but not always, lethal. Most mutations are deleterious but are often retained in the population because they also tend to be recessive and can thus be carried in the genotype without affecting the viability of the organism. The natural rate of mutation is low, and varies with different gene loci. The mutation frequency can be increased by MUTAGENS. *See also* polyploidy.

mutual antagonism   The situation in which two species have reciprocal negative effects on each other, for example, they may both compete for the same resources, or they may prey on each other.

mutual interference Interference between PREDATORS or CONSUMERS that leads to a reduction in the consumption rate of individual predators/consumers that increases with predator/consumer density, even when resources are not limiting, for example, where aggressive behavior (often territorial) reduces the time and energy available for predation or foraging.

mutualism The close relationship between two or more species of organisms in which all benefit from the association. There are two types of mutualism: *obligatory mutualism*, in which one cannot survive without the other, for example the algal–fungal partnership found in lichens; and *facultative mutualism* (*protocooperation*), in which one or both species can survive independently, for example, marine crabs and their associated invertebrate fauna of sponges, cnidarians, etc., that attach to the crab shell and act as camouflage. In the association between the bacterium *Rhizobium* and members of the family Fabaceae, which leads to the formation of root nodules in which nitrogen fixation occurs, the bacterium can also live independently of the plant. In lichens, often the alga can live independently of the fungus. In many mutualistic associations, the relationship can change from a parasitism to a mutualism or vice versa. *Compare* amensalism; commensalism; symbiosis.

mya Abbreviation for million years ago.

mycelium *See* Fungi.

mycobiont *See* lichens.

mycorrhiza (*pl.* mycorrhizae) The association between the hyphae of a fungus and the roots of a higher plant. Two main types of mycorrhiza exist, *ectomycorrhizae* (ec-totrophic mycorrhizae), in which the fungus forms a mantle around the smaller roots, as in many temperate and boreal trees, and *endomycorrhizae* (endotrophic mycorrhizae), in which the fungus grows around and within the cortex cells of the roots, as in many herbaceous plants. Orchids and heathers have further specialized types of mycorrhizae, in which there is both a mantle and cell penetration.

The fungus benefits by obtaining carbohydrates and possibly B-group vitamins from the roots. The tree benefits in that mycorrhizal roots absorb nutrients more efficiently than uninfected roots, and it is common forestry practice to insure the appropriate fungus is applied when planting seedling trees.

The mycorrhizal relationship is somewhat flexible. In seedling trees and orchids and sometimes as adults, the plant is in effect parasitic on the fungus, but the relationship changes, and in the fungal fruiting season the fungus takes a lot from the plant, which is transporting nutrients to its roots. Between times the relationship is effectively mutualistic.

mycotrophic Describing an association between an organism and a fungus in which the organism derives nutrients from the fungus (*see* mycorrhiza). The term is also used in a more restricted sense to refer only to associations where the fungus extends into the aerial parts of a plant, as in some heathers (Ericaceae) and orchids (Orchidaceae). *See* mutualism; symbiosis.

myxomatosis *See* myxoma virus.

myxoma virus The causative agent of *myxomatosis*, an often fatal disease of rabbits that is sometimes used to control their numbers. Some individuals have genetic resistance to the virus.

N

NAD *See* nucleotide; respiration.

nanoplankton *See* plankton.

natality rate (*b*; birth rate) The number of births per year in a population, usually shown as the number of births per 1000 individuals.

native *See* indigenous.

natural gas Gas containing 50–90% methane found trapped in certain types of rock formations. It is classified as a CLEAN FUEL, widely used for domestic heating and in industry. It can be compressed for use in vehicles. Liquefied petroleum gas (LPG) is produced during natural gas production and petroleum refining. It gives a similar performance to gasoline. In the United States reformulated gasoline is being developed to emit lower levels of pollutants than traditional gasoline.

natural resource *See* resource.

natural selection The process, which Darwin called the 'struggle for survival', by which organisms less adapted to their environment tend to perish, and better-adapted organisms tend to survive to reproduce and pass on their genes to the next generation. It may be summed up as 'the *survival of the fittest*', fitness in this context being a measure of an organism's ability to contribute to the next generation (*see* fitness). There has been much debate about the level at which natural selection acts: traditionally it is regarded as the action of the environment on individual organisms that determines which genotypes survive and reproduce, but there are strong arguments for natural selection acting at the level of the gene (the selfish gene hypothesis) and some authors argue that it happens in the long term at higher levels such as population or species. In the end the less successful types will die out. According to DARWINISM, natural selection acting on a varied population results in evolution. *See* evolution.

nature and nurture The interaction between inherited and environmental factors (*nature* and *nurture*, respectively) in determining the observed characteristics of an organism. It is often applied in a discussion of behavioral characteristics, such as intelligence, in which the relative importance of inherited and environmental factors, including such factors as social background, are a matter of great controversy. The term HERITABILITY is sometimes used as an alternative, meaning the proportion of the total variation caused by genetic influences alone.

nature conservation The management of the Earth's natural resources and environment to ensure they are used wisely and not depleted or degraded.

nature reserve An area set aside primarily for the preservation and protection of certain species of animals or plants or certain habitats.

Nearctic One of the six zoogeographical regions, including North America from the Central Mexican Plateau in the south to the Aleutian islands and Greenland in the north. The fauna include mountain goats, pronghorn antelopes, caribou, and muskrat.

nearest-neighbor distance   The average distance between individuals in a population. Assuming that individuals are distributed at random with respect to one another, the expected distance between neighbors is given by:

$\sqrt{(\pi/n)}/2$ m, where n is the density of individuals per square meter. Ratios of less than 1 indicate clumping. Ratios greater than 1 indicate relatively even spacing.

nectarivore   An animal that feeds mainly on nectar.

negative feedback   *See* feedback mechanism.

neighborhood size   The number of individuals in a population that are within the dispersal distance of a single individual.

nekton   Animals of the pelagic zone of a sea or lake that are free-swimming and independent of tides, currents, and waves, such as fish, whales, squid, crabs, and shrimps. Nekton are limited in distribution by temperature and nutrient supply, and decrease with increasing depth.

Nematoda   A large phylum of marine, freshwater, and terrestrial invertebrates, the roundworms. Most are free-living, e.g. *Anguillula* (vinegar eel), but many are parasites, e.g. *Heterodera* (eelworm of potatoes) and *Ascaris* (found in pigs' and human intestines). Some cause serious diseases in humans, e.g. *Wuchereria* (causing elephantiasis). Nematodes are bilaterally symmetrical with an unsegmented smooth cylindrical body pointed at both ends and covered with a tough cuticle. The body cavity is not a true coelom and there are no blood or respiratory systems. The muscular and excretory systems and embryonic development are unusual. Nematodes are not closely related to any other phylum.

neoDarwinism   Darwin's theory of evolution through NATURAL SELECTION, modified and expanded by genetic studies arising from the work of Mendel and his successors. This fusion of Darwin's theory of natural selection with Mendel's genetics was called the *neoDarwinian synthesis* in the 1930s. The inclusion of genetics in evolutionary studies answered many questions that Darwin's theory raised but could not adequately explain because of lack of knowledge at the time it was formulated. Notably, genetics has revealed the source of variation on which natural selection operates, namely MUTATIONS of GENES and CHROMOSOMES, and provided mathematical models of how ALLELES fluctuate in natural populations, thereby quantifying the process of evolution. More recent discoveries in molecular biology have added to the understanding of the causes of variation and the nature of evolution at the molecular level.

neon   *See* atmosphere.

Neotropical   One of the six main zoogeographical regions of the earth. It includes South and Central America, the West Indies, and the Mexican lowlands. The characteristic fauna includes sloths, armadillos, anteaters, cavies, vampire bats, llama, alpaca, peccary, rhea, toucan, curassows, and certain hummingbirds.

NEP   *See* net ecosystem production.

nephelometry   A method of measuring turbidity in water by passing light through the sample and measuring the amount of light that is deflected by the suspended particles.

neritic   The marine environment from low water level to a depth of about 200 m, a zone that in many areas corresponds to the extent of the continental shelf. It makes up less than 1% of the marine environment. Nutrients are relatively abundant in this zone and it is penetrated by sunlight.

net assimilation rate   A measure of plant productivity based on net assimilation by leaves. The net assimilation rate is the rate of increase in dry weight divided by the area of the leaf.

net ecosystem production (NEP)   The difference between the gross primary production of all the primary producers in an ecosystem (plants, algae, and bacteria) and the loss of nutrients and/or energy through their respiration; in other words, the NET PRIMARY PRODUCTION of the ecosystem. *See* producer.

net photosynthesis   The rate of carbon dioxide uptake by a plant minus the rate of carbon dioxide evolution by respiration.

net plankton   *See* plankton.

net primary production (NPP)   The GROSS PRIMARY PRODUCTION of an organism, population, or community (i.e. the total energy or nutrients accumulated) minus the loss of energy or nutrients by respiration.

net production efficiency   The percentage of the food that is assimilated by an organism that is used in its growth and reproduction. *Compare* gross production efficiency.

net reproductive rate (R_0)   The number of offspring a female may be expected to produce during her lifetime.

neutral theory   The theory that most evolutionary changes are caused by random GENETIC DRIFT (random changes in the frequencies of alleles in a population) rather than by NATURAL SELECTION. Genetic MUTATIONS arise at random and clearly some will have increased or decreased fitness and be subject to selection. But many, perhaps a majority, are neutral – they do no affect the fitness or survival of the carrier. They therefore survive in the GENOME and changes in their frequencies are due to the BREEDING SYSTEM and chance rather than to natural selection. *See* evolution.

neutral variation   Genetic diversity that confers no obvious selective advantage. *See* natural selection; variation.

niche (ecological niche)   The functional role of an organism, population, or species in a community. It includes the habitat in which the organism lives, the resources it

uses, and the periods of time in which it is active there and its interactions with other organisms. The range of conditions tolerated and resources used by an individual, population, or species is called its *niche breadth*. A statistical measure of niche breadth is *niche width*: the STANDARD DEVIATION of the distribution of resource used.

If two species occupy the same niche, competition may occur until either one has replaced the other or the two divide the niche between them to some extent. Thus coexisting species tend to differ, if only slightly, in their niche requirements, a phenomenon known as *niche differentiation*. If coexisting species occur in a similar position along one niche dimension (e.g. an altitude or vegetation gradient), they will differ along another (e.g. food preferences), a phenomenon known as *niche complementarity*. Thus coexisting species will tend to specialize to occupy between them all the available 'space' within a given set of niche parameters; this is known as *niche packing*. The overlap in resources used or conditions tolerated by two or more different species is their *niche overlap*. During ecological SUCCESSION, colonizing species tend to monopolize a proportion of the available resources, leaving less choice and a narrower potential niche breadth for the next species in the succession, a behavior termed *niche preemption*.

A similar niche may be occupied by different species in different areas, for example the giant anteater (*Myrmecophaga tridactyla*) of South America occupies the same type of niche as the aardvark (*Orycteropus afer*) of southern Africa. Conversely one type of organism may evolve by ADAPTIVE RADIATION to fill several different niches, as in the finches of the Galápagos Islands.

If a niche is considered to be a space with *n* dimensions, with the different axes representing resources and environmental factors, then the *niche space* occupied by a community is the 'volume' into which the niches of all the species fit. Thus the number of species in the community depends on the total available niche space and the average size and dimensions of each species' niche. The size of niche that a given species would occupy in the absence of competition is termed its *fundamental niche*, whereas the minimum niche space in which it can persist, even in the presence of competitors and predators, is called its *realized niche*.

nicotine A poisonous alkaloid that is the main active component of tobacco and is also used as an insecticide.

nitrate A salt of nitric acid. Nitrates are present in the atmosphere and in the soil; they are highly soluble and when present in significant concentrations in drinking water can harm humans (especially babies) and livestock. High levels of nitrates that enter water supplies from the overuse of fertilizers and from livestock waste, sewage systems, and garbage dumps cause health problems and EUTROPHICATION.

nitrification The oxidation of ammonia compounds to nitrites or nitrites to nitrates, carried out by certain *nitrifying bacteria* in the soil. The chemosynthetic soil bacteria *Nitrosomonas* and *Nitrobacter*

carry out the first and second stages, respectively, of this conversion. The process is important in the NITROGEN CYCLE because many plants assimilate nitrates as their source of nitrogen. *Compare* denitrification.

nitrogen An essential element found in all amino acids and therefore in all proteins, and in various other important organic compounds, e.g. nucleic acids. Gaseous nitrogen (N_2) forms about 78% by volume of the ATMOSPHERE but is unavailable in this form except to a few nitrogen-fixing bacteria (*see* nitrogen fixation). Nitrogen is incorporated into plants as the nitrate ion NO_3^- or, especially in acid soils, the ammonium ion NH_4^+, absorbed in solution from the soil by roots. Plants suffering from nitrogen deficiency tend to develop chlorosis and become etiolated, the effects being seen first in the oldest parts. In animals, the nitrogen compounds urea and uric acid form the main excretory products. *See also* nitrogen cycle.

nitrogen cycle The circulation of nitrogen between organisms and the environ-

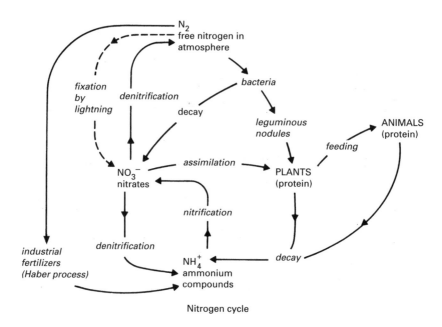

Nitrogen cycle

ment. Atmospheric gaseous nitrogen can be used directly only by certain nitrogen-fixing bacteria (e.g. *Clostridium*, *Nostoc*, *Rhizobium*). They convert nitrogen to ammonia, nitrites, and nitrates, which are released into the soil by excretion and decay. Some are free-living, whereas others form symbiotic associations with plants (*see* nitrogen fixation). Another method by which atmospheric nitrogen is fixed is by lightning, which causes nitrogen and oxygen to combine. The oxides so produced dissolve in rain to form nitrous and nitric acids; in the soil these acids combine with mineral salts to form nitrites and nitrates. This is an insignificant process when compared with microbial NITROGEN FIXATION. When plants and animals die, the organic nitrogen they contain is converted back into nitrate in the process termed NITRIFICATION. Apart from uptake by plants, nitrate may also be lost from the soil by DENITRIFICATION and by LEACHING. The use of nitrogen fertilizers in agriculture and the emission of nitrogen oxides in car exhaust fumes have influenced the nitrogen cycle and contributed to urban air pollution (*see* photochemical smog) and acid precipitation.

nitrogen fixation   The formation of nitrogenous compounds from atmospheric nitrogen. In nature this may be achieved by electric discharge in the atmosphere or by the activities of certain microorganisms. For example, symbiotic bacteria of the genus *Rhizobium* are associated with leguminous plants, causing the root cortex to form nodules, which house the bacteria. These bacteria contain the nitrogenase enzyme that catalyzes the fixation of molecular nitrogen to ammonium ions, which the plant can assimilate. In return the legume supplies the bacteria with carbohydrate. Many other nitrogen-fixing symbioses are known, e.g. the nitrogen-fixing root nodules of bog myrtle (*Myrica gale*) and alder (*Alnus glutinosa*) appear to contain unicellular protoctists called plasmodiophorans, while water ferns (*Azolla*) have CYANOBACTERIA in their roots. Free-living bacteria that can fix nitrogen include members of the genera *Azotobacter*, *Klebsiella*, and *Clostridium*, some sulfur bacteria (e.g.

Chlorobium), and most cyanobacteria (e.g. *Anabaena*). Some yeast fungi have also been shown to fix nitrogen. Industrially, nitrogen fixation is mainly achieved using the HABER PROCESS. *See* nitrogen cycle; symbiosis.

nitrogenous wastes   Organic wastes that contain significant amounts of nitrogen.

noise pollution   Harmful or unwanted sounds in the environment.

nomenclature   An internationally recognized system of naming of living organisms and fossils that forms the basis of their scientific classification. *See* classification; International Codes of Nomenclature; taxonomy.

nominal scale   *See* measurement scales.

nonequilibrium models   Models of POPULATION DYNAMICS that take into account spatial and temporal variation to explain the behavior of populations that are not in a state of equilibrium.

nonequilibrium theory   *See* equilibrium theory.

nonionizing electromagnetic radiation   Radiation that has no effect on the structure of atoms, but which heats tissues and may have harmful effects. Examples include radio waves, microwaves, and low-frequency electromagnetic fields from high-voltage electric power cables.

nonpoint source   A source of POLLUTION that cannot be recognized at a single location, for example, photochemical smog caused by vehicle exhausts.

nonpotable   Undrinkable. The term may refer to water that is unsafe to drink due to contamination with bacteria or pollutants, or unpalatable water.

nonrandom sampling   *See* sampling.

nonrenewable resources   RESOURCES that

have been built up over a long period of time (on a geological time scale) that cannot be renewed within a human lifespan, either because they take too long to accumulate or because the conditions that favored their accumulation no longer exist. Examples include FOSSIL FUELS, water in AQUIFERS that is not being replenished fast enough (see fossil water), and many mineral resources. See sustainable development.

normal distribution A symmetrical *bell-shaped curve* produced on a GRAPH or histogram of distribution. A normal distribution is used to model the variation in a set of experimental results and allows comparisons with other sets of results. For the whole range of experimental values it shows the number of results that have each value. There is an even distribution of values on either side of the mean. See standard deviation.

North Atlantic Deep Water A body of cold saline water formed by winter cooling as the Norwegian Current flows out from the Norwegian Sea over the rim of the ridge that stretches from Scotland and Iceland to Greenland, and plunges down to the of the Atlantic Ocean, where it spreads slowly south along the western edge of the North Atlantic. It is joined by water from the Labrador Sea, and from highly saline water formed by evaporation in the Mediterranean Sea, which flows into the North Atlantic through the Strait of Gibraltar. It takes about 200 years for water in the NADW to reach the Antarctic Circumpolar Current and spread into the Indian and Pacific Oceans.

northern coniferous forest See boreal forest.

NPK Nitrogen, phosphorus, and potassium, used together in a fertilizer.

nuclear energy Energy released by processes that disrupt the nuclei of atoms, especially the breakdown (*nuclear fission*) in specially designed reactors of naturally unstable elements, resulting in the alter-

ation of atomic nuclei and thus the formation of different elements. The alternative term *atomic energy* is not accurate, as it implies the rearrangement of atoms rather than of nuclear particles. In nuclear fission the nucleus of an atom breaks up into two lighter nuclei either naturally or when excited by particles such as alpha particles, deuterons, neutrons, or protons. The reaction released more neutrons, which stimulate further reactions, setting up a chain reaction that releases vast amounts of energy. In a nuclear reactor such chain reactions are controlled and the heat evolved used to raise steam to drive turbines for generating electricity. The main element involved in the production of nuclear energy by fission is uranium, which splits to form two lighter elements (*fission fragments*). Other by-products of the fission of uranium and plutonium include strontium-90, iodine-131, and cesium-137. The persistence of radioactive products of fission is measured as the half-life – the time taken for half of a given mass of an element to be changed into another isotope or element by radioactive decay. The early products of fission are unstable and rapidly decay, emitting beta particles (electrons). Later products are more stable and have longer half-lives. Cesium-137, for example, has a half-life of 30 years. These radioactive fission products are dangerous by-products of the generation of nuclear energy, and their storage and containment imposes high costs on the nuclear energy industry. When not contained, they can cause serious contamination of the environment, especially when released into the atmosphere or the sea. Another drawback is the possibility of catastrophic accidents such as those at Chernobyl (*see* Chernobyl nuclear accident) and THREE-MILE ISLAND. Supporters of nuclear energy point to the fact that it does not contribute to global warming.

Another form of nuclear energy that has not yet been harnessed for industrial use is *nuclear fusion*, in which light elements react together to form heavier ones. Nuclear fusion reactions are the main source of energy for the Sun, involving the 'burning' of hydrogen to form helium. The heat from nuclear fission or fusion reactions can be used to produce steam or other hot gases to drive electricity-generating turbines. Energy released during radioactive decay can be stored in semiconductor materials to produce batteries for heart pacemakers, remote-controlled exploration instruments, and other applications. *See* high-level radioactive waste; low-level radioactive waste; radioactive decay.

nuclear fallout   *See* radioactive fallout.

nuclear fission   *See* nuclear energy.

nuclear fusion   *See* nuclear energy.

nuclear membrane   *See* nucleus.

nuclear pore   *See* nucleus.

nuclear power   Electricity produced by power plants in which NUCLEAR ENERGY is converted into heat, which is used to produce steam to drive turbines, which power electricity generators.

nuclear reaction   A reaction that involves a change in the nucleus of an atom, for example radioactive decay, nuclear fission or fusion, or neutron capture. It differs from chemical reactions, which are limited to changes in the number and/or distribution of electrons surrounding the nucleus. *See* nuclear energy; radioactive decay.

nuclear reactor   A device that initiates and controls nuclear fission reactions to produce heat. *See* nuclear energy.

nuclear winter   The significant cooling of the global climate suggested to be a possible consequence of nuclear war, due to the resulting high-altitude dust clouds caused by the nuclear explosions and the burning of forests, fossil fuels, and plastics set alight by firestorms triggered by exploding nuclear warheads. Such a large accumulation of dust would block sunlight for weeks afterward, perhaps reducing surface temperatures by as much as 22°C (40°F). The low temperatures, frosts, and dim light, as well as the radioactive fallout, would destroy plant and animals life.

nucleoplasm   *See* nucleus.

nucleotide   A compound formed by condensation of a nitrogenous base (a purine, pyrimidine, or pyridine) with a sugar (ribose or deoxyribose) and phosphoric acid. ATP is a *mononucleotide* (consisting of a single nucleotide), the coenzymes NAD and FAD are *dinucleotides* (consisting of two linked nucleotides), and the nucleic acids (*see* DNA: RNA) are *polynucleotides* (consisting of chains of many linked nucleotides).

nucleus (*pl.* nuclei)   **1.** An organelle of EUKARYOTE cells containing the genetic information (DNA) and hence controlling the cell's activities. It is the largest organelle, typically spherical and consisting of *nucleoplasm* bounded by a double membrane, the *nuclear membrane*, which is perforated by many pores (*nuclear pores*) that allow exchange of materials with the cytoplasm. The outer nuclear membrane is an extension of the *endoplasmic reticulum*, a network of membrane compartments distributed throughout the cytoplasm, in which glycoproteins and other complex compounds are synthesized. In the nondividing (interphase) nucleus the genetic material is irregularly dispersed as chromatin; during nuclear division (MITOSIS or MEIOSIS) this condenses into densely staining CHROMOSOMES, and the nuclear envelope. **2.** The central positively charged 'core' of an atom.

NUE   *See* nutrient use efficiency.

null hypothesis   A statistical hypothesis that can be tested experimentally. It proposes that any observed difference (for example, between the means of two samples) is due to chance alone and not to the effect of an internal or external factor. To test how significant the difference between the expected result and that proposed by the null hypothesis, a statistical test such as Student's T-TEST may be used.

numerical response   A change in the size of a predator population as a result of a change in the density of its prey. *Compare* functional response.

numerical taxonomy (taxometrics) The assessment of similarities between organisms by mathematical procedures. It involves statistical analysis of many measurable characteristics and uses phenetic rather than phyletic evidence. The results are usually expressed as one or more DENDROGRAMS.

nutrient A chemical substance that an organism must obtain from its environment in order to sustain life and grow. Nutrients are used to build up the body's cells, and to release energy for vital processes.

nutrient cycle *See* biogeochemical cycle.

nutrient enrichment The addition of nutrients to natural waters through run-off from land treated with fertilizers, effluent from sewage treatment works and livestock manure, urban drainage systems and industrial sources, or deforestation. *See* eutrophication.

nutrient foraging The regulation of nutrient uptake. Nutrient foraging strategies include changes to root architecture (patterns of branching and spacing of roots) in response to soil conditions, and to stem growth. Roots or stems may become thinner and longer in poor nutrient conditions. Symbiotic associations with microbes or other plants (e.g. mycorrhizae and root nodules) may also be affected, and also processes that change the availability of ions in the soil, such as the exudation of mucilages from roots.

nutrient recycling *See* biogeochemical cycle.

nutrient use efficiency (NUE) The ratio of productivity to the uptake of a specific essential nutrient.

nutrification The addition of nutrients to an ecosystem that increases the production of BIOMASS. This may lead to a decrease in the oxygen concentration in soil or water.

nutrition The way in which an organism assimilates and utilizes NUTRIENTS. Organisms may be classified on the basis of the way in which they derive nutrients and energy. *See* autotroph; chemoautotroph; chemoheterotroph; heterotroph; lithotroph.

obligate   Describing an organism that requires specific environmental conditions for its survival and cannot adopt an alternative mode of living. For example, an obligate aerobe is an organism that can grow only under aerobic conditions and cannot survive in anaerobic conditions. An obligate parasite cannot live in the absence of its host. *Compare* facultative.

occupational exposure standards (OES) Limits for exposure to chemical, biological, or physical hazards, beyond which there will be a risk to health. Such limits may be defined by manufacturers of products such as chemicals and equipment, or by industry regulators.

occupational health   An area of legal responsibility for employers and employees, concerning protection from physical hazards, stress, and occupational diseases due to poor ventilation and lighting, risks related to equipment, machinery, or building design, shift patterns, and chemicals in the workplace. It involves planning to prevent or minimize such risks, safety training, and monitoring safety and emergency procedures.

ocean   The continuous body of more or less salty water that occupies some 71% of the Earth's surface, excluding lakes and inland seas.

ocean current   A net movement of ocean water along a definable path due to gravity, wind friction, or gradients in water density.

oceanic   The marine environment beyond the continental shelf, which is usually deeper than 200 m. It makes up about 99%

of the total marine environment. *Compare* sublittoral.

oceanic crust   *See* crust.

oceanography   The study of the oceans, their distribution and mapping, the physics, chemistry, biology and ecology of their waters, and the exploitation of their resources.

odor threshold   The lowest concentration of a substance in air or water that has a detectable odor.

OECD   *See* Organization for Economic Cooperation and Development.

OES   *See* occupational exposure standards.

oil   1. A triacylglycerol that is liquid at or near room temperature. The commonest fatty acids in oils are oleic and linolenic acids, which are unsaturated. Storage oils, synthesized in the endoplasmic reticulum, make up as much as 60% of the dry weight of certain seeds, such as the castor bean (*Ricinus communis*). After oil is extracted from seeds, the residual 'cake' is used to feed livestock or, if poisonous (as in castor bean), as fertilizer.
   Essential oils are volatile oils secreted by aromatic plants; they are the source of characteristic odors and flavors. Some essential oils repel insects or grazing animals, while others (*allelochemicals*) deter encroaching neighboring plants. Resinous oils help prevent loss of sap and protect against the entry of pathogens and parasites. Essential oils are common in XERO-PHYTES, where in hot conditions the evaporating oils increase the density of the

boundary layer at the leaf surfaces, so reducing TRANSPIRATION.
2. *See* petroleum.

oil field   An area rich in petroleum deposits, especially one that is already being exploited.

oil fingerprinting   A detailed chemical analysis that distinguishes between oils from different sources. It is used to trace the sources of oil spills.

oil refinery   A processing unit that separates crude petroleum or petroleum mixtures into their component parts, producing fractions suitable for different purposes such as fuel or lubrication.

oil shale   A brown or black SEDIMENTARY ROCK that yields liquids or gaseous hydrocarbons on heating.

oil slick   A layer of mineral oil on the surface of water. It has the effect of making the surface of the water smooth. The oil prevents oxygen and light penetrating into the water below, so may suffocate marine life. Spillages from oil tankers are a major pollutant of coastline.

okta   A unit of cloud cover, defined as cloud covering one-eighth of the sky area. It is used especially in relation to aviation.

old-growth forest   In North America, a forest that has remained undisturbed since before European settlement. It is therefore at a late stage of succession and probably resembles the climax vegetation (*see* climax community).

Oligocene   An epoch of the TERTIARY, 38–25 mya. It is characterized by the gradual disappearance of earlier mammal groups, including primitive insectivores and primates, and their replacement by more modern forms. The world's climate ranged from temperate to subtropical, and the epoch was marked by an expansion of grasslands at the expense of forests, a change that promoted the evolution of large herbivores. Around the great Tethys

Sea there were tropical swamps, which gave rise to extensive deposits of lignite in Germany and neighboring countries. The angiosperms were evolving and diversifying rapidly, and overtook the gymnosperms in their abundance. Flowers and insects were evolving together, and complex COEVOLUTION was taking place as pollination mechanisms diversified. *See also* geological time scale.

Oligochaeta   *See* Annelida.

oligotrophic   Describing lakes and ponds that are low in nutrients and consequently low in productivity. *Compare* eutrophic.

omnivore   An animal that eats both plant and animal material, e.g. some species of bears. *Compare* carnivore; herbivore.

ontogeny   The course of development of an organism from fertilized egg to sexual maturity with the production of gametes and the next generation.

OP   *See* osmosis.

opacity   The degree to which light is able to penetrate a substance. For example, clear glass has an opacity of zero; a brick wall has an opacity of 100%. Opacity is used to indicate the degree of particulate pollution in air or water.

open community   *See* community structure.

open-pit mining   The extraction of minerals or metal ores that lie near the surface by removing the overlying rocks and soil to expose the ore, which is then broken up and removed. Such mining techniques can produce unsightly scars on the landscape.

operational sex ratio   *See* sex ratio.

operon   A genetic unit found in prokaryotes and comprising a group of closely linked genes acting together and coding for the various enzymes of a particular biochemical pathway. They function as a unit, since either all are transcribed or none. Transcription is controlled by regulator genes and proteins, and inducer molecules. Such systems are usually controlled by negative feedback mechanisms. An example is the *lac* operon in the bacterium *Escherichia coli*, which is involved in the metabolism of lactose. In the absence of an inducer molecule (the substrate lactose or one of its derivatives), no transcription occurs. But if lactose binds to the regulator protein, transcription can proceed and enzyme synthesis is initiated. *See also* feedback mechanism.

opportunist species *See* fugitive species.

optimal foraging theory The theory that organisms attempt to balance the costs and benefits of feeding by adopting strategies that maximize their food intake per unit of time or minimize the time required to acquire sufficient food. Other factors, such as predator avoidance, may also be taken into account when simulating such strategies.

optimum similarity The degree of similarity between competing species at which both species show optimum FITNESS. Any change would result in the lowering of fitness of individuals from at least one of the species. *See also* competitive exclusion principle.

order A TAXON that consists of a collection of similar families. Plant orders generally end in *ales* (e.g. Liliales) but animal orders do not have any particular ending. Orders may be divided into suborders. Similar orders constitute a class.

ordinal scale *See* measurement scales.

ordinate *See* graph.

Ordovician The second oldest period of the PALEOZOIC era, some 510–440 million years ago. It is characterized by an abundance of marine invertebrates (e.g. brachiopods and echinoderms) but an almost total absence of vertebrates apart from some jawless fish. Many of the inverte-brates were primitive forms of life that have no living representatives. *See also* geological time scale.

ore A mineral that is mined and worked to extract a valuable constituent, such as a metal.

organ A part of an organism that is made up of a number of different tissues specialized to carry out a particular function. Examples include the lung of an animal and the stem of a plant.

organelle *See* cell.

organic 1. Derived from or relating to living organisms.
2. (*Chemistry*) Describing a compound that contains carbon.

organic farming Farming that does not involve the use of chemical pesticides or artificial (inorganic) fertilizers.

organic loading The amount of organic matter in water. This arises naturally by the decomposition of plants and animal remains, and from organic matter washed into rivers and lakes from the surrounding land, but it may be greatly increased by runoff and seepage of livestock manure, sewage, and industrial effluents into waterways and groundwater. Deforestation causes a significant increase in organic load due to the decomposition of the felled timber and other vegetation and to the soil erosion that often follows. High organic loads may lead to EUTROPHICATION.

organic matter The organic component of an ecosystem, especially the organic material present in soils.

Organization for Economic Cooperation and Development (OECD) An international organization founded in 1961 to promote world trade, economic growth and employment, and an increase in living standards in member countries. It attempts to boost this by liberalizing international trade and facilitating the movement of capital, including development aid, between countries. It also researches and compiles economic data relating to natural resources, pollution and environmental degradation, farming, capital markets and taxation, scientific research, and education.

organochlorines Organic compounds containing chlorine used in pesticides. Organochlorines are not readily broken down and persist in animal tissues, affecting breeding success and causing other problems. They are banned from use in most countries, but are still used in some, especially where insect-borne diseases such as malaria are prevalent.

organomercury compounds *See* methylmercury compounds.

organophosphates Organic phosphate-containing chemicals, such as malathion and parathion, used to control pests such as aphids and mites. Many are systemic insecticides, being taken up by the roots of plants and affecting any insect that bites into or sucks sap from them. These insecticides are nonselective and highly toxic, destroying natural predators of pest species and affecting animals even farther up the FOOD CHAIN. Insects also rapidly evolve resistance to them. Organophosphates, which are chemically similar to some nerve gases, are also toxic to humans, affecting the nervous system. Strict precautions against skin contact and inhalation must be taken. Some organophosphates persist in the soil.

Oriental One of the six zoogeographical regions of the earth. It includes the southern Asian countries of India, Southeast Asia, and the western Malay archipelago. The characteristic fauna include the Indian elephant, rhinoceros, macaque, gibbon, orang-utan, jungle fowl, and peacock. The boundary between this region and the Australasian region has been the subject of contention in the past.

origin of life Geological evidence strongly suggests that life originated on Earth about 4600 million years ago. The basic components of organic matter – water, methane, ammonia, and related compounds – were abundant in the atmosphere, which had much greater reducing properties than today's atmosphere. Until recently, it was widely accepted that life evolved in warm lagoons or hot springs, where energy from the sun (cosmic rays) and lightning storms caused simple molecules to recombine into increasingly complex organic molecules that eventually showed the characteristics of living organisms. However, recent discoveries suggest that life may have evolved deep underground in fissures in hot rocks – a habitat in which today vast numbers of THERMOPHILE bacteria live. Life probably had its origins in organic self-replicating molecules that consumed the chemicals around them to duplicate themselves. These molecules were probably a form of RNA, nicknamed the *ur-gene* – a combined replicator and catalyst. RNA is a relatively unstable molecule, and eventually the more stable DNA evolved, together with a system for making RNA copies of it – a kind of primitive RIBOSOME. This was the *Last Universal Common Ancestor (Luca)*.

oscillation A regular cyclical change in climate, especially in global climate. *See* Southern Oscillation; Pacific Decadal Oscillation.

osmometer An instrument for measuring osmotic pressure (*see* osmosis).

osmoregulation The process by which animals regulate their internal osmotic pressure by controlling the amount of water and the concentration of salts in their bodies, thus counteracting the tendency of water to pass in or out by osmosis.

osmosis The movement of solvent from a dilute solution to a more concentrated solution through a *selectively permeable membrane* (or *semipermeable membrane*), i.e. one that allows the passage of some kinds of molecule and not others. For example, if a concentrated sugar solution (in water) is separated from a dilute sugar solution by a selectively permeable membrane, water molecules can pass through from the dilute solution into the concentrated one by DIFFUSION but sugar molecules will not cross the membrane.

Osmosis between two solutions will continue until they have the same concentration. If a certain solution is separated from pure water by a membrane, osmosis also occurs. The pressure necessary to stop this osmosis is called the *osmotic pressure* (OP or π) of the solution. The more concentrated a solution, the higher its osmotic pressure. Osmosis is a very important feature of both plant and animal biology. Cell membranes act as differentially permeable membranes and osmosis can occur into or out of the cell. It is necessary for an animal to have a mechanism of OSMOREGULATION to stop the cells bursting or shrinking. In the case of plants, the cell walls are slightly elastic – the concentration in the cell can be higher than that of the surroundings, and osmosis is prevented by the pressure exerted by the cell walls (*wall pressure*).

Where the solvent is water, physiologists now describe the tendency for water to move in and out of cells in terms of *water potential*, which is the difference between the energy of the system and that of pure water. A solution that contains the same concentration of osmotically active solutes as the cell, so there is no net movement of water between the solution and the cell, is said to be *iso-osmotic*. A solution that has a higher concentration of osmotically active solutes than the cell is termed *hyperosmotic*, and water moves out of the cell. Conversely, a solution with a lower concentration than the cell is *hypo-osmotic*, and water moves from it into the cell.

osmotic desert *See* physiological drought.

osmotic pressure *See* osmosis.

outbreeding Breeding between individuals that are not closely related. In plants the term is often used to mean cross-fertilization, and various methods exist to promote it naturally, e.g. stamens or ovaries maturing before pistils. In animals behav-

ioral mechanisms often promote outbreeding. The most extreme form – crossing between species – usually results in sterile offspring and there are various mechanisms to discourage it. Outbreeding increases heterozygosity (*see* heterozygous), giving more adaptable and more vigorous populations. *Compare* inbreeding.

outbreeding depression In species that show ecotypic variation over relatively small distances, OUTBREEDING may reduce FITNESS because the resulting individual will not be so closely adapted to local environmental conditions as either of its parents.

outfall The point at which effluent is discharged from a conduit, drain, or sewer into receiving waters.

overcompensation An apparent increase in the growth, FITNESS, or reproductive success of plants resulting from increased herbivory. The effects are often due to changes in plant architecture caused by grazing; for example, where removal of a flowering stalk results in the production of several more. Such effects are not necessarily advantageous, for example the seeds may be smaller or there may be fewer produced per flower.

overdispersion *See* dispersion.

overexploitation The excessive use of natural resources, resulting in environmental degradation and long-term deleterious ecological effects.

overpopulation A population density that exceeds the capacity of the environment to support it in a healthy state.

overturn *See* fall bloom.

oxidant A substance containing oxygen that reacts chemically with other materials to produce new substances. Oxidants are important contributors to PHOTOCHEMICAL SMOG.

oxidation An atom, an ion, or a molecule is said to undergo oxidation or to be oxidized when it loses electrons. The process may be effected chemically, i.e. by reaction with an *oxidizing agent*, or electrically, in which case oxidation occurs at the anode. For example,

$$2Na + Cl2 \rightarrow 2Na^+ + 2Cl^-$$

where chlorine is the oxidizing agent and sodium is oxidized, and

$$4CN^- + 2Cu^{2+} \rightarrow C2N2 + 2CuCN$$

where Cu^{2+} is the oxidizing agent and CN^- is oxidized.

oxidation ditch (oxidation pond) An artificial body of water in which organic wastes are decomposed by bacteria, e.g. a sewage lagoon.

oxygen An element essential to living organisms both as a constituent of carbohydrates, fats, proteins, and their derivatives, and in aerobic RESPIRATION. It enters plants in both carbon dioxide and water, the oxygen from water being released in gaseous form as a by-product of photosynthesis. Plants are the main, if not the only, source of gaseous oxygen and as such are essential in maintaining oxygen levels in the air for aerobic organisms.

oxygen demand *See* biological oxygen demand; chemical oxygen demand.

oxygen method *See* dark bottle.

ozone (trioxygen; O_3) A pale blue gas with a pungent odor, and a powerful oxidizing agent. It is unstable and decomposes to oxygen on warming. It can seriously impair the respiratory system. *See* ozone layer; photochemical smog.

ozone depletion A reduction over time of the concentration of ozone in the OZONE LAYER. *See* Farman; Molina; Rowlands.

ozone hole *See* ozone layer.

ozone layer (ozonosphere) A layer consisting of OZONE (O_3) molecules scattered

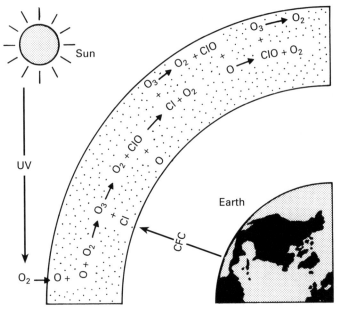

Ozone layer: chlorine from CFCs combine with ozone to form chlorine monoxide (ClO), which react with atomic oxygen (O) to form O_2 and chlorine atoms. The presence of the chlorine atoms has a significant effect on the amount of ozone present.

through the stratosphere roughly 15–50 km above the Earth's surface. The ozone is formed by the action of ultraviolet solar radiation on oxygen molecules. The ozone layer absorbs about 99% of the harmful ULTRAVIOLET radiation entering the Earth's ATMOSPHERE, and hence provides a shield for living organisms without which life on Earth would be impossible. This absorption of ultraviolet radiation has two consequences: it warms the stratosphere, helping maintain the temperature of the Earth's surface; and it stabilizes the stratosphere, preventing it from mixing with other layers, so reducing winds in the lower atmosphere.

In the early 1980s it was discovered that the ozone layer was becoming thinner, with seasonal 'holes' (*ozone holes*) appearing over Antarctica and elsewhere. This de-pletion is believed to be caused by increasing atmospheric concentrations of chlorine compounds, especially CFCS – very stable chemicals used as refrigerants and aerosol propellants – which disrupt the delicate balance between ozone production and breakdown in the atmosphere. A single atom of chlorine or fluorine can destroy about 10^5 molecules of ozone. Other ozone-destroying chemicals include nitrogen oxides from aircraft engine exhausts and HALONS – gases widely used in industry and in certain kinds of fire extinguishers. Such findings led to the introduction in 1987 of restrictions on the use of CFCs, with a complete ban proposed to take effect by 2000. However, even with a ban in force it may take many years before the ozone layer is stabilized. *See* Montreal Protocol.

P

Pacific Decadal Oscillation (PDO) An EL NIÑO-style shift in the temperature of the Pacific Ocean that has two components, a 15- to 20-year oscillation and a second oscillation of about 70 years. It affects sea surface temperature, sea-level pressure, and surface wind stress. Although the temperature difference is only 1 to 2°C, it has a significant effect on the climate of North America, altering the path of the JET STREAM, and hence of storms. During warm phases there is increased PRODUCTIVITY in marine coastal ecosystems in Alaska and decreased productivity off the remainder of the west coast of the USA. The PDO is found in both north and south Pacific, but not in equatorial regions.

Palearctic One of the six zoogeographical regions, including Europe, Russia and the former Soviet states, northern Arabia, and the Mediterranean coastal strip of Africa. The fauna include the hedgehog, wild boar, and fallow and roe deer.

paleobotany *See* paleontology.

Paleocene The oldest epoch of the TERTIARY PERIOD, 65–55 million years ago. It was a period of warm humid climate with large areas of temperate and subtropical forests, but only limited grasslands, because the grasses were of very recent origin. The dinosaurs had become extinct, and there were various primitive mammals, including the first insectivores, rodents, and primates. During the Paleocene there was a great deal of COEVOLUTION of plants and animals, the plants evolving adaptations to animal pollination and fruit and seed dispersal. In the latter part of the Paleocene, large angiosperm fruits and seeds evolved as fruit- and seed-eating mammals also

evolved and diversified. *See also* geological time scale.

paleoecology *See* paleontology.

paleontology The study of extinct organisms, including their fossil remains and impressions left by them. Sometimes the subject is divided into *paleobotany*, the study of fossil plants, *paleozoology*, the study of fossil animals, and *paleoecology*, the study of past life and environments and the interactions between them.

Paleozoic The first and oldest era in which multicellular life became abundant, about 590–230 million years ago. It is divided into six main periods: the Cambrian, Ordovician, Silurian, Devonian, Carboniferous, and Permian. Beginning with aquatic invertebrates and algae, the era ended with the invasion of land by tree ferns and reptiles. *See also* geological time scale.

paleozoology *See* paleontology.

palynology (pollen analysis) The study of living and fossil pollen. It is mainly used as a means of obtaining information about the composition and extent of past floras, and may give reliable quantitative information on the vegetative cover many thousands of years go. Study of living and fossil pollen can also be used as a character to determine plant PHYLOGENY and the nature and adaptation of pollen grains.

pampas Temperate GRASSLAND in South America, found especially in Argentina. Originally dominated by tussock grasses, much of the wetter northeastern pampas is cultivated, while the drier areas remain under short grasses and xerophytic shrubs.

Pangaea A vast landmass or supercontinent, which once contained much of the Earth's land. It broke up during the Mesozoic era into two smaller landmasses, GONDWANALAND in the south and LAURASIA in the north. *See* continental drift; plate tectonics.

parallel evolution (parallelism) The development of similar features in closely related organisms as a result of strong selection in the same direction (*see* directional selection). This may occur between species of the same genus that are widely separated geographically but which live in similar environments. There are few examples of this phenomenon and some authorities deny its existence. A commonly quoted example is the similarity between Australian marsupials and placental mammals, which share a common ancestor in the distant past. Both groups have species that look and behave like wolves, cats, mice, moles, and anteaters. *Compare* convergent evolution.

paralytic shellfish poisoning *See* red tide.

parapatric speciation *See* speciation.

paraquat A nonselective contact HERBICIDE that causes severe and often irreversible damage to lungs, liver, and kidneys if ingested. Paraquat becomes inert on contact with the soil.

parasite *See* parasitism.

parasitism An association between two organisms in which one, the *parasite*, benefits at the expense of the other, the *host*. The tolerance of the host varies from being almost unaffected to serious illness and often death. An *obligate parasite* such as potato blight (*Phytophthora infestans*) can live only in association with the host, whereas a *facultative parasite* such as the damping off fungus (*Pythium*) can exist in other ways, for example as a saprophyte.

Some parasites – *ectoparasites* – such as fleas (Order Siphonaptera) live on the outside of their hosts, whereas others – *endoparasites* – such as the pig tapeworm (*Taenia solium*) are internal parasites. In *brood parasitism*, e.g. the European cuckoo (*Cuculus canorus*) and South American brown-headed cowbird (*Molothrusater*), the female birds lay their eggs in the nests of other species, leaving the other species to rear their young.

Parasites can also be classified into *microparasites* (microscopic organisms such as bacteria, fungi, viruses, and protozoans), and *macroparasites* (larger animals such as arthropods and worms).

parasitoid An animal (usually a wasp or a fly) that spends part of its life as a parasite, and part as a predator. For example, parasitic braconid wasps (family Braconidae) undergo their larval development inside the body of another insect, eventually killing and consuming it, but are free-living as adults. Such wasps are important agents of BIOLOGICAL CONTROL, used for killing aphids and similar pests. *See also* parasitism.

parathion *See* organophosphates.

parental investment The energy and resources that a parent organism puts into caring for its offspring, providing it with resources, so enhancing the survival chances of the offspring.

parthenocarpy The development of fruit in unfertilized flowers, resulting in seedless fruits. It may occur naturally, as in the banana (*Musa*), which is triploid. It may also be induced artificially by the application of the plant hormone auxin, as in commercial tomato growing.

parthenogenesis The development of unfertilized eggs to form new individuals. It occurs regularly in certain plants, e.g. dandelion (*Taraxacum*), in which meiosis is defective, and in some animals (e.g. aphids). Animals produced by parthenogenesis are always female and, if diploid, look exactly like the parent. *See also* apomixis.

partial parasite (hemiparasite) A parasitic plant that has chlorophyll and can photosynthesize, but depends on another plant for its water and mineral salts. An example is mistletoe.

particle size distribution The proportion of particles of different sizes in a sample, usually expressed as weight percentages. There are standard terms for particles of different sizes. However, the same terms may be assigned to different size ranges by soil scientists and geologists. One of the commoner scales in use is the Udden–Wentworth particle size scale.

particulates 1. Substances made up of separate particles.
2. Particles of pollutants. Two types are recognized: primary particulates, which occur in smoke and industrial emissions; and secondary particulates, which are formed in the air by chemical reactions such as those involved in producing PHOTOCHEMICAL SMOG. *See also* PM_{10}.

pascal (Pa) The SI unit of pressure. 1 pascal is a pressure of 1 newton per square meter. (A newton is the force needed to give an acceleration of one meter per second to one kilogram of mass.)

passerine A songbird with a perching habit. More than half of all living species of birds are passerines.

pasture Land covered in grass or other small plants, used as grazing by livestock.

patch dynamics The POPULATION DYNAMICS associated with migration between discrete patches of habitat. It provides the basis for some models of competition.

patchiness 1. The degree to which a HABITAT shows spatial variation in its suitability for a particular species, or the degree to which the distribution of a population shows spatial variation within the habitat.

2. The distribution of different types of HABITAT within a landscape. Habitat quality may be affected by the presence of different habitat types nearby, providing such resources as nesting sites or pollinators. Landscape patchiness affects population dynamics and the success and survival of particular species.

patch-use model An optimal foraging model concerned with how foraging animals choose between patches containing prey.

pathogen Any organism that is capable of causing disease or a toxic response in another organism.

PDO *See* Pacific Decadal Oscillation.

peat Partially decomposed plant material that accumulates in waterlogged anaerobic conditions in temperate humid climates, often forming a layer several meters deep. Peat varies from a light spongy material (*sphagnum* moss) to a dense brown humidified material in the lower layers. If mineral salts are present in the waterlogged vegetation, neutral or alkaline *fen peat* is formed (the salts neutralize the acid produced by decomposition). If there are no mineral salts in the water (as in rain), acid *bog peat* is formed. Peat is used as a fuel and is the first step in COAL formation. It is also used to improve soil and as a component of potting compost in horticulture. These uses have led to concern about the destruction of natural peat bogs. *See also* bog; fen.

peck order *See* dominance.

ped *See* crumb structure.

pedogenesis The process of SOIL formation from unconsolidated parent material.

pedology The study of the formation, structure, distribution, and classification of SOILS.

pelagic Inhabiting the open upper waters rather than the bed of a sea or ocean.

Compare benthic zone. *See also* photic zone; plankton.

Pennsylvanian   The US name for the Upper CARBONIFEROUS period.

peppered moth   *See* industrial melanism.

peptide   *See* protein.

per capita   Per unit or individual of the population or per person.

percolation   The trickling of a liquid through a substance composed of solid particles; for example the trickling of sewage through a filter bed, or of rainwater through soil. *See also* infiltration.

perennating organ   In BIENNIAL or PERENNIAL plants, a storage organ that enables the plant to survive an adverse season, such as winter or a dry season. Most such organs are underground, e.g. rhizomes, corms, bulbs, swollen taproots, and tubers. Perennating organs are also used by many plants in asexual reproduction (VEGETATIVE PROPAGATION).

perennial   A plant that may live for several years. Perennials may reproduce in their first growing season or may have to attain a certain age before seed production commences. *Herbaceous perennials* die back each year and survive until the next growing season as tubers (e.g. *Dahlia*), bulbs (e.g. daffodil), rhizomes (e.g. *Iris*), etc. *Woody perennials*, such as trees and shrubs, persist above ground throughout the year but may show adaptations (e.g. leaf fall) to survive unfavorable seasons. A few plants exhibit *semelparity*, living for several years without flowering and then flowering once and dying after setting fruit, e.g. the century plants, *Agave* (*compare* iteroparity). *See also* deciduous; evergreen; perennating organ.

period   *See* geological time scale.

periphyton   A community of microscopic organisms, such as diatoms, algae, and cyanobacteria, that live on underwater surfaces.

permafrost   *See* tundra.

permanent wilting point   The point at which soil has dried to the extent that plants can no longer remove the remaining water held in the soil particles, and begin to wilt and will not recover unless more water is added to the soil. At the permanent wilting point the water potential (*see* osmosis) of the soil is equal to or lower than the water potential of the plant. *See also* field capacity.

permeability   The ease with which a substance (e.g. rock or soil) will allow a liquid or gas to pass through it.

Permian   The most recent period of the PALEOZOIC era, some 280–250 million years ago. Life became dominated by a few types of reptiles, while amphibians were greatly reduced in number and size. Insect groups appeared. As the climate warmed and dried during the Permian, the ferns, seed ferns, and clubmosses that dominated Permian marshes and swamps were joined by the evolving conifers and some other gymnosperms, whose evolution was probably driven by the need to cope with drier environments. These groups gradually began to replace the older floras dominated by nonseed-bearing plants. *See also* geological time scale.

persistence   The length of time an introduced substance stays in the environment before it is decomposed, degraded, or washed away. Persistence has a big effect on the potential toxicity of POLLUTANTS: less toxic pollutants that are persistent, such as organochlorines, may cause more harm than highly toxic pollutants, such as cyanides, that have low persistence.

persistent organic pollutants (POPs) Introduced organic compounds that are harmful to the environment or to human health and that remain for along time in the environment without being broken down – longer than one growing season or one year after introduction. *See* persistence.

Peru Current   A cold-water ocean current that flows north along the coast of South America in the southeast Pacific Ocean, especially along Chile, Peru, and Ecuador, causing fog in coastal areas, but keeping the rainfall very low. The flow is reinforced by upwelling of deep water due to the effects of surface winds and the CORIOLIS FORCE, making the waters rich in nutrients, supporting a rich fishing industry. During EL NIÑO episodes, the current is forced farther south, leading to a sudden decline in plankton, fish, and bird populations and heavy storms.

pest   Any species considered undesirable in its present location, especially one that competes with humans for food, fiber, or shelter, feeds on people, contaminates food and goods, transmits pathogens, and generally threatens human health or comfort.

pest control   Limiting the numbers of pest populations by killing them, preventing them from breeding, or preventing them from attacking people, livestock or crops, or from any other undesirable activities. *See* biological control.

pesticide   A chemical that kills PESTS. *See also* botanical insecticide; contact pesticide; contact pesticide; fungicide; insecticide; pesticide tolerance; systemic pesticide.

pesticide tolerance   1. The development of tolerance of (resistance to) PESTICIDES. During successive generations individuals may arise whose enzymes or other metabolic processes can tolerate the presence of pesticides or break them down in the body to harmless substances. Such resistant individuals are more likely to survive in the population and pass on their genes for resistance, so resistance will gradually spread through the population. For example, some rats have become resistant to warfarin. One way to overcome this is to administer more than one poison or pesticide at once, because very few individuals will have resistance to more than one type. *See also* pesticide treadmill.
2. A legal limit, enforced by the United States Environmental Protection Agency, for the amount of a chemical residue that is permitted to remain in or on a harvested food or feed crop as a result of the application of a chemical for pest control purposes. Such limits are designed to be well below the point at which the chemical would pose a health risk to consumers.

pesticide treadmill   The cycle of events in which the application of pesticides leads to pests evolving resistance and secondary pests arising that were originally preyed on or outcompeted by organisms affected by the pesticides, requiring the application of more and different pesticides, leading to resistance to these new chemicals and more secondary pests, and so on. *See also* pesticide tolerance.

pest pressure hypothesis   The idea that consumers readily locate and feed on abundant species, so reducing their densities, allowing other, less common species to grow and multiply, so that herbivores can actually promote high diversity in plants.

petrochemical   A chemical manufactured from the products of refining crude oil and coal carbonization, or produced as a fuel oil residue.

petrol   *See* petroleum.

petroleum   A mixture of hydrocarbons formed originally from marine animals and plants, found beneath the ground trapped between layers of rock. It is obtained by drilling (also called *crude oil*). Different oilfields produce petroleum with differing compositions. The mixture is separated into fractions by fractional distillation in a vertical column. The main fractions are:
Diesel oil (gas oil) in the range 220–350°C, consisting mainly of C13–C25 hydrocarbons. It is used in diesel engines.
Kerosene (paraffin) in the range 160–250°C, consisting mainly of C11 and C12 hydrocarbons. It is a fuel both for domestic heating and jet engines.

Gasoline (*petrol*) in the range 40–180°C, consisting mainly of C5–C10 hydrocarbons. It is used as motor fuel and as a raw material for making other chemicals.
Refinery gas, consisting of C1–C4 gaseous hydrocarbons.

In addition lubricating oils and paraffin wax are obtained from the residue. The black material left is bitumen tar.

petroleum gas   *See* natural gas.

pH   A measure of the acidity or alkalinity of a solution on a scale 0–14. A neutral solution (such as pure water) has a pH of 7. Acid solutions have a pH below 7; alkaline solutions have a pH above 7. The lower the pH value, the higher the concentration of H^+. The pH is the logarithm to base 10 of the reciprocal of the hydrogen-ion concentration of a solution; i.e. $\log_{10} 1/[H^+]$.

phage   *See* bacteriophage.

phagoplankton   *See* plankton.

phanerophyte   *See* Raunkiaer's life-form classification.

Phanerozoic   *See* Precambrian.

phenology   The study of biological phenomena in relation to the seasons (e.g. migration and seasonal breeding).

phenotype   The observable characteristics of an organism, which are determined by the interaction of the GENOTYPE with the environment. Many genes present in the genotype do not show their effects in the phenotype because they are masked by dominant alleles (*see* dominant). Genotypically identical organisms may have very different phenotypes in different environments, an effect particularly noticeable in plants grown in various habitats.

phenotypic variance (V_p)   A statistical measure of the variation in value of a phenotypic character between individuals of a population. *Compare* phenotypic variation.

phenotypic variation   Variation in the appearance of individuals in a population. This is due to the effects of both the GENOTYPE and the environment. *Compare* phenotypic variance.

phloem   Plant vascular tissue in which food is transported from areas where it is made to where it is needed or stored. It consists of *sieve tubes* (*sieve elements*), which are columns of living cells (though without nuclei) with perforated end walls that allow passage of substances from one cell to the next; and companion cells, full of mitochondria, which provide energy for the transport. Phloem also contains fibers (sclerenchyma) and parenchyma.

phosphates   *See* phosphorus cycle.

phosphorescence   1. The absorption of energy by atoms followed by emission of electromagnetic radiation. Phosphorescence is a type of *luminescence*, and is distinguished from *fluorescence* by the fact that the emitted radiation continues for some time after the source of excitation has been removed. In phosphorescence the excited atoms have relatively long lifetimes before they make transitions to lower energy states. However, there is no defined time distinguishing phosphorescence from fluorescence.
2. In general usage the term is applied to the emission of 'cold light' – light produced without a high temperature. The light comes from excited atoms produced directly in the reaction – not from the heat produced. It is thus an example of *chemiluminescence*. In biological systems, the light emitted is often light that has been previously absorbed, but which is re-emitted at a different wavelength. Phosphorescent seas are due to the bioluminescence of DINOFLAGELLATES, in particular *Noctiluca* species. Mechanical disturbance of the water causes them to emit flashes of blue light, which is produced by enzyme-catalyzed reactions in special vesicles called *scintillons*. Phosphorescence may sometimes be seen on rotting wood from certain fungi (known as 'fox fire'). *See also* bioluminescence.

phosphorus One of the essential elements (macronutrient) in living organisms. Phosphorus has an important role in the metabolic activities in plants and animals and is essential for plant growth and skeleton formation in vertebrates. Phosphorus may be added to soils as inorganic basic slag or as phosphate fertilizers, or in organic fertilizers such as GUANO and bonemeal.

phosphorus cycle The circulation of phosphorus between organisms and the environment. Phosphorus occurs in the environment as (ortho)phosphates (PO_4^{3-}) derived from the WEATHERING of rocks, especially of the mineral apatite, as organic phosphates released during the DECOMPOSITION of dead organisms, and as colloidal phosphate, a form not available to organisms. High levels of aluminum and iron may cause phosphates to precipitate out of the soil, becoming unavailable to living organisms. Phosphorus occurs in natural waters in the form of soluble phosphates derived from runoff and water percolating through the soil and into lakes and waterways. Its concentration is often boosted by runoff from land treated with fertilizers and from sewage effluents. Rivers carry phosphates into the oceans. Phosphates are assimilated from the soil by soil organisms and plant roots, and from water by bacteria and phytoplankton. In lakes the cycling of phosphorus is affected by STRATIFICATION and seasonal TURNOVER. *See also* biogeochemical cycle; eutrophication.

photic zone In lakes, seas, and oceans, the zone of surface water penetrated by light. It is divided into two further zones by the *compensation depth* (the depth at which the rate of carbon fixation equals the rate of carbon loss by respiration). The zone above this is termed the *euphotic zone* and the zone below the *dysphotic zone*. The depth of the euphotic zone depends on the turbidity of the water. In the dysphotic zone the light penetration or temperature is so low that oxygen consumption by respiration exceeds oxygen production by photosynthesis. *Compare* aphotic zone.

photoautotroph *See* autotroph.

photobiont *See* lichens.

photochemical air pollution *See* ozone layer; pollution.

photochemical smog A type of SMOG that occurs most commonly in urban areas and is caused by the photochemical reactions of the exhaust emissions of vehicles with sunlight. These reactions produce the highly toxic gas OZONE, as well as nitrogen dioxide. Photochemical smog has a brownish tinge, and causes reduced visibility, aggravation of the respiratory tract, irritation of the eyes, and damage to plants.

photoheterotroph A phototrophic organism that uses organic compounds as its main source of carbon. *See* phototroph.

photoinhibition 1. The reversible inhibition of PHOTOSYNTHESIS in certain leaves by light of a particular quality.
2. The stopping or slowing of a process by light, for example, germination of some seeds is inhibited by light.

photoperiodism The response of an organism to changes in day length (*photoperiod*). In plants, leaf fall and flowering are common responses to seasonal changes in day length, as are migration, reproduction, molting, and winter-coat development in animals. Many animals, especially birds, breed in response to an increasing spring photoperiod, a long-day response. Some animals (e.g. sheep, goats, and deer) breed in autumn in response to short days so that offspring are born the following spring. The substance *melatonin*, produced by the pineal gland, is thought to play a role in regulating such changes. Diapause and seasonal changes in form, as in aphids, are photoperiodically induced in insects. Such responses are especially common in species of high and mid-latitudes, where day length is variable and provides a good indicator of the time of year. The length of day that triggers a photoperiodic response is called the *critical day length*.

Plants are classified as *short-day plants* (SDPs), e.g. cocklebur, or *long-day plants* (LDPs), e.g. cucumber and barley, according to whether they flower in response to short or long days. *Day-neutral plants*, e.g. pea and tomato, have no photoperiodic requirement. However, the critical factor is not the length of the day, but the length of the dark period, because flowering of SDPs is inhibited by even a brief flash of red light in the dark period, and an artificial cycle of long days and long nights inhibits flowering in LDPs. The response is mediated by a pigment called *phytochrome*, a protein that can absorb red light, which converts it to a form (P_{FR}) that absorbs far-red light. Absorption of far-red light causes the phytochrome to revert to the red light-absorbing form (P_R). This change is linked to sequences of metabolic reactions that trigger the physiological or morphological response of the plant. *Compare* circadian rhythm.

photophosphorylation *See* photosynthesis.

photorespiration A light-dependent metabolic process of most green plants that resembles true (or 'dark') RESPIRATION only in that it uses oxygen and produces carbon dioxide. Carbon dioxide production during photorespiration may be up to five times greater than in dark respiration. Photorespiration wastes carbon dioxide and energy, using more ATP than it produces (*see* C_3 plant). Photorespiration rates in C_4 PLANTS are negligible. Attempts are being made to introduce the C_4 photosynthetic system into important C_3 crop plants by means of genetic engineering. This may be an advantage for crops in the tropics, because the rate of photorespiration is much greater at higher temperatures.

photosynthesis The synthesis of organic compounds using light energy absorbed by the green pigment CHLOROPHYLL. With the exception of a small group of bacteria, organisms photosynthesize from inorganic materials. All green plants photosynthesize, as well as algae and certain bacteria. In green plants, photosynthesis takes place in CHLOROPLASTS, mainly in leaves and stems. Directly or indirectly, photosynthesis is the source of carbon and energy for all except chemoautotrophic organisms. The mechanism is complex and involves two sets of stages: *light reactions* followed by *dark reactions*. The overall reaction in green plants can be summarized by the equation:

$$CO_2 + 4H_2O \rightarrow ?[CH_2O?] + 3H_2O + O_2$$

The light reactions of photosynthesis involve reaction centers in the chloroplasts and involves chlorophyll *a* molecules. The whole unit is called a photosystem. There are two kinds of photosystems in green plants, *Photosystems I and II*, which use slightly different forms of chlorophyll *a*. Both are usually involved in the light reactions of photosynthesis. Light energy is passed to the reaction center chlorophyll molecules, where it excites electrons to leave the photosystems and pass along an *electron transport chain*. The electrons are replaced by electrons from the *photolysis of water*, releasing molecular oxygen and protons:

$$2H_2O \rightarrow O_2 + 4H^+ + 4e^-$$

These reactions are collectively called the light reactions.

Energy released as the electrons flow along the electron transport chain is used to generate ATP (adenosine triphosphate), a process called *photophosphorylation*, and reducing agents (NADPH). ATP and NADPH are used in the stroma (non-membrane portion) of the chloroplast to reduce carbon dioxide to carbohydrate and for other syntheses – the so-called dark reactions. Carbon dioxide is 'fixed' by combination with the 5-carbon sugar *ribulose bisphosphate* (RuBP), forming two molecules of phosphoglyceric acid (PGA). This reaction is catalyzed by the enzyme *ribulose bisphosphate carboxylase* (rubisco). In a series of reactions using NADPH and ATP from the light reactions, PGA is converted to a succession of 3-, 4-, 5-, 6- and 7-carbon sugar phosphates in a series of reactions collectively termed the *Calvin cycle* or *Calvin–Benson cycle*. These products are then used in the synthesis of carbohydrates, fats, proteins, and other compounds, and RuBP is regenerated.

In determining photosynthesis rates in a whole plant, a distinction is made between *gross photosynthesis* (GP, the total carbon dioxide uptake or total oxygen released in the light) and *net photosynthesis* (NP, which is GP minus the carbon dioxide released or oxygen consumed by respiration): $NP = GP - R$. Net photosynthesis is the actual net carbon dioxide uptake or oxygen evolution that is measured in a controlled experiment.

photosynthetic bacteria A group of bacteria able to photosynthesize through possession of chlorophyll pigments. They include the CYANOBACTERIA, green sulfur bacteria (Chlorobia), green nonsulfur bacteria (Chloroflexa), purple sulfur bacteria, and purple nonsulfur bacteria (both in the phylum Proteobacteria). The photosynthetic bacteria fix carbon using the Calvin cycle, but use a variety of hydrogen sources. The Cyanobacteria derive hydrogen from water, the sulfur bacteria use hydrogen sulfide or reduced sulfur compounds, and the nonsulfur bacteria obtain hydrogen from organic sources, such as ethanol, lactate, or pyruvate.

phototroph An organism that uses light energy to synthesize its organic requirements. *Compare* chemotroph.

phycobiont *See* lichens.

phylogeny The evolutionary history of groups of organisms, in particular the relationships between groups of organisms based on their past evolutionary history. It provides a basis for classification.

phylum (*pl.* phyla) One of the major groups into which a kingdom of organisms is classified. Phyla may be divided into subphyla. In some plant classifications (especially older ones) the term 'division' is used instead of phylum.

physiognomy The study of the structure of natural communities.

physiographic factors Environmental factors other than climatic, biotic, and edaphic (soil) factors, which affect the prevailing conditions in a habitat and the distribution of plants and animals. They include factors relating to topography, such as altitude, slope, drainage, and degree of erosion of the land.

physiological drought A situation in which plants are unable to take up sufficient water even though there is water in the soil. Physiological drought occurs in salt marshes and other brackish coastal habitats, and in deserts where there are high concentrations of salts in the upper layers of the soil. It also occurs in cold weather because the permeability of the endodermal cells decreases rapidly below $5°C$, so the root resists the entry of water. *See* osmosis.

physiological ecology The study of the physiological processes in organisms in relation to their environments.

physiology The way in which organisms or parts of organisms function. *Compare* MORPHOLOGY.

phytoalexin A chemical produced by a plant that inhibits the growth of a pathogenic fungus.

phytochrome *See* photoperiodism.

phytogeography (plant geography) The study of the geographical distribution of plant species.

phytophagous Describing an organism that feeds on plants.

phytoplankton *See* plankton.

phytoremediation *See* bioremediation.

phytosociology The classification of plant COMMUNITIES based on their species composition and distribution. Modern phytosociology uses quantitative methods based on computer analysis of data. It encompasses not only the geographical distribution and classification of plant communities, but also their organization, development, and interdependence.

phytostimulation *See* rhizodegradation.

phytotoxic Describing a substance that is toxic to plants.

phytotron A large chamber or greenhouse with a precisely controlled environment in which plants may be grown for experimentation. Such factors as temperature, light, humidity, and photoperiod can be varied or held constant as desired.

picoplankton *See* plankton.

piezometer A device for measuring the hydraulic head of groundwater at a particular level. It consists of a narrow observation well that allows groundwater to enter at a particular depth.

pine forest *See* boreal forest.

pioneer species A species that colonizes a new physical environment, for example, land exposed by retreating glaciers, felling of forest, or bare mud on coasts or estuaries. Pioneer species are the first stage in plant SUCCESSION. Most are fast-growing and fairly short-lived, with small seeds and good powers of seed dispersal. They may modify the environment so that new species can move in and perhaps replace them.

Pisces A term that includes the two classes of fish – Osteichthyes (bony fish) and Chondrichthyes (cartilaginous fish). Fish are cold-blooded aquatic vertebrates with a streamlined body, a powerful muscular finned tail for propulsion, and paired pectoral and pelvic fins for stability and steering. There is usually a body covering of scales. They breathe by means of internal *gills*, which open to the outside through gill slits or pores. The bony fish have a skeleton of bone, only one external gill opening each side, which is covered by a flap called an operculum, and a body covering of overlapping scales. The cartilaginous fish,

which include the sharks, rays, and skates, are mostly predatory fish with a cartilaginous skeleton, separate gill openings not covered by an operculum, and a skin covering of small scales that are modified in the mouth as rows of teeth.

pitfall trap A device for the live capture of ground-dwelling invertebrates, which consists of a jar containing bait, which is buried in the ground with its mouth at ground level. A raised cover protects the jar from rain and also prevents larger animals from exploring it. Small invertebrates crawl under the cover and fall into the jar.

plagioclimax *See* climax community.

plankton A varied collection of aquatic organisms that drift freely, not being attached to any substrate and not possessing any organs for locomotion. The most important components of the plant plankton (*phytoplankton*) are the DIATOMS upon which the planktonic animals (*zooplankton*) (e.g. crustaceans) feed. Other members of the phytoplankton include microscopic algae and CYANOBACTERIA. The larvae of many species (e.g. cod) make up a large part of the plankton, especially in early summer. The plankton form the basis of the FOOD CHAIN in the sea. Plankton may be classified according to the size of the organisms: *picoplankton* are less than 2 μm in diameter, *nanoplankton* between 2 and 20 μm, *microplankton* between 20 and 200 μm, and *mesoplankton* over 200 μm in diameter. Plankton larger than 25 μm in diameter are sometimes called *net plankton*, because they are large enough to be caught in a plankton net. *Compare* nekton.

Plantae The kingdom that contains plants. It contains organisms that can (usually) make their own food by taking in simple inorganic substances and building these into complex molecules by PHOTOSYNTHESIS. This process uses light energy, absorbed by the green pigment CHLOROPHYLL, which is found in all plants but no animals. There are a few exceptions, in the form of certain parasitic plants. Most plants have cellulose cell walls, and have starch (*see* carbohydrates) as a storage polysaccharide, whereas animals have no cell walls, store glycogen, and do not have plastids or chlorophyll; and fungi usually lack cellulose and also do not form plastids. Plants lack motility, with the exception of the gametes of many species. One major characteristic that distinguishes plants from other plantlike organisms, such as algae or fungi, is the possession of an embryo that is retained and nourished by maternal tissue. Fungi and algae lack embryos and develop from spores. Plants also differ from fungi in having a regular alternation of diploid and haploid generations. *Compare* algae; Animalia; Fungi; Protoctista.

plant breeding The selection and improvement of plants for economic or aesthetic purposes. At its simplest, it involves choosing individuals with desirable characteristics, controlling their mating, and selecting suitable individuals from their progeny for further breeding. It may just consist of choosing the best seeds to plant for the next year's crop. Breeding can be controlled by hand-pollination, development of hybrids (and induction of POLYPLOIDY to render the hybrid progeny fertile), and vegetative propagation to get identical progeny. Additional variation may be introduced by chemicals or radiation that induce mutations. Genetic engineering now enables scientists to produce highly specific combinations of genetically determined characteristics, and tissue culture allows the rapid propagation of selected varieties and individuals (clones). *See* hybridization; vegetative propagation.

plant-derived pesticide *See* botanical insecticide.

plasma membrane *See* cell.

plasmid A genetic element found within bacterial cells that is not part of the bacterial DNA and replicates independently of it. Plasmids are widely used as cloning vectors (agents that transfers pieces of 'foreign' DNA) in GENETIC ENGINEERING.

plasmolysis Loss of water from a walled cell (e.g. of a plant or bacterium) due to OSMOSIS to the point at which the protoplast shrinks away from the cell wall.

plasticity The ability of an organism to change its form in response to changing environmental conditions. For example, some plants develop hairier leaves if moved to a windy site; leaves growing in shade have thinner broader laminae and less densely packed mesophyll than leaves growing in full sunlight. These changes can occur only in parts that develop and grow after the environmental change. They are phenotypic changes that reflect changes in the activation of genes, but not in the genes themselves.

plate tectonics A theory formulated in the 1960s to explain the phenomenon of CONTINENTAL DRIFT and the formation of major features on the Earth's surface. It sees the Earth's crust as a mosaic of rigid major and minor plates up to 100 km thick, which move relative to each other as a result of convection currents in the mantle below. The plates bearing the continents are thicker and less dense than those bearing the ocean floor.

Where two plates are moving apart, molten rock (*magma*) from the mantle wells up in the space between them, hardening to new crust, forming an underwater ridge of mountains called a *mid-oceanic ridge* (e.g. the mid-Atlantic ridge, on which Iceland sits). As new rock wells up in the middle of the ridge it pushes the mountain flanks apart, causing the sea floor to spread (*sea-floor spreading*). On land, such upwellings form *rift valleys*, deep gorges with mountains on either side, such as the East African Rift Valley.

Where a continental and an oceanic plate moving toward each other meet, as is happening along the west coast of the North America, a *subduction zone* is formed, where the denser plate is forced under the lighter one. The descending plate melts to form magma, which may then surface through cracks and faults, forming volcanoes. Where two plates of continental crust meet, the result is a crumpling of the rocks to form mountains such as the Himalayas and the European Alps.

Where two plates slide past each other, friction may build up, and sudden movements releasing it cause earthquakes, e.g. along the San Andreas Fault in California.

Platyhelminthes A phylum of primitive wormlike invertebrates, the flatworms, including the classes Turbellaria (aquatic free-living planarians), and the parasitic Trematoda (flukes) and Cestoda (tapeworms). Flatworms are unsegmented animals with no coelom or blood system. The flat body provides a large surface area for gaseous exchange. The gut, when present, has only one opening (the mouth) and a sucking pharynx. Tapeworms, which live in the guts of other animals, have no gut, but simply absorb dissolved nutrients all over their body surface.

pleiomorphism The occurrence of different morphological stages during the life of an organism. Examples are the larval, pupal, and adult forms of an insect. *Compare* polymorphism.

Pleistocene The first epoch of the QUATERNARY PERIOD, which started with a GLACIATION about 2 million years ago and ended with the last glaciation about 10 000 years ago. Several Ice Ages drove many organisms toward the Equator while others (e.g. mammoth) became extinct. For example, TUNDRA covered temperate parts of the USA and Central Europe where today deciduous forests are the natural vegetation. Many present-day mammals of South America and Africa resemble pre-Ice Age mammals of Europe. Modern humans (*Homo sapiens*) evolved during this period. *See* geological time scale; Ice Age.

Pliocene The epoch of the TERTIARY PERIOD, about 7–2 million years ago, which followed the MIOCENE. In the Pliocene the hominids, such as *Australopithecus* and *Homo*, became clearly distinguishable from the apes. Grasses became more abundant, perhaps because the climate was becoming drier. Modern conifers were spreading, and leptosporangiate ferns underwent adaptive radiation. *See also* geological time scale.

plutonium (Pu) A radioactive silvery metallic element found only in minute quantities in uranium ores but readily obtained, as ^{239}Pu, by neutron bombardment of natural uranium. The readily fissionable ^{239}Pu is a major nuclear fuel and nuclear explosive. Plutonium is highly toxic because of its radioactivity; in the body it accumulates in bone.

PM$_{10}$ The concentration (in mg m^{-3}) of particles in the atmosphere that have a diameter equal to or less than 10 μm. It is pollutant PARTICULATE level standard, used to measure compliance with the CLEAN AIR ACT. Such particulates are believed to be the most damaging to human health.

podzol (podsol) The type of SOIL found under heathland and coniferous forests in temperate and boreal climates where there is heavy rainfall and long cool winters, such as the BOREAL FORESTS of northern North America and Eurasia. It is strongly acid and often deficient in nutrients, especially iron compounds and lime, as a result of leaching by the heavy rain or snow-melt in spring. Beneath the humus layer lies a bleached horizon (A horizon) composed mainly of quartz sand. In the clay-rich B horizon below the leached layer iron compounds accumulate, staining the layer brown and forming an impermeable HARDPAN that prevents drainage. Where this layer is sufficiently impermeable to maintain the A horizon in a waterlogged condition, the soil is termed a GLEY podsol. The type of vegetation that grows on podsols tends to produce highly acidic leaf litter, which further increases the acidity of the soil.

pogonophores Tube worms that grow up to 3 m long, found in large numbers in the communities that live around HYDROTHERMAL VENTS on the ocean floor. They contain symbiotic bacteria that fix carbon using energy from the oxidation of hydrogen sulfide, forming the basis of the FOOD CHAIN.

poikilothermy The condition of having a body temperature that varies approximately with that of the environment. Most animals other than birds and mammals are poikilothermic ('cold-blooded'). *Compare* homoiothermy.

point quadrat *See* cover.

point source A single identifiable source from which pollutants are discharged, e.g. an outfall pipe or chimney.

Poisson distribution A kind of data distribution in which the variance is equal to the area. It occurs when very large numbers of samples or measurements are taken, so that even when an event or value has a low probability of occurring, it will occur eventually. For example, it is typical of the measurement of plant DENSITY using QUADRATS.

polar vortex A region of extremely cold air that forms over Antarctica in winter, which is surrounded by strong westerly winds that isolate the cold air from surrounding air masses in lower latitudes. Temperatures in the vortex may fall as low as −200°C in the lower stratosphere. These particular conditions are ideal for chemical reactions between sulfuric acid aerosols, chlorine, and nitrogen oxides, causing severe depletion of the OZONE LAYER.

polder An area of flat low-lying land, drained and reclaimed from the sea and protected from inundation by embankments, dams, or levees. Polders are often below sea level. They are particularly common in the Netherlands and Belgium.

pollarding A method of pruning that involves cutting off the branches to leave about 2 m of trunk above ground. This produces new bushy growth out of reach of grazing animals. It is commonly seen in riverside willows, in street trees to provide shade, and in certain woodland species to provide a source of small poles for fencing or firewood out of reach of grazing mammals. *See also* coppice.

pollen The male spores of seed plants, produced in large numbers in the pollen sacs, or stamens. *See* palynology.

pollination The transfer of pollen from the anther to the stigma in angiosperms, and from the sporangiophores to the micropyle in gymnosperms. Plants may be self-pollinating, e.g. barley (*Hordeum vulgare*), thus ensuring that seed will be set. *Cross pollination* (the transfer of pollen from one individual to another), also takes place and is brought about by wind, water, insects, or other animals.

pollutant A contaminant in a concentration that adversely changes the environment. A pollutant may be chemical (e.g. CFCs), of biological origin (e.g. sewage), physical (e.g. RADIOACTIVE WASTE, dust from soil erosion, heat emissions, noise), or pathogenic in nature (e.g. *Cryptosporidium* bacteria in swimming pools). *Primary pollutants* are those that are emitted into the environment from a specific source, e.g. sulfur dioxide from power plants. *Secondary pollutants* are formed in the environment itself, e.g. photochemical smog.

pollutant standard index (PSI) An indicator of one or more pollutants used to warn people about the current potential health risks, for example, an indicator of the sulfur dioxide concentration in an urban area.

polluter-pays principle The principle that those who cause pollution should meet the costs of clearing it up. This is sometimes extended to include the costs of making good the damage it causes.

pollution Any damaging or unpleasant change in the environment that results from the physical, chemical, or biological side effects of human activities. Pollution can affect the atmosphere, rivers, seas, and the soil.

Air pollution is caused by the domestic and industrial burning of carbonaceous fuels, by industrial processes, and by vehicle exhausts. Industrial emissions of sulfur dioxide cause ACID RAIN, and the release into the atmosphere of CFCS, used in refrigeration, aerosols, etc., leads to the depletion of the OZONE LAYER in the stratosphere. Carbon dioxide, produced by burning fuel and by motor vehicle exhausts, is slowly building up in the atmosphere, and is one of the factors leading to an overall increase in the temperature of the atmosphere (*see* greenhouse effect). Vehicle exhausts also contain carbon monoxide, nitrogen oxides, and other hazardous substances, such as fine particulate dusts. Lead was formerly a major vehicle pollutant, but the widespread introduction of lead-free gasoline has eliminated this problem in most countries. PHOTOCHEMICAL SMOG, caused by the action of sunlight on hydrocarbons and nitrogen oxides from vehicle exhausts, is a problem in many major cities, and may contain significant amounts of ozone. In plants, gaseous pollutants may inhibit stomatal action, damage leaf surfaces, or inhibit enzymes. Sulfur dioxide causes chlorosis between the leaf veins, and the death of leaves in forest trees. Oxides of nitrogen can cause black necrotic lesions on leaves. Acid rain caused by sulfur dioxide and nitrogen oxides leads to the release of toxic aluminum in soils, and inhibition of nutrient uptake. Fine particulate dust blocks stomatal pores and prevents light from reaching the leaf cells. Not all air pollution results from human activity. For example, volcanic activity discharges vast quantities of ash and toxic gases into the atmosphere. Also, natural sandstorms and vegetation fires contribute dust and smoke, but the incidence of these events may be exacerbated by DEFORESTATION and DESERTIFICATION brought about by humans.

Water pollution and *soil pollution* is caused by dissolved chemicals, suspended particulates, and floating substances, such as oil. Such pollutants include substances that are biodegradable, such as sewage effluent and nitrates leached from agricultural land, which if allowed to enter water courses can lead to EUTROPHICATION and algal BLOOMS. Nonbiodegradable pollutants, such as certain chlorinated hydrocarbon pesticides (e.g. DDT) and HEAVY METALS, such as lead, copper, and zinc in some industrial effluents, accumulate in the environment. Heavy metals, even when present at low concentrations, are toxic to plants, inhibiting water or nutrient uptake, damaging cell membranes, and inhibiting enzymes. Soil contaminated with fuel ash may contain increased levels of boron, which at high concentrations causes chlorosis and necrosis to tissues. Soil contaminated with mining waste may have high levels of copper, which damages root cell membranes and inhibits growth. Any contamination leading to high concentrations of magnesium or manganese impedes calcium uptake.

Overcultivation of marginal arid regions and the use of irrigation water in areas where evaporation exceeds precipitation can lead to SALINATION of soils. Deforestation and intensive agriculture can also cause lowering of the water table, again leading to accumulation of salts near the soil surface. This has several effects on plants. Sodium competes with potassium for uptake and osmoregulation, affecting the osmotic balance of the soil, and upsetting stomatal regulation. Chloride also has osmotic effects and it competes with other anions, preventing their uptake and leading to deficiency. Other forms of pollution are noise from airplanes, traffic, and industry and the disposal of RADIOACTIVE WASTE.

pollution credit *See* Clean Air Act.

polyandry *See* polygamy.

polycentric distribution The occurrence of a population, species or other taxonomic group in several widely separated places.

Polychaeta *See* Annelida.

polychlorinated biphenyl (PCB) A type of compound based on biphenyl ($C_6H_5C_6H_5$), in which some of the hydrogen atoms have been replaced by chlorine atoms. They are used in certain polymers used for electrical insulators. PCBs are highly toxic and concern has been caused by the fact that they can accumulate in the food chain.

polyclimax *See* climax community.

polygamy A mating system in which a male pairs with more than one female at a time (polygyny) or a female pairs with more than one male at a time (polyandry). *Compare* monogamy.

polygenic inheritance *See* quantitative variation.

polygyny *See* polygamy.

polymorphism A distinct form of VARIATION in which significant proportions of different types of individuals exist within a population of a species at the same time and in the same place, such that the frequency of the rarest form cannot be explained on the basis of recurring mutation. The term is generally used where the frequency of the least common morph exceeds 1 in 20, less frequent occurrences being usually due to rare mutation or recombination events.

If the differences persist over many generations there is a *balanced polymorphism*. This may happen where the heterozygotes (*see* heterozygous) have an advantage not possessed by either of the homozygotes (*see* homozygous), or where different forms are successful in different MICROHABITATS or in different years. For example, carriers of the recessive allele for sickle-cell anemia (i.e. those who are heterozygous for the allele) enjoy greater protection from malaria than noncarriers, so the incidence of sickle-cell anemia (due to the possession of two recessive alleles) is more common in areas where malaria is prevalent. An example of success in different habitats are the water striders, freshwater insects of the genus *Gerris*, which hunt on the water surface film. Species of large permanent lakes have short wings or none at all, and do not travel from lake to lake. Those of temporary ponds have long functional wings and disperse regularly as ponds dry up. Species of more persistent small ponds that nevertheless may dry in summer often have both long- and short-winged forms, a situation known as *alary polymorphism*.

Polymorphism usually results from the occurrence of different alleles of a gene. The caste system in social insects results, in some cases, from differences in nutrition rather than genotype and is thus an *environmental polymorphism* rather than a genetic polymorphism.

polyp *See* Cnidaria; corals.

polypeptide A compound that contains many amino acids linked together by bonds formed between the carboxylate (–COOH) group on one acid and the amino (–NH$_2$) group on another. The resulting –CO.O.NH– linkages are known as *peptide bonds*. They are the linkages present in proteins.

polyploidy The condition in which a cell or organism contains three or more times the HAPLOID number of CHROMOSOMES. It occurs when chromosomes fail to separate during MEIOSIS, giving rise to gametes that are DIPLOID instead of haploid. Fertilization results in *triploid* or *tetraploid* individuals. Polyploidy is far more common in plants than in animals and very high chromosome numbers may be found; for example in octaploids and decaploids (containing eight and ten times the haploid chromosome number). Some 60% of monocotyledons and 40% of dicotyledons are polyploid.

Polyploids are often larger and more vigorous than their diploid counterparts and the phenomenon is therefore exploited in plant breeding, in which the chemical colchicine can be used to induce polyploidy.

Polyploids may contain multiples of the chromosomes of one species (*autopolyploids*) or combine the chromosomes of two or more species (*allopolyploids*). Chromosome pairing problems during meiosis may prevent hybrid individuals from reproducing sexually. However, a subsequent doubling of the chromosome number restores the ability to pair and these new polyploids may be fertile. Once formed, a polyploid individual may be incapable of reproducing with its parents, and immediately constitutes a new species.

Polyploidy thus contributes to evolution. Polyploids have a greater store of genetic variability than nonpolyploids, and harmful recessive alleles are more easily masked by normal dominant alleles, so further adding to their evolutionary potential. Polyploidy is rare in animals because the sex-determining mechanism is disturbed.

polysaccharides   *See* carbohydrates.

pond   A body of still fresh water in a small surface depression. A pond is smaller than a lake, but larger than a pool.

ponding   A condition in a filter bed used to treat SEWAGE in which, as a result of excessive fungal growth, the surface of the filter becomes covered in settled sewage, reducing air flow, so that the bed becomes anaerobic and treatment stops.

pool   A body of still fresh water in a small surface depression, such as the surface of a marsh. A pool is smaller than a POND.

pooter   A device used to collect small animals. It consists of a jar sealed with a bung through which pass two tubes pointing in opposite directions. The operator sucks on one tube, which ends inside the jar, where its opening is covered in muslin. This draws up the small animals through the other tube and into the jar.

POP   *See* persistent organic pollutants.

population   A group of interbreeding organisms of the same species (or other groups within which individuals may exchange genetic information) occupying a particular space. A population is continually modified by increases (birth and immigration) and losses (death and emigration), and is limited by the food supply and the effects of environmental factors such as disease. The concept is easier to apply to animals than to plants. Many plants cannot be divided strictly into populations because they may reproduce vegetatively, with the result that one individual can occupy a large area and parts can detach.

Some other species are almost entirely self-fertilizing or reproduce by agamospermy, so do not interbreed with any other individual. The number of individuals in a population per unit area (or sometimes per unit volume or other suitable parameter) is the *population density*. In some species or under some conditions the population size may fluctuate between extremes over a fairly regular period – this is known as a *population cycle*. The *minimum viable population* (MVP) is the smallest population that can persist for a long time, usually taken to be 1000 years.

The sum of the characteristics of a population – its density, dispersion, the movement and age classes of its members, genetic VARIATION, and size and distribution pattern of its habitats – is known as the *population structure*.

population dynamics   The study of the factors influencing the fluctuations in numbers in a population or of its GENE POOL. These factors include those affecting birth and mortality rates, immigration and emigration, and reproductive potential (for example, seed size, dispersal, and dormancy).

population ecology   The study of the variations in the size and density of populations in space and time and their interactions with their environment.

population equilibrium   A state in which numbers of individuals in a population remain stable because the number of deaths equal the number of births.

population genetics   The study of inherited VARIATIONS in populations of organisms, and their distribution in time and space. It involves the quantification of GENE FREQUENCIES, and the effects of GENETIC DRIFT, immigration and emigration, NATURAL SELECTION, and MUTATION on their distribution. *See* gene pool.

population matrix   *See* Lewis–Leslie matrix.

population pyramid   A diagram representing the age structure of a population, in which the youngest age class is placed at the bottom and successive age classes are stacked above it. It is usually shown as a series of horizontal bars, the length of which represents the numbers of individuals of each age.

population regulation The tendency for a population to regulate its own numbers at a fairly constant level. For example, factors such as abundant food resources may cause density to increase when it is low, and other factors, perhaps grazing exceeding the CARRYING CAPACITY of the habitat, cause it to decrease when it is high.

population vulnerability analysis (PVA; population viability analysis) An analysis of the populations of a species, a community, or ecosystem to assess the chances that it may become extinct. It encompasses study of ecological, economic, and political issues and other factors relating to its conservation.

Porifera A phylum of primitive multicellular animals, the sponges, that probably evolved a multicellular structure independently of the other multicellular animals. All are sessile and almost all are marine. The body of a sponge is a loose aggregation of cells, with minimal coordination between them, forming a vase-like structure. Flagellated cells (choanocytes) line the vase, and cause water currents to flow in through apertures (ostia) in the body wall and out through one or more openings (oscula) at the top. Sponges have an internal skeleton of chalk or silica spicules or protein fibers (as in the bath sponge).

positive feedback See feedback mechanism.

potable water Treated or untreated water that is considered safe to drink.

potassium (K) One of the essential elements in plants and animals. It is absorbed by plant roots as the potassium ion, K^+, and in plants is the most abundant cation in the cell sap. Potassium is important in OSMOREGULATION and other meta-

bolic activities in animals and plants. A deficiency of potassium in plants leads to a characteristic purple coloration of the leaves, and poor root growth, flower and fruit formation. Potassium is usually added to soils in compound fertilizers.

potassium–argon dating See isotopic dating.

prairie An extensive area of relatively level GRASSLAND in North America, with very few trees. The plant communities consist of various dominant grass species mixed with a variety of perennial herbs. Different communities occur on different soils, drainage, and topography. There are two main types – tall-grass prairie (long-grass prairie) in the east, where the grass varieties are mainly tall, and short-grass prairie in the west. Large tracts of former prairie are now urbanized or under wheat cultivation.

preadaptation An adaptation that evolved in one habitat but happens to be advantageous to survival in an adjacent habitat, so allowing the organism to radiate into it. See adaptation; adaptive radiation.

Precambrian The time in the Earth's geological history that precedes the CAMBRIAN, i.e. from the origin of the Earth, nearly 5 billion years ago, to the start of the Cambrian, around 570 million years ago. The term 'Precambrian' is now used mainly descriptively, and has been largely discarded as a geological term in the light of greater knowledge of the early evolution of life. Precambrian time is now divided into three eons: Hadean, from the Earth's origin to about 3900 mya; Archean, 3900–2390 mya; and Proterozoic, 2390–570 mya (the Cambrian marks the start of the Phanerozoic eon, which extends to the present day).

The oldest fossils discovered so far are remains of bacterialike organisms, dating from about 3500 mya. Indeed, there is abundant evidence of flourishing colonies of CYANOBACTERIA and other bacteria throughout the Archean and Proterozoic eons. This takes the form of stromatolites, rock structures representing the remains of sediment trapped or precipitated by cyanobacteria communities. Stromatolites are still found in a few special locations today. The earliest remains of single-celled EUKARYOTES are much later, dating from about 1400 mya, while the first appearance of multicellular animals is in the so-called Ediacara fauna, in rocks dated to the last 100 million years of Precambrian time.

precautionary principle A strategy that if there is a strong suspicion that certain activities that cause changes to the environment may prove to be harmful or undesirable, action should be taken to prevent or ameliorate those effects, even if the scientific basis for such suspicions is not absolutely certain. The Rio Declaration (*see* Earth Summit) states that where there are threats of serious or irreversible damage, lack of full scientific certainty shall not be used as a reason for postponing cost-effective measures to prevent environmental degradation.

precipitation 1. The formation of an insoluble solid as a result of a chemical reaction that occurs in solution. Precipitation is used to separate suspended material from sewage by adding chemicals to cause it to precipitate.
2. Moisture that falls onto the Earth's surface from clouds, for example rain, hail, sleet, or snow. There are two main methods of measuring precipitation. A *nonrecording gauge* is a simple cylinder that collects rain through a funnel of standardized diameter. The volume of water collected over a given period is measured. Snow or hail is melted before measuring. A *recording gauge* is linked to a rotating drum that provides detailed data on the time and intensity of the precipitation, which is recorded as mm h^{-1}.

precipitation enhancement The increasing of precipitation that results from changes in the colloidal stability of clouds. This may be due to intentional seeding of clouds with condensation nuclei, or the effect of pollution.

precocity The occurrence of reproduction at an early stage of the life cycle of an organism relative to other individuals of the same or related species.

predation The capturing and killing by a PREDATOR of another animal (the PREY) in order to eat it.

predation curve A graphical representation of the interaction between the numbers and behavior of PREDATOR and PREY populations. It may show, for instance, if predator or prey behavior changes with prey population density, and indicate whether prey switching is going on, or whether there are external factors limiting the predator population.

predator An animal that captures and kills another animal (the PREY) in order to eat it. A predator that is at the top of the food chain is often called a *top predator*. Top predators are often KEYSTONE SPECIES: if their populations are healthy, it is a good indication that the ecosystem as a whole is in a healthy state, because the rest of the FOOD CHAIN must be in a healthy enough state to support it.

predator–prey ratio The ratio of the number of species of predators to the number of prey species in a given community. When this is modeled assuming that there is competition between predators for the most prey-rich areas and competition between prey for the areas with the fewer predators, the ratio remains fairly constant. In practice, feeding relationships are extremely complex with interactions between many different levels in the FOOD WEB. Ratios are also complicated by omnivores, where the same species may be counted as both predator and prey. *See also* Lotka–Volterra model.

predator–prey relationships The functional and numerical interactions between predator and prey populations. Such studies may include the relationship between a predator's consumption rate and the density of its prey – a *functional response*. Or they may explore changes in the size of a predator population in response to changes in density of its prey population – a *numerical response*. A predator may opt to hunt mainly the most common species of prey, even though other species are readily available – this is called *apostatic prey selection*. *See also* predator–prey ratio; predation curve.

prescribed burning Deliberate controlled setting of surface fires in forests or grasslands to remove dry dead vegetation and prevent more destructive fires, or to kill off unwanted species that compete for nutrients with commercial species.

preservationism A conservation movement that seeks to preserve natural environments in their pristine state, not allowing any kind of exploitation, from hunting and fishing to logging or tourism.

press experiment An experiment in which variables in a the system are changed (e.g. one or more environmental factors are changed or individuals are added to or removed from a population) continuously, so that the system is in a state of constant perturbation. *Compare* pulse experiment.

prevalence The proportion of potentially inhabitable sites in which a particular species is present. It is usually expressed as a percentage.

prey An organism that is likely to be killed and eaten by a PREDATOR.

prey switching The switching of a predator's preference from one prey species to another, often in response to changes in prey density or accessibility, or to competition from other predators.

primary data Direct observations or values of variables obtained in the field or laboratory, before any data processing has taken place.

primary drinking water regulation A regulation that applies to public water systems, specifying the maximum permitted levels of particular contaminants deemed not to have adverse effects on health.

primary forest Forest that is believed not to have been affected by any human activity throughout its history. *Compare* secondary forest.

primary production The total assimilation (*see* gross primary production) or accumulation (*see* net primary production) of nutrients or energy by plants and other autotrophs.

primary productivity The rate of production of BIOMASS per unit area by plants and other autotrophs. *Compare* secondary productivity.

primary sex ratio *See* sex ratio.

primary succession *See* succession.

Primates The order of mammals that contains the monkeys, great apes, and humans. Most primates are relatively unspecialized arboreal mammals with a very highly developed brain, quick reactions, and large forward-facing eyes allowing binocular vision. The opposable thumb and (usually) big toe are used for grasping and the digits have nails. The young undergo a long period of growth and development, during which they learn from their parents. The New World monkeys have prehensile tails; the more advanced Old World monkeys lack prehensile tails, and great apes are larger tailless primates that typically swing from trees by their long arms.

prisere *See* sere.

probability The chance that a given event will occur, or that over a series of observations a particular kind of observation will occur regularly as a given proportion of the total number of observations. Statistical probability is usually based on an infinite number of observations. For example, in genetics, if a heterozygous plant is selfed, the probability of finding the double recessive is 1 in 4, or a 25% chance. The greater the number of offspring the better the chance that this actual percentage will be achieved.

producer The first TROPHIC LEVEL in a FOOD CHAIN. Producers are those organisms that can build up foods from inorganic materials, i.e. green plants, algae, and photosynthetic and chemosynthetic bacteria. Producers are eaten by herbivores, which are primary CONSUMERS.

production efficiency The percentage of the energy assimilated by an organism that becomes incorporated into new BIOMASS.

productivity The rate of production of biomass per unit area by a specific organism or group of organisms.

profundal The deepwater zone of a lake beyond a depth of 10 m. Little light penetrates this zone and thus the inhabitants are all heterotrophic, depending on the littoral and sublittoral organisms for basic food materials. Commonly found inhabitants include bacteria, fungi, mollusks, and insect larvae. Species found in the profundal zone are adapted to withstand low oxygen concentration, low temperatures, and low pH. *Compare* littoral; sublittoral; photic zone.

prokaryote (procaryote) An organism whose genetic material (DNA) is not enclosed by membranes to form a nucleus but lies free in the cytoplasm. Organisms can be divided into prokaryotes and EUKARYOTES, the latter having a true nucleus. This is a fundamental division because it is associated with other major differences. Prokaryotes constitute the domain BACTERIA, with two kingdoms, or even domains, the ARCHAEA and the EUBACTERIA. All other organisms are eukaryotes. Prokaryote cells evolved first and gave rise to eukaryote cells. *See* cell; endosymbiont theory.

protein A large, complex molecule made up of one or more polypeptide chains, i.e. it consists of AMINO ACID molecules joined together by peptide links. The molecular weight of proteins may vary from a few thousand to several million. About 20 amino acids are present in proteins. Simple proteins contain only amino acids. In conjugated proteins, the amino acids are joined to other groups. Proteins are vital compounds found in all living organisms.

Proterozoic *See* Precambrian.

Protista (protists) In some classifications, a kingdom of simple single-celled eukaryotic organisms including the algae and protozoans. It was introduced to overcome the difficulties of assigning such organisms, which may show both animal and plantlike characteristics, to the kingdoms Animalia or Plantae. Today the grouping is considered artificial, and protists are included in the kingdom PROTOCTISTA.

Protoctista (protoctists) A kingdom of simple eukaryotic organisms that includes the algae, diatoms, slime molds, fungus-like oomycetes, and the organisms traditionally classified as protozoa, such as flagellates, dinoflagellates, ciliates, and sporozoans (*see* eukaryote). The members of this kingdom do not share clear phylogenetic links, but are grouped together simply because they do not belong in any of the other four kingdoms (*see* Five Kingdoms classification). Most are aerobic, some are capable of photosynthesis, and most possess UNDULIPODIA at some stage of their life cycle. Protoctists are typically microscopic single-celled organisms, such as the amebas, but the group also has large multicellular members, for example the seaweeds and other large algae, some of which are sometimes classified as plants. Protoctists show a wide range of nutritional habits, including photoautotrophs, heterotrophs, phagotrophs, and mixotrophs.

protoplasm The living contents of a cell, comprising the cytoplasm plus nucleoplasm.

protoplast The protoplasm and plasma

membrane of a cell after removal of the cell wall, where present.

protozoa   An old name for animal-like *Protoctista* that also included some photosynthetic organisms such as *Euglena* and *Volvox*.

pruning   The removal of part of a plant. This may be done to remove injured or diseased parts, or to train the plant into a particular shape. *See also* coppice; pollarding.

psammosere   A SUCCESSION that develops on sand dunes.

PSI   *See* pollutant standard index.

psu   *See* salinity.

psychrometer   An instrument for measuring HUMIDITY.

psychrophilic   Describing microorganisms that can live at temperatures below 20°C. *Compare* mesophilic; thermophilic. *See also* extremophile.

pteridophyte   A general term, now largely obsolete, that includes any vascular nonseed-bearing plant. Pteridophytes include the clubmosses (phylum Lycophyta), horsetails (Sphenophyta), ferns (Filicinophyta), and whisk ferns (Psilophyta or included in Filicinophyta). *See* alternation of generations.

pulse experiment   An experiment in which one or more variables are changed (e.g. one or more environmental factors are altered or individuals are added to or removed from a population or community) at one single moment, and the progress of the experiment or the population dynamics are monitored before and after the perturbation. *Compare* press experiment.

purine   A nitrogenous organic molecule with a double ring structure. Members of the purine group include adenine and guanine, which are constituents of the nucleic acids, and certain plant alkaloids, e.g. caffeine and theobromine.

PVA   *See* population vulnerability analysis.

pyramid of biomass   A type of ecological pyramid based on the total amount of living material at each TROPHIC LEVEL in the community, which is normally measured by total dry weight or calorific value per unit area or volume, and shown diagrammatically. The biomass depends on the amount of carbon fixed by green plants and other producers. The pyramid of biomass usually has a more gentle slope than the PYRAMID OF NUMBERS because organisms at successively higher levels in the pyramid tend to be larger than those below.

pyramid of energy   A type of ecological pyramid in which the energy contained in the organism at each stage of a food chain is depicted diagrammatically. It shows that the energy flux decreases at progressively higher trophic levels, energy being lost be-

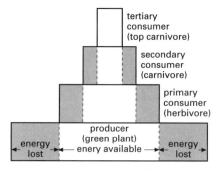

Pyramid of energy

tween levels due to inefficiencies of digestion and assimilation, respiration, heat loss, energy of movement, nervous transmission, and so on.

pyramid of numbers   A type of ecological pyramid in which the number of individual organisms at each stage in the FOOD CHAIN of the ecosystem is depicted diagrammatically. The producer level forms the base, and successive TROPHIC LEVELS the tiers. The shape of the pyramid of numbers depends upon the community considered; generally, the organism forming the base of

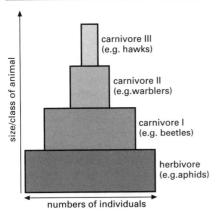

Pyramid of numbers

a food chain is numerically very abundant, and each succeeding level is represented by fewer individual organisms, culminating with the final large predator. The pyramid of numbers may be partly inverted (base smaller than one or more of the upper tiers) if the producing organisms are large.

Theoretically the higher the level in the pyramid, the fewer organisms it can support. This has important implications regarding human food supplies because it illustrates that more people can be supported in a given area if their diet is mainly vegetarian – more land is needed per head if meat forms a large part of the diet.

pyrethrum *See* botanical insecticide.

pyrimidine A nitrogenous organic molecule whose ring structure is contained in the pyrimidine bases cytosine, thymine, and uracil, which are constituents of the nucleic acids, and in thiamine (vitamin B_1).

Q

Q_{10} A coefficient that shows by what factor the rate of a chemical reaction increases for every 10°C rise in temperature. For most metabolic reactions $Q_{10} = 2$, i.e. the reaction rate doubles for every 10°C rise in temperature.

quadrat A square area (standard size is one meter square) taken at random, within which the composition of organisms is noted. The quadrat sampling technique is mostly used in plant ecology to study plant communities but quadrats are also used as a sampling unit to count and weigh animals for an estimate of density or to discover animal distribution in a selected area. Permanent quadrats can be established that are examined at given intervals as a means of assessing changes in species composition in an area over a period of time.

qualitative variation (discontinuous variation) A form of variation in which a character has two or more distinct forms. Examples are human blood groups and Mendel's pea characters. It generally occurs when there are two or more allelic forms of a major gene in a population.

quantitative variation (continuous variation) A form of variation in which a character shows a continuously varying range of values, for example height in humans or grain yield in cereal crops. It occurs where traits are controlled by many different genes (*polygenes*). The pattern of inheritance of such traits is called *polygenic inheritance* or *multifactorial inheritance*.

quartile The 25th, 50th, or 75th percentile (data value equal to 25, 50, or 75% of the highest value) of a set of data or a distribution.

quartz (SiO_2) A mineral composed of crystalline silica that is widespread in rocks of all kinds, the second most abundant mineral in the Earth's crust.

Quaternary The most recent period of the CENOZOIC era from about 2 million years ago to the present day and composed of PLEISTOCENE and HOLOCENE epochs. Literally, the 'fourth age,' it is characterized by the emergence of humans. *See also* Ice Age.

R

race In classification, a TAXON below the rank of species, sometimes being placed between subspecies and variety. The term is sometimes used instead of ecotype for groups of individuals that have uniform ecological preferences or physiological requirements. *See* ecotype.

radiation The dissemination of energy from a source. In the absence of absorption, the energy decreases as the inverse square of the distance from the source. Types of radiation include electromagnetic radiation (*see* electromagnetic energy) and the emission of particles such as alpha and beta radiation and gamma rays (*see* radioactive decay). *See also* ionizing radiation.

radiation sickness The collection of symptoms shown by persons suffering from radiation injury resulting from excessive exposure of part or all the body to IONIZING RADIATION. These usually include fatigue, nausea, vomiting, and diarrhea, and later hemorrhage, inflammation of the mouth and throat, and hair loss.

radiation standards Regulations imposing maximum exposure limits for radioactive material in order to protect the health of workers and the general public.

radical A group of atoms in a molecule, for example, the carbonate radical is CO_3^-. Where the radical has a single unpaired electron, it is highly reactive, and is called a *free radical*. They are often formed in light-induced reactions, as in the photochemical reactions with CFCS that cause damage to the OZONE LAYER.

radioactive dating *See* isotopic dating.

radioactive decay The spontaneous breakdown of a radioactive isotope. Such isotopes break down with the emission of alpha particles, beta particles, or gamma waves, forming other isotopes or even different elements. A useful measure of the rate of decay is the half-life – the time required for half of the atomic nuclei in the sample to decay. *See also* ionizing radiation; isotope.

radioactive fallout (nuclear fallout) The radioactive material that falls to the ground or into the Earth's atmosphere or water at some distance from the site of a nuclear explosion.

radioactive isotope (radioisotope; radionuclide) Any of several different forms of the same element that have different masses and whose nuclei are unstable, spontaneously breaking down with the emission of particles, e.g. alpha, beta, or gamma radiation. *See also* isotope; radioactive decay.

radioactive waste (nuclear waste) Any waste material that contains radioactive material. Radioactive waste can come from the mining, extraction, and processing of radioactive ores, the normal operation of nuclear reactors, decommissioning nuclear reactors, the manufacture of nuclear weapons, and from radioactive materials used in hospitals, laboratories, and industry.

Since radioactive waste can be very harmful to all living matter and since it can contain radionuclides that have half-lives of many thousands of years the problem of how to dispose of radioactive waste is very important. Different levels of radioactivity have to be dealt with. *High-level waste* has to be stored until its radioactivity has been reduced to a level at which it can be processed. *Intermediate-level waste* can be contained by burial in deep mines or below the seabed. *Low-level waste* can be stored in steel drums at special sites. Spent nuclear fuel is an example of high-level waste. Reactor components and processing plant sludge are examples of intermediate level waste. Solids or liquids slightly contaminated by radioactive substances are examples of low-level waste.

radioisotope   *See* radioactive isotope.

radiometric dating   *See* isotopic dating.

radionuclide   *See* radioactive isotope.

radon   *See* atmosphere.

rainfall   Water formed by condensation when moist air is cooled below its dew point. The droplets form clouds, and coalesce until they become heavy enough to overcome the frictional resistance of the air and fall to the ground. Carbon dioxide in the atmosphere dissolves in rainwater to form a very weak solution of carbonic acid (H_2CO_3), sufficient to cause slow weathering of rocks and building stone. *See also* acid rain.

rainforest (equatorial forest; tropical rainforest)   A TROPICAL FOREST in climates with regular heavy rain, which supports prolific plant growth. Tropical rainforests also require constantly high temperatures. Rainforest is a species-diverse broad-leaved forest, with tall canopy trees forming a closed canopy penetrated occasionally by even taller trees (EMERGENTS). There is a lower layer of smaller trees, and dense undergrowth, as well as climbing vines and lianas, strangling figs, and many epiphytes. With the constant climate, flowering and fruiting take place all year round, allowing for the evolution of specialist herbivores. Most rainforest trees are evergreen. Temperate rainforest (*see* temperate deciduous forest) forms in moist coastal regions, such as the northwest coast of North America and Tasmania. Growth here, while lush, is not so prolific as in the tropics, but there is a dense growth of mosses and ferns.

rain gauge   *See* precipitation.

rain shadow   A geographical area that receives less rain than neighboring areas because higher land in the direction of the prevailing wind causes clouds to rise and cool, thus shedding their rain before reaching the lower land on the far side of the hills. Examples include the deserts of Nevada and eastern California, which are in the rain shadow of the Sierra Nevada and the Coast Range.

rain-wash   The movement of material over the land surface and down slopes as a result of rainfall. There are usually two processes involved: rain-wash, the loosening and detachment of small soil particles by the impact of raindrops; and soil-wash, the downhill movement of material by surface water flow (runoff). *See also* solifluction. *Compare* creep.

ramet   An offshoot or other module formed by vegetative growth that is capable of an independent existence, for example, the runners of strawberry or the polyps of colonial hydroids (*see* Cnidaria). Ramets are clones of their parents, and their production is a form of ASEXUAL REPRODUCTION.

Ramsar Convention   *See* Convention on Wetlands of International Importance.

random error   A deviation of an observed value from a true value that occurs as if selected at random from a probability distribution of such errors.

randomization   A form of experimental design in which experimental units are allocated to random positions in the experiment, usually by means of random number tables. Randomization is used when similar units are being observed under controlled conditions, such as pure-breeding plants being grown in a phytotron. It increases the number of degrees of freedom, so reducing the error variance and making it easier to detect significant differences between treatments. *Compare* Latin square.

random niche model (broken stick model) A model of SPECIES DIVERSITY in which species abundance is determined by a random partitioning of resources between different species. It was first conceived by imagining a stick that was broken at random points along its length. The length of the different pieces represented the proportion of the resources utilized by one species. *See also* dispersion.

random sample A sample in which each individual organism or QUADRAT sampled is independent of all other individuals and also independent of any obvious physical features of the environment. In an ecological survey, this is usually achieved by laying out a grid over the sample area, and using random number tables to select coordinates for points at which to place quadrats. *See also* randomization.

range 1. The difference between the largest and smallest data values in a sample. The size of the range is determined by the DISPERSION of the data about the mean. 2. (geographic distribution) The geographic area over which a species occurs. 3. A large area of open grassland used for grazing or browsing.

range management The management of livestock production or the grazing of mammals on large areas of open grassland. It involves controlling stocking density so that the CARRYING CAPACITY of the land is not exceeded, and ensuring the population density and other aspects of the grazing management do not lead to environmental degradation.

rank 1. The hierarchical status of a TAXON in a classification scheme. For example, the taxon Annelida has the rank of phylum, while the taxon Oligochaeta has the rank of class.
2. The position of a data value when all the data are arranged in ascending order of value.

rarefaction A method of determining the relationship between species diversity and sample size by progressively reducing the sample size by randomly deleting individuals.

ratio scale *See* measurement scales.

Raunkiaer's life-form classification A classification of plant growth forms based on the persistence of the shoots and the position of the resting buds. The system simplifies assessment of the percentages of different plant forms in any given type of vegetation, especially in temperate communities. A *cryptophyte* is a plant in which the resting buds are below the soil surface or in water. Within this group, plants are categorized according to their habitat as land plants (*geophytes*), marsh plants whose overwintering buds are under water (*helophytes*), and aquatic plants (*hydrophytes*). A *hemicryptophyte* is a perennial plant, usually nonwoody, with its overwintering buds at or just below soil level, and are common in cold, moist climates, and include many rosette plants. *Chamaephytes* include small bushes and herbaceous perennials with buds at or near the soil surface and are commonly found in cold or semiarid climates. *Phanerophytes* are perennial plants with persistent shoots and buds on upright stems well above soil level, e.g. herbs, shrubs, trees, and climbing plants. *Therophytes* survive part of the year as seeds and complete their life cycles during the remainder of the year.

raw sewage Untreated domestic or industrial waste water.

reafforestation *See* reforestation.

realized niche The portion of its potential NICHE (*see* fundamental niche) occupied by a species. This will be less than the full potential niche if predators or competitors are present.

Recent *See* Holocene.

recessive Describing an allele that is expressed in the PHENOTYPE only when it is present in the homozygous condition (i.e. both alleles are recessive, a condition called *double recessive*). *Compare* dominant.

recharge area   An area of land in which there is a net annual movement of water from the surface to the groundwater. It determines the maximum sustainable rate at which an aquifer can be exploited.

reciprocal predation   A situation in which two species or individuals prey on each other, a kind of competitive interaction.

reciprocal transplant experiment   An experiment in which individuals from different environments are moved to the other environment to compare their genotype–environment interactions.

reclamation   1. The conversion of land that is periodically inundated by the sea or which is waterlogged for other reasons, its drainage and management for farming, urban, or industrial development. *See also* polder.
2. The cleaning up of contaminated land to make it suitable for farming, urban, or industrial development.
3. The transformation of substances found in waste into useful materials.

recombinant DNA technology   The technique by which foreign DNA, whether from another organism or genetically engineered, is inserted into another DNA fragment or molecule. The product – *recombinant DNA* – is fundamental to many aspects of GENETIC ENGINEERING, particularly the introduction of foreign genes to cells or organisms. There are now many techniques for creating recombinant DNA, depending on the nature of the host cell or organism receiving the foreign DNA.

recombination   The regrouping of genes that regularly occurs during MEIOSIS as a result of the INDEPENDENT ASSORTMENT of chromosomes into new sets, and the exchange of pieces of chromosomes (crossing over). Recombination results in offspring that differ both phenotypically and geno-

typically from both parents and is the most important means of producing VARIATION in sexually reproducing organisms. *See* crossing over; independent assortment; meiosis.

recruitment   Additions to a population as a result of births or immigration. *Net recruitment* is the difference between these additions and the losses due to deaths and emigration.

recycling   The processing of waste material so that it can be used again. For example, the making of new paper and glass from waste paper and glass, or the extraction of metals from waste cans. This saves raw materials and some of the energy needed to make these products from raw materials, as well as saving on landfill space.

red-and-yellow forest soil   An acid SOIL formed in wet subtropical regions. The heavy rain causes LEACHING and accumulation of salts in the B horizon. Iron oxides in the A horizon give the soil its red color, but if the soil is sandy, it is usually yellow. Such soils have a low nutrient value, contain little humus, and are easily eroded.

Red Data Book   A set of information on rare, threatened, and endangered species worldwide, compiled by the World Conservation Monitoring Center, and published at regular intervals by the Species Survival Commission of the World Conservation Union. The *Red List of Threatened Plants* is incorporated with the Red Data Book enabling continuous updating and monitoring of data concerning these species.

red desert soil   A type of coarse SOIL formed in hot deserts, which is rich in salts and lime but poor in humus. Such soils may be cultivated if irrigated.

Red List of Threatened Plants   *See* Red Data Book.

Red Queen effect   A theory of evolution that considers that most of a lineage's evolution consists of keeping up with environmental changes rather than adapting to new environments. The name comes from the Red Queen in Lewis Carroll's *Through the Looking Glass*, who had to keep running as fast as she could just to stay in the same place.

red tide A BLOOM of marine plankton that is toxic and often fatal to fish and humans who consume the shellfish (*paralytic shellfish poisoning*). Shellfish concentrate these toxins when they feed on the plankton. The water may be colored red, green, or brown by the organisms. Red-colored tides are especially toxic, and are due to blooms of DINOFLAGELLATES. Such blooms may be natural, or they may be caused by nutrient enhancement of the water, for example by phosphates and nitrates from fertilizer runoff or wastewater, or untreated sewage discharged into the sea from coastal resorts.

reduction The gain of electrons by such species as atoms, molecules, or ions. It often involves the loss of oxygen from a compound, or addition of hydrogen. Reduction can be effected chemically, i.e. by the use of *reducing agents* (electron donors), or electrically, in which case the reduction process occurs at the cathode. For example,

$$2Fe^{3+} + Cu \rightarrow 2Fe^{2+} + Cu^{2+}$$

where Cu is the reducing agent and Fe^{3+} is reduced, and

$$2H_2O + SO2 + 2Cu^{2+} \rightarrow$$
$$4H^+ + SO4^{2-} + 2Cu^+$$

where SO2 is the reducing agent and Cu^{2+} is reduced.

reforestation (reafforestation) The planting of trees on land from which trees have previously been removed.

refugia (*sing.* refugium) Isolated areas in which plant and animal species were able to survive a period of substantial climatic change that made the surrounding areas unsuitable for them to live in, for example an ice-free area during a glaciation, or a cool mountain top during a period of warming. Such refugia serve as centers of

DISPERSAL if the climate becomes more favorable again. *See also* geographical isolation.

regolith A general term for loose unconsolidated material such as rock fragments, mineral grains, soil, and other deposits that rest on solid bedrock. Much is derived from weathering of the rock below, or from nearby areas.

regression 1. A return to a former state, or to a more primitive state, for example when cultivated land reverts to natural vegetation.
2. A retreat of the sea from a coast as a result of a fall in sea level.
3. A statistical method for investigating the interdependence of variables. *Regression analysis* attempts to find a mathematical model to explain how changes in the independent variable(s) affect the response of the dependent variable(s). Such models may be used to predict values of the responses. Where there is more than one dependent or independent variable, this is called *multiple regression*.

regular distribution *See* dispersion.

rehabilitation ecology (restoration ecology) Where an area has been contaminated or degraded, and it is impossible to restore it to its original wild condition, the establishment there of as similar a community as possible. *Compare* remediation.

relative frequency In probability theory, the proportion of events in which a variable has a particular value in a sample of infinite size is its relative frequency. For example, in tossing a coin, with a large enough sample size the proportion of heads will approach 50%, a relative frequency of one-half. *See also* probability.

relict population A population with a very restricted range that is all that remains of a former larger and more widely distributed population.

relief The shape of the land – the difference in height above or below sea level across the landscape. A *relief map* usually has contours or colors showing different elevations.

relief map   *See* relief.

remediation   The cleaning up of a damaged or degraded site and restoration of the environment to its former state, so far as is possible, or to remove, contain, or in other ways make safe pollutants, toxic spills, or other contamination. *Compare* rehabilitation ecology.

remote sensing   The detection, identification, location and/or analysis of land, water, or living organisms by the use of distant sensors/recording devices. Examples include satellite imaging, aerial photography, radar, sonar, and radio-tracking.

rendzina   A type of brown earth SOIL typical of humid to semiarid grasslands over calcareous rocks. The rendzina is rich in lime and has an upper humus-rich horizon. *See* brown earth.

renewable energy   Energy from a source that is not in limited supply, for example, geothermal energy, sunlight, wind, tides, and the burning of waste.

renewable resource   A resource that is not in limited supply, or whose supply rate is likely to exceed any proposed rate of consumption, for example an aquifer in an area of high rainfall.

replication   The mechanism by which exact copies of the genetic material are formed. Replicas of DNA are made when the double helix unwinds as a result of helicase enzyme action and the separated strands serve as templates along which complementary NUCLEOTIDES are assembled through the action of the enzyme DNA polymerase. The result is two new molecules of DNA each containing one strand of the original molecule, and the process is termed *semiconservative replication*.

reproduction   The production of new individuals by existing individuals. *See* asexual reproduction; sexual reproduction.

reproductive barrier   Any factor that prevents two species or populations from producing viable fertile hybrids. They may live in different habitats or locations and seldom meet, or use different attraction signals during courtship or have flowers designed to attract different pollinators; they may breed at different times; their reproductive organs may be incompatible (e.g. different lengths of stigma and stamens in primroses *Primula vulgaris*); the gametes may be chemically incompatible, or genetic incompatibility (for example, between different polyploids) which may prevent the zygote from dividing successfully or prevent the resulting hybrid forming viable gametes. *See also* ecological isolation; geographical isolation; reproductive isolation.

reproductive isolation   The prevention of gene flow between members of a population due to REPRODUCTIVE BARRIERS that prevent them interbreeding, thus separating distinct breeding groups. Reproductive isolation may also be a consequence of gradual genetic divergence due to other isolating factors. It may eventually lead to SPECIATION, as in the monkeyflowers *Mimulus cardinalis*, which is pollinated by hummingbirds, and *M. lewisii*, which is bee-pollinated. The differences in flower color and structure involve very few genes, but the flowers attract different pollinators, so do not normally interbreed in the wild. *Compare* ecological isolation; geographical isolation.

reproductive rate   The number of offspring produced by an individual per unit time or over a particular period of time. The *average reproductive rate* is the number of young produced per mother or per breeding pair averaged over the population. The *net reproductive rate* is the average number of offspring an individual would produce in its lifetime. It depends on age-specific mortality rates. The average number of offspring that each individual in a population gives rise to in one unit of time is the *fundamental net reproductive* *rate* (R), a factor expressing the relationship between the size of a population to its size one time unit later. If $R=0$ the population is stable, if $R>1$ it is growing, and if $R<1$ it is declining.

Reptilia The class of vertebrates that contains the first wholly terrestrial tetrapods, which are adapted to life on land by the possession of a dry skin with horny scales, which prevents water loss by evaporation. Fertilization is internal and there is no larval stage. The young develop directly from an amniote egg that has a leathery shell and is laid on land, i.e. it is cleidoic. Respiration is by lungs only and the heart has four chambers, although oxygenated and deoxygenated blood usually mix. Other advanced features are the clawed digits and the metanephric kidney. Like amphibians, but unlike birds and mammals, reptiles are poikilothermic.

Reptiles, notably the dinosaurs, were the dominant tetrapods in the Mesozoic period. Modern forms include the predominantly terrestrial lizards and snakes (order Squamata), as well as the aquatic crocodiles and turtles. Reptiles evolved from primitive Amphibia and Mesozoic reptiles included aerial members (e.g. *Pteranodon*) and aquatic forms (e.g. *Ichthyosaurus*), as well as the terrestrial dinosaurs (e.g. *Tyrannosaurus*). Some groups gave rise to birds and mammals. Primitive reptiles had a pineal eye but this is lost in most modern forms.

reservoir 1. An artificial or natural lake or pond managed as a store of water, which is released in a controlled manner. 2. An underground rock with sufficient interstitial space to store water, oil, or natural gas. 3. A store of a nutrient in a BIOGEOCHEMICAL CYCLE. *Abiotic reservoirs* include rock, sediments, soil, water, and the atmosphere. *Biotic reservoirs* include nutrients contained in the bodies of living organisms. Reservoirs are important when considering nutrient and biogeochemical cycles.

resistance The ability of a population to avoid displacement from its present state by a disturbance.

resource Any item in the environment that can be used by a living organism, for example water or nutrients. Items that may be exploited by humans, such as timber, coal, and minerals, are termed *natural resources*. *See also* essential resource; limiting resource.

resource partitioning *See* differential resource utilization.

respiration The oxidation of organic molecules to provide energy in plants and animals. In animals, food molecules are respired, but autotrophic plants respire molecules that they have synthesized by photosynthesis. Respiration occurs in all cells. The energy from respiration is used to attach a high-energy phosphate group to ADP (adenine diphosphate) to form the short-term energy carrier ATP (adenine triphosphate), which can then be used to power energy-requiring processes within the cell. ATP is not transported between cells, but is made in the cell where it is required.

The chemical reactions of respiration normally require oxygen from the environment (*aerobic respiration*). Some organisms are able to respire, at least for a short period, without the use of oxygen (*anaerobic respiration*), although this process produces far less energy than aerobic respiration. A few bacteria can survive indefinitely in anaerobic conditions.

The complex reactions of cell respiration fall into two stages, *glycolysis* and the *Krebs cycle*. The first stage, glycolysis, occurs within the CYTOPLASM, but the Krebs cycle enzymes are localized within the MITOCHONDRIA of EUKARYOTES; cells with high rates of respiration (e.g. insect flight muscles) have many mitochondria. Glycolysis results in partial oxidation of the respiratory substrate to the 3-carbon compound pyruvate. It does not require oxygen, so can occur in anaerobic conditions, when the pyruvate is converted to ethanol in plants and most bacteria or to lactic acid in animals and certain bacteria. The yield of glycolysis is two molecules of ATP plus two molecules of the reduced coenzyme $NADH_2$ for each molecule of glucose respired. In the presence of oxygen, the pyruvate is further oxidized in the Krebs cycle.

In the Krebs cycle, which requires free oxygen, pyruvate is converted into the 2-carbon acetyl group, which becomes attached to a coenzyme forming acetyl coenzyme A. This then enters a cyclic series of reactions during which carbon dioxide is evolved and hydrogen atoms are transferred to the coenzymes NAD and FAD. The energy released by the Krebs cycle is transferred via the reduced coenzymes $NADH_2$ and $FADH_2$ to an electron transport chain embedded in the inner mitochondrial membrane, which drives the formation of more ATP. Overall, 38 molecules of ATP are generated for each molecule of glucose oxidized during aerobic respiration, compared with only two molecules of ATP during anaerobic respiration. At the end of the electron transport chain the electrons react with protons and oxygen to give water. In aerobic respiration, therefore, pyruvate is completely oxidized to carbon dioxide and water.

response   A change in an organism or in part of an organism that is produced as a reaction to a stimulus.

restoration ecology   *See* rehabilitation ecology.

retrogression   A reversion to an earlier stage of a SUCCESSION, with less species diversity and complexity, due to a disturbance, often the effects of a pollutant.

Revelle, Roger   (1909–91) American oceanographer. Revelle was responsible for directing much of the work at the Scripps Institute of Oceanography, La Jolla, California, that eventually led to the discovery of sea-floor spreading and magnetic reversals. He also turned in the 1950s to what was then the far from fashionable topic of GLOBAL WARMING. The issue had first been raised by ARRHENIUS in 1895. Revelle lobbied for real measurements and

as a result gas recorders were set up in 1957 at Mauna Loa, Hawaii, and at the South Pole. By 1990 the carbon dioxide concentration had risen to 350 parts per million, an increase of 11%.

reverse osmosis   A method for purifying water in which the water is forced under pressure through a membrane that is not permeable to the impurities it is desired to remove.

rhizodegradation   (phytostimulation) The breakdown of contaminants in the soil by microbial activity that is enhanced in the presence of the *rhizosphere* (the area of soil around the plant roots). Both plants and microorganisms may be involved. For example, hydrocarbon pollutants are broken down into derivatives that are less toxic and less persistent in the environment. *See also* bioremediation; rhizofiltration.

rhizofiltration   The decontamination of polluted water by passing it through the roots of certain plants that extract the toxins. The pollutants become adsorbed or precipitated onto the plant roots and the plants are harvested and disposed of when the roots become saturated with contaminants. *Compare* bioremediation; rhizodegradation.

rhizosphere   *See* rhizodegradation.

ribonucleic acid   *See* RNA.

ribosome   A small organelle that is the site of protein synthesis in the CELL.

Richter scale   *See* earthquake.

rift valley   *See* plate tectonics.

Rio Conference   *See* Earth Summit.

risk factor   Any factor that is associated with an increased risk to the wellbeing of an individual or environment. For example, prey animals have to compromise between gathering food and keeping a lookout for predators. Risk factors such as duration and intensity of exposure to pesticides can be quantified and managed. Analysis of risk factors is an early step in planning for the conservation of species.

risk management The process of taking and acting on decisions about the level of risk that constitutes a threat to public health and the best method of controlling or minimizing that risk.

river continuum The concept of a river as a single continuous ECOSYSTEM, in which there is a gradient of environmental factors and of trophic relationships. While there are predators throughout the system, in the upper reaches of a river, where considerable amounts of organic matter enter the system from the river banks and overhanging vegetation, there are many organisms that break it down, supporting communities of detritus-feeders farther down the river. As the river becomes wider and more light enters it, there are more autotrophic organisms, so more primary production and an increase in grazers. When much of the river becomes too deep for aquatic plants, heterotrophic feeders again predominate.

riverine forest (fringing forest; gallery forest) A narrow strip of forest similar to a RAINFOREST that occurs along river banks in drier regions than those usually colonized by rainforests.

RNA (ribonucleic acid) A nucleic acid comprising a single polynucleotide chain similar in composition to a single strand of DNA except that the sugar ribose replaces deoxyribose and the pyrimidine base uracil replaces thymine. RNA is synthesized in the nucleus, using DNA as template, and exists in three main forms: *messenger RNA* (mRNA) is synthesized during transcription from template DNA in the nucleus, and passed out through the nuclear envelope to the ribosomes in the cytoplasm, which decode it in the process of translation to synthesize peptides; *transfer RNA* (tRNA), which when activated carries specific amino acids to the ribosome to be joined together; and *ribosomal RNA* (rRNA), which is the main component of the ribosome itself. In certain viruses, RNA is the genetic material. *See* transcription; translation.

rock Any mineral matter making up part of the Earth's CRUST. Ecologists tend to use the term to mean mainly consolidated solid material, while geologists include in it unconsolidated material such as sand, mud, clay, peat, and gravel (*see* regolith). *Biogenic rocks* are derived from the structures or activities of living organisms, e.g. limestones that represent fossilized coral reefs. *Chemical rocks* are formed by the precipitation of chemicals or the mineralization of organic remains. *Fragmental rocks* (*clastic rocks*) are formed by the consolidation of rock fragments that have been weathered or eroded from pre-existing rocks. *See also* igneous rock; metamorphic rock; sedimentary rock.

rock cycle The cycle of formation, degradation, and disintegration of ROCKS. Rocks are uplifted, then weathered and eroded, transported by rivers or ice, deposited somewhere else, consolidated into rocks and possibly metamorphosed, then uplifted again to start a new cycle.

Rodentia The largest and most successful order of mammals, including *Rattus* (rat), *Mus* (mouse), *Sciurus* (squirrel), and *Castor* (beaver). Rodents are herbivorous or omnivorous mammmals with one pair of chisel-like incisor teeth projecting from each jaw at the front of the mouth and specialized for continuous gnawing. Rodents are found universally and are mostly nocturnal and terrestrial. They are noted for their rapid breeding.

root The organ that anchors a plant to the ground and that is responsible for the uptake of water and mineral nutrients from the soil. *See* mycorrhiza.

root nodule *See* nitrogen fixation.

root pressure The pressure that may build up in a plant root system due to the osmotic potential of the root cells, which is thought to help force water upward in the XYLEM vessels. Root pressure tends to build up at night, when the rate of transpiration is low, and is a cause of GUTTATION in some species, especially grasses. *See* osmosis.

root–shoot ratio The ratio of the weights of the roots to the weight of the shoots of a plant. It indicates the ability of the plant's roots to compete for nutrients and water (a high ratio indicates competitive success), or the ability of the shoots to compete for light for photosynthesis (a low ratio indicates success). A high ratio is characteristic of plants in the early stages of a succession, whereas success in competing for light becomes more important in the later stages as the vegetation becomes taller and denser.

rosette plant Any plant whose leaves radiate outward from a short stem at soil level, e.g. daisy (*Bellis perennis*). This growth form helps plants withstand trampling or grazing and in exposed habitats avoids exposure to strong winds.

rotation of crops The planting of different crops on the same land in successive years to ensure that species-specific pests do not get a hold, and that the land is not depleted of the same combination of minerals. Leguminous crops are frequently included in crop rotation for their nitrogen-fixing abilities, and plowed back in to the soil to replenish the supply of nitrates (*see* nitrogen fixation).

roundworms *See* Nematoda.

Rowland, Frank Sherwood (1927–) American chemist. In 1973, Mario MOLINA took to Rowland, his postdoctoral adviser, some calculations suggesting that CFCS (chlorofluorocarbons), which were widely used in aerosol propellants, rise to the upper atmosphere and destroy the ozone layer, located 8 to 30 miles above the Earth. As the layer protects us from harmful ultraviolet rays, its destruction could have disturbing consequences. Rowland and Molina published their preliminary results in June 1974. Rowland discovered that 400 000 tons of CFCs had been produced in the United States in 1973, and that the bulk of this was being discharged into the atmosphere. He calculated that at the then current production rate there would be a long-term steady-state ozone depletion of 7–13%. The National Academy of Sciences published a report in September 1976 supporting the work of Rowland and Molina, and in October 1978 CFC use in aerosols was banned in the United States. Final confirmation came when Joe FARMAN discovered in late 1984 a 40% ozone loss over Antarctica. For his work on CFCs Rowland shared the 1995 Nobel Prize for chemistry with Mario Molina and Paul CRUTZEN.

ruderal Describing a plant that grows on wasteland, agricultural land, or around or on human dwellings. Ruderals include many weeds that cannot tolerate much competition and have relatively high nutrient demands.

ruminants Hoofed herbivorous mammals (order Artiodactyla) that chew the cud, e.g. cattle, deer, goats, and sheep. A ruminant's stomach has a series of chambers that house symbiotic microorganisms capable of digesting CELLULOSE, a major constituent of plant cell walls that most mammals have difficulty digesting. Food is regurgitated from the first chamber (the rumen) and chewed again while the animal is resting before being swallowed again for complete digestion.

Ruminants have their own internal nitrogen cycle. Microbial fermentation of protein in their food releases ammonia, which is converted into urea in the liver, some of which is excreted into the saliva. When the saliva passes with the food into the rumen, the microbes convert the urea into protein. This conserves water, because less is needed to dilute the urine. Desert ruminants such as camels drink rarely compared to nonruminants such as donkeys.

runoff The flow of rainwater or snowmelt from the land surface into streams and rivers.

S

Sagan, Carl Edward (1934–96) American astronomer. Sagan's main area of research was the solar system, particularly the physics and chemistry of planetary atmospheres and surfaces. In 1984 Sagan coauthored, with R. Turco, O. Toon, T. Ackerman and J. Pollock, an influential paper, *Nuclear Winter: Global Consequences of Multiple Nuclear Explosions*, referred to since as the TTAPS paper. The authors argued that even a relatively small-scale nuclear bomb of 5000 megatons would create enough atmospheric smoke (300 million tons) and dust (15 million tons) to produce a temperature drop of 20–40°C, which would persist for many months. This prolonged NUCLEAR WINTER would destroy much of the world's agriculture and industry. The impact of the paper on politicians and the public was dramatic although the paper and the nuclear-winter argument itself was heavily criticized by many other influential scientists.

salination (salinization) The accumulation in water or soil of salts of magnesium, potassium, and sodium. Salination of soils occurs in arid and semiarid regions, where water is drawn to the surface by CAPILLARITY due to high rates of evaporation. As the water evaporates, salts accumulate in the upper horizons of the soil. Plants cannot grow well in such soils and thus salination contributes to DESERTIFICATION. It can also occur where salty GROUNDWATER seeps into land adjacent to the coast, often when excess water has been drawn off for irrigation.

salinity The concentration of dissolved salts or ions in a given volume of water, measured when all the organic matter has been completely oxidized and all bromide and iodide converted to chloride. It is usually measured as parts per thousand by weight: 1 part per thousand (ppt or $^0/_{00}$) is equal to 1 *practical salinity unit* (psu).

Fresh water has zero salinity, while the average salinity of sea water is about 35 psu, and that of estuarine water varies from 0 to 35 psu. Inland lakes and seas such as the Red Sea may have higher salinities. Increased salinity raises the relative density, lowers the freezing point, and greatly influences marine ecosystems in that the flora and fauna have to adapt to saline conditions.

salinization *See* salination.

Salisbury, Sir Edward James (1886–1978) British botanist. Salisbury's first substantial work was his *Plant Form and Function* (1938) written in collaboration with Felix Fritsch, a widely used textbook. However Salisbury was primarily a plant ecologist and did much work both on the effects of soil conditions and on the seed-producing capacity of British species. This was presented in his *The Reproductive Capacity of Plants* (1942). He also carried out a long-term study of sand-dune ecology, the results of which were published in his *Downs and Dunes* (1952).

salt Chemically, a compound formed when an acid is neutralized by a base. Most salts are crystalline ionic compounds, and most minerals are composed of salts. In everyday language, the term salt usually means common salt, sodium chloride (NaCl).

saltation An important method of transport of sediment, such as sand grains or river sediments, in which the particles are lifted steeply off the land or river bed and swept along and then, as the wind speed or current slackens, gently descend again.

salt dome (salt plug) An intrusion of salt (halite) that pushes up existing sediments into a dome below which natural gas, oil, or water may accumulate in porous rocks or in the porous salt deposits.

salt gland A specialized secretory structure involved in the removal of excess salt from the leaves of certain HALOPHYTES, such as sea lavenders (*Limonium*). The salt is exuded in solution, and washed away from the plant by rain or by the next tide. Certain marine birds and reptiles that drink salt water have a nasal gland just above each eye (birds) or between the eye and nostril (marine reptiles, e.g. sea turtles), which remove salt from the blood and excrete it as brine.

salt marsh Vegetation growing in saline or brackish marshy places, such as river estuaries and sheltered muddy coasts. Salt marshes consist mainly of HALOPHYTES. They occur on land regularly inundated by the tides, and represent an early stage in vegetation succession (*see* halosere). The dominant species include salt-tolerant grasses such as *Puccinellia* spp. and sometimes sea lavenders (*Limonium*). In wetter places, pioneer species such as glasswort (*Salicornia*), cordgrass (*Spartina*), sea aster (*Aster tripolium*), and dwarf shrubs such as seablite (*Suaeda*) are found. Salt marshes can be productive grazing land, and serve as important barriers against inundation of the hinterland by the sea. Soil accumulates steadily and a salt marsh can eventually rise above the level of inundation and lose its salinity.

salt pan A basin or pool in a semiarid region in which salts are precipitated and saline water evaporates, e.g. Death Valley, California. Such salts are called *evaporites*.

salt plug *See* salt dome.

salvage The reuse of waste materials.

sampling There are many different methods of sampling. *Random sampling* collects information in an unbiased way. It involves the use of random numbers to select individuals or locations to be sampled. For very large populations *systematic random sampling* is used: the total number of individuals is divided by the number of individuals to be collected in each sample, giving a *sampling interval*. A random number is used to select the first individuals to be sampled, after which the sampling interval is added to that number to locate the next sample, and so on, so that samples are taken at regular intervals.

Various methods are used for sampling populations of living organisms and their habitats. In *point sampling*, sample points are selected by random numbers. This may involve using a QUADRAT of selected size. Alternatively, a *transect* may be taken – a line from two specific points, often across an environmental gradient or slope – and an area of defined width sampled on either side of it or on one particular side.

sampling estimate A method of estimating the size of an animal population by counting in sample plots. It assumes that all members of the population have an equal chance of being sampled, and that there is negligible recruitment and mortality during the sampling period.

sand Mineral particles consisting mainly of quartz, feldspar, and mica, and measuring between 2.00 and 0.05 mm in diameter. *Sandy soils* contain at least 85% sand and not more than 10% clay. They are light and drain well. However, retention of nutrients and water is poor. *See also* soil structure; soil texture.

sand dune *See* dune; xerophyte; xerosere.

sand filter A container of sand that filters suspended solids from SEWAGE. Sand particles are relatively large, and air can pass between them, aiding bacterial decomposition.

sand sea *See* erg.

sandstone   A sedimentary ROCK consisting of sand particles bound by cementing minerals such as silica or calcium carbonate.

sandy soil   *See* sand.

saprobe (saprophyte)   An organism that derives its nourishment by digesting and then absorbing the products or remains of other organisms. Many fungi and bacteria are saprobes, and are important DECOMPOSERS in food chains, returning nutrients to the soil by putrefaction and decay.

saprophyte   *See* saprobe.

saprotrophic   The mode of nutrition of a SAPROBE. *See* decomposer.

sapwood   *See* wood.

saturation index   *See* Langelier index.

saturation point   The level of light intensity above which PHOTOSYNTHESIS no longer increases with increasing light intensity.

savanna   Tropical GRASSLAND with tall coarse tufted grasses and often scattered trees or tall bushes. In Africa, baobab (*Adansonia digitata*) and *Acacia* and *Euphorbia* species are common. The climate alternates between cool dry winters and hot summers with heavy rains. Some savannas have many grazing animals that affect the vegetation structure; these and other savannas have periodic or irregular fires in the dry season. Savannas are found in large areas of South America (*campos*), East and South Africa, South-east Asia, and northern Australia. *See* climax community.

scatter diagram   *See* graph.

scavenger   *See* detritivore.

Schimper, Andreas Franz Wilhelm (1856–1901) German plant ecologist. During travels to the West Indies in 1881 and 1882–83, Brazil (1886), Ceylon (Sri Lanka) and Java (1889–90), and the Canary Islands, Cameroons, East Africa, Seychelles, and Sumatra (1898–99) with the *Valdivia* deep-sea expedition, he made ecological studies of tropical vegetation. His results led to publication of important papers on the morphology and biology of epiphytes and littoral vegetation, culminating with his masterpiece, *Pflanzengeographie auf physiologischer Grundlage* (1898; Plant Geography Upon a Physiological Basis), which relates the physiological structure of plants to their type of environment.

sclerophyllous   Describing vegetation dominated by trees or shrubs with small or leathery leaves resistant to drought. *See* garrigue; maquis.

screening   The removal of coarse floating and suspended solids from SEWAGE by the filtering through racks or screens (sieves).

sea-floor spreading   *See* plate tectonics.

sea grasses   Any of various angiosperms that live in the intertidal or immediate subtidal zone, e.g. eelgrass (*Zostera noltii*), tapeweeds (*Posidonia*). The term is sometimes extended to include seaweed communities of green, brown, and red macroscopic algae.

sea level   The level of the ocean surface, usually taken to be the mean level halfway between low and high tide averaged over all stages of the tide over a long period of time (*mean sea level*). Where this is estimated on a global scale it is called the *global mean sea level*. Sea level is used as a baseline for measuring the elevation of land or the depth of the seabed.

season   A subdivision of the year consisting of a period of supposedly uniform or similar climatic conditions. In tropical areas, temperatures are uniform throughout the year and a distinction is made between wet and dry seasons only. In temperate latitudes, the seasons are based on the equinoxes and solstices. In the northern hemisphere the four seasons are spring (March, April, May); summer (June, July, August); fall (September, October, November); and winter (December, January, February). In polar regions there is a two- season year: summer and winter.

seaweed Any large marine alga (macro-alga). Seaweeds grow on rocky coasts from high-water mark to well below low-water mark, where some species (e.g. kelps) may form underwater forests. A few species are free-floating, e.g. *Sargassum* weed, which may form large floating masses of seaweed, especially in the Sargasso Sea of the North Atlantic, supporting an entire ecosystem of highly adapted organisms.

secondary forest A forest growing on the site of a former forest that has been cleared. It consists mainly of PIONEER SPECIES that will be replaced as succession proceeds.

secondary productivity The rate of accumulation of BIOMASS per unit area by HETEROTROPHS. *Compare* primary productivity.

sediment Solid particles derived from existing rocks and deposited by wind, water, or ice or precipitated out of solution onto land or onto the beds of rivers, lakes, and seas.

sedimentary rock A rock formed by consolidation and compression of sediments laid down by wind or water.

sedimentation A process used in water treatment to remove solid particles before FILTRATION. It takes place in a sedimentation (settling) tank.

seed *See* Angiospermophyta; gymnosperms.

seed bank A collection of seeds kept for research purposes or for the conservation of rare, threatened, or endangered species.

seed dressing A chemical applied to seeds to protect them against attack by fungi or insects. *See* fungicide.

seed plants *See* spermatophyte.

segregation The separation of the two ALLELES of a gene into different gametes, brought about by the separation of homologous chromosomes at anaphase 1 of MEIOSIS. *See* independent assortment.

seismic wave *See* earthquake.

seismograph *See* earthquake.

seismology The study of EARTHQUAKES.

seismometer *See* earthquake.

selection differential (S) The difference between the mean value of a phenotypic character of a selected group of individuals and that of the population from which they came.

selection pressure The degree to which the environment tends to change the balance of ALLELES in a population or the frequency of a particular allele in the population. It is not quantifiable. Comparison of survival rates of individuals with different alleles gives a measure of the FITNESS of a genotype relative to others. Strong selection pressures result in evolutionary change. *See* evolution; natural selection.

selective breeding *See* artificial selection.

selectively permeable membrane *See* osmosis.

selective pesticide A PESTICIDE that targets only selected groups of pests, leaving other species unharmed.

selenium (Se) A highly toxic micronutrient that occurs in meat, nuts, oranges, and in bread made from wheat grown in soils containing selenium. Deficiency of selenium has been linked to disorders such as prostate cancer. In higher amounts it is toxic, and livestock grazing on selenium-rich soils, as in South Dakota, may suffer selenium toxicity.

self-limitation A situation where as population density increases, intraspecific competition causes the reproduction rate or survival rate to decrease.

self-purification The reducing of the biological oxygen demand (BOD) in a stream following an inflow of organic effluent that has raised BOD. As microbes decompose the organic matter and there is therefore less of it, microbial populations and BOD decline, and increased growth of algae and plants, together with natural turbulence, raise the oxygen concentration.

self-thinning The reduction in numbers of plants as density increases, due to COMPETITION for resources and other plant interactions.

semelparity *See* perennial.

senescence The phase of the aging process of an organism or part of an organism between maturity and natural death. It is usually characterized by a reduction in capacity for self-maintenance and repair of cells, and hence by deterioration.

sensor A device that detects the value of or change in a physical entity such as pressure, temperature, pH, light, or sound, and converts it into a signal from which meaningful data can be derived.

septic tank An underground storage tank for domestic waste in an area not connected to sewers. The waste is partially or wholly broken down by anaerobic microorganisms, and the final effluent may be allowed to soak away, or the tank may be emptied at regular intervals.

sequestration *See* chelating agent.

seral stage *See* sere.

sere Any plant COMMUNITY in a SUCCES-

SION in which each community itself effects changes in the HABITAT that determine the nature of the following stage. The successive stages are known as *seral stages*. Seres result eventually in a CLIMAX COMMUNITY. Stages in a secondary succession, which appear when the biotic components of a primary sere are destroyed, e.g. by fire, are called *subseres*; seres in microhabitats are called *microseres*. The initial (pioneer) community in a succession is termed a *prisere*. *See also* halosere; hydrosere; xerosere.

set-aside Land that is taken out of cultivation or other usage and left to revert to its natural state, often incurring a government subsidy, so as to prevent overproduction or to protect the environment from overexploitation or soil degradation in areas of poor soils. The term is also used for land that is set aside in national parks, nature reserves, and other protected areas.

sewage Liquid-borne waste that contains organic matter in solution or suspension, especially that produced by domestic and commercial premises.

sewage farm An area of land on which sewage or other wastewater is spread as manure or to purify it by oxidation and microbial DECOMPOSITION.

sewage fungus A slime or woolly growth in sewage and on objects in waters polluted by sewage, which is formed by communities of filamentous bacteria, fungi, and protozoa.

sewage sludge (sludge) A slimy semisolid substance produced by the sedimentation and precipitation of solid material in sedimentation tanks during sewage treatment.

sex The quality of being either male or female, with the ability to produce one of two types of gamete – sperm in the male and eggs (ova) in the female. An individual's sexuality may be determined by genetic or environmental factors or both, according to the species, or by whether or not an egg has been fertilized.

sex ratio The ratio or males to females (or vice versa) in a population. A distinction is made between the *primary sex ratio* – the ratio of males to females among zygotes – and the *operational sex ratio*, the ratio of males to female at the time of mating, because in many species the sexes have different mortality rates.

sexual dimorphism A difference between the phenotypes of males and females of the same species, e.g. a difference in size, color, or courtship display.

sexual reproduction The formation of new individuals by fusion of two HAPLOID nuclei or sex cells (gametes) to form a diploid *zygote*. In unicellular organisms whole individuals may unite but in most multicellular organisms only the gametes combine. In organisms showing sexuality, the gametes are of two types: male and female (e.g. in animals, spermatozoa and ova). They may be produced in special organs (e.g. testis and ovary in animals, anther and carpel in angiosperms, antheridia and archegonia in lower plants and some algae). The gametes may be derived from the same parent (*autogamy*) or from two different parents (*allogamy*). Individuals producing both male and female gametes are termed *hermaphrodite*. If a flower produces both male and female gametes, the plant is said to be *monoecious* if there are separate male and female flowers; those in which male and female gametes are borne on different individuals are termed *dioecious*. See alternation of generations; natural selection; recombination.

sexual selection The selection by one sex of particular characteristics in the other when choosing a mate, usually mediated by courtship behavior.

shade plant A plant that can tolerate and thrive in low light intensity. Some, but not all, shade plants are sensitive to very bright light and cannot live in open habitats. Shade plants are found where trees or shrubs form a canopy that cuts out much of the light, especially red light, leaving a higher proportion of far-red to red light than in sunlight. *Compare* sun plant.

Shannon-Weaver index (*H*) A logarithmic measure of SPECIES DIVERSITY that is weighted by the abundance of each species.

shelter belt A strip of trees planted upwind of a site to shelter the land or buildings from the prevailing wind.

shifting cultivation *See* slash-and-burn.

shoot The aerial photosynthetic portion of a plant that generally consists of a stem upon which leaves, buds, and flowers are borne.

short-day plant *See* photoperiodism.

shrub A woody PERENNIAL plant smaller than a tree that branches very close to the ground so that there is no obvious main trunk.

shrub layer In the layered structure of a forest, a layer of low-growing woody plants that includes shrubs and saplings.

sievert (Sv) A unit of radiation dose: 1 Sv is a dose equivalent to the gamma radiation received from 1 mg of radium enclosed in platinum 0.5 mm thick delivered for 1 hour at a distance of 1 cm, or 100 rems. For most purposes a millisievert (1/1000th of a sievert, mSv) is more useful.

sieve tube *See* phloem.

sigmoid curve (S-shaped curve) *See* exponential growth.

silage Cattle feed produced by the anaerobic microbial decomposition (FERMENTATION) of green plant matter (such as freshly harvested grass) mixed with diluted molasses and sometimes chemical additives in a pit or storage tower (silo).

silicon A micronutrient found in many animals and plants, especially grasses, where it can be a useful taxonomic feature.

It forms the skeleton of certain marine animals, e.g. the siliceous sponges.

silt   Mineral particles between 0.05 and 0.002 mm in diameter. A silt SOIL has more than 80% silt and less than 12% clay, and has a smooth soapy texture. *Compare* sand.

Silurian   The period, some 440–405 million years ago, between the Ordovician and the Devonian periods of the Paleozoic. It was a relatively warm period in the Earth's history, during which the first land plants appeared. It is characterized by simple plants such as liverworts; fossils of plants similar to vascular plants, e.g. *Cooksonia*, in the later part of the period, some of which had simple strands of tracheids; primitive jawless fish; and many invertebrates. *See also* geological time scale.

silviculture   The management of forest or woodland for the production of timber and/or other wood crops.

Simpson's index (*D*)   A measure of species diversity weighted by the relative abundance (*see* abundance) of each species.

sink   1. A natural reservoir that can take in and store energy or materials without itself undergoing change, e.g. the deep oceans are sinks for carbon dioxide 2. An ecosystem, habitat, community, or population that receives an input of materials or individual organisms.

SI units (Système International d'Unités)   The internationally adopted system of units used for scientific purposes. It has seven base units (the meter, kilogram, second, kelvin, ampere, mole, and candela) and two supplementary units (the radian and steradian). Derived units are formed by multiplication and/or division of base units; a number have special names. Standard prefixes are used for multiples and submultiples of SI units. *See Appendix.*

slash-and-burn (shifting cultivation)   A small-scale agricultural system typical of semi-nomadic people, in which a small area of natural vegetation is cleared by cutting down and burning, cultivated for up to 5 years, then abandoned as the soil becomes less fertile and yields fall. The abandoned area is soon colonized by PIONEER SPECIES from the surrounding natural vegetation, and undergoes succession back to climax forest. Slash-and-burn is now practiced mainly in tropical rainforests where there is no significant population pressure on the land.

sling psychrometer   *See* humidity.

sludge   *See* sewage sludge.

smog   Naturally occurring fog contaminated with smoke and/or invisible pollutants. *See* photochemical smog.

smoker   *See* hydrothermal vent.

social dominance   *See* dominance.

society   A minor plant COMMUNITY within a larger community, characterized by a specific dominant species. For example, a *Trillium*-dominated society in a community such as an oak woodland.

sociobiology   The study of social behavior, based on the assumption that all behavior has evolved as an adaptation to the environment.

sodium   An element essential in animal tissues, and often found in all terrestrial plants, although it is believed not to be essential in most, with the exception of some salt-tolerant C_4 PLANTS. It is implicated in CRASSULACEAN ACID METABOLISM. Sodium occurs widely in nature as sodium chloride (NaCl) in seawater and as deposits of halite from dried-up seas.

sodium chloride   *See* salt.

soft detergents   BIODEGRADABLE detergents.

soft water   Water that does not contain

appreciable amounts of dissolved minerals such as calcium and magnesium salts.

softwood *See* wood.

soil The accumulation of mineral particles and organic matter that forms a superficial layer over large parts of the Earth's surface. It provides support and nutrients for plants and is inhabited by numerous and various microorganisms and animals. A section down through the soil is termed a *soil profile* and this can characteristically be divided into three main layers or *horizons*. Horizon A, the topsoil, is darker than the lower layers due to the accumulation of organic matter as humus. It is the most fertile layer and contains most of the soil population and a high proportion of plant roots. Horizon B, the subsoil, contains materials washed down from above and may be mottled with various colors depending on the iron compounds present. Horizon C is relatively unweathered parent material from which the mineral components of the layers above are derived. The depth and content of the horizons are used to classify soils into various types, e.g. podsols and brown earths. The texture, structure, and porosity of soil depends largely on the sizes of the mineral particles it contains and on the amount of organic material present. Soils also vary depending on environmental conditions, notably rainfall.

soil color Color is an important feature of different soil types and horizons. For example, it can give an indication of the amount of humus in a horizon, or the degree of oxidation of iron minerals. *See also* Munsell soil color chart system.

soil conservation The management of soil to prevent or reduce soil erosion by wind and water, and to avoid overuse, leading to mineral depletion and loss of soil structure, i.e. to ensure the sustainable use of soil.

soil erosion The loosening and removal of topsoil or soil from upper horizons by the action of wind, excessive runoff or other moving water, and excessive downhill movement of loosened material under the influence of gravity. Erosion is often initiated or exacerbated by human activity, such as overgrazing and plowing across contours instead of along them on steep slopes, so that water funnels down the furrows, removing natural vegetation from steep slopes in areas of heavy rainfall. The cultivation of grassland has led to much soil erosion – the loss of the close mesh of grass roots following plowing renders soils susceptible to erosion if it is a time of high winds or heavy rainfall. *See also* Dust Bowl.

soil moisture Water stored in soils. It has three components. *Hygroscopic water* clings to the surface of soil particles by molecular attraction sufficiently powerful that it is unavailable to plants. *Capillary water* forms thicker films and fills small spaces between soil particles. It is rich in nutrients and available to plants. *Gravitational water* is soil water that is neither hygroscopic nor capillary water, but free to flow under the influence of gravity. *See* field capacity.

soil pH The acidity or alkalinity of the water in soil (soil solution). This depends on the soil's parent rock, the rate of weathering, the nature of any solutions percolating through it, the amount of rainfall, the structure of the soil, and the nature of the overlying vegetation. pH has a major effect on the availability of nutrients to plants. An increase in acidity can interfere with root osmoregulation, affect the root's ability to exchange gases, increase the concentration of heavy metals, and may affect the balance of nutrient ions in the soil. In alkaline soils iron, phosphate, and trace elements such as manganese may be trapped in insoluble compounds.

soil profile *See* soil.

soil skeleton The physical structure of the soil – its mineral particles. *See* soil structure; soil texture.

soil structure The arrangement of the mineral particles in the soil – whether they

---

are free or bound into aggregates by decomposing organic matter

soil survey The systematic investigation and mapping of soils in the field.

soil texture The sizes of the different mineral particles in a soil, which also influence its structure, drainage and aeration, and the retention of water by CAPILLARITY. The size of the particles depends in part on the parent rock and how it weathers. The main size classes of mineral particles are gravel and stones (greater than 2 mm in diameter), sand (0.05 to 2 mm), silt (0.002 to 0.05 mm), and clay (smaller than 0.002 mm). Clay soils, which have small particles, tend to be wet and sticky, with poor drainage and aeration. However, they tend to be electrostatically charged and attract humus, forming a clay–humus complex that can attract and retain minerals. The humus sticks the clay particles together, forming larger aggregates and improving drainage and aeration. Clay soils can be further improved by adding lime (*liming*), which promotes clumping of particles. Sandy soils have large particles, and therefore larger spaces between them for drainage and aeration. They may need addition of HUMUS or MULCHES to improve water and mineral retention and prevent leaching.

soil type The most widely used classification system for soils is that of the United States Soil Survey, which classifies soils into 11 orders on the basis of their soil profiles, acidity, and moisture content.

solar energy (solar radiation) Energy transmitted from the Sun as electromagnetic radiation (*see* electromagnetic energy). It covers the spectrum from ultraviolet radiation through visible light to infrared radiation (heat). As well as reaching the surface of land and sea directly as radiation, it may also be transformed into other forms of energy, driving the atmospheric circulation, setting up thermal gradients in oceans and lakes, and becoming incorporated into biomass (chemical energy).

solar power The use of SOLAR ENERGY to generate electricity.

solar radiation *See* solar energy.

solid edition *See* map.

solid waste Any solid material discarded as WASTE, e.g. garbage, sludge from sewage treatment plants, commercial and domestic refuse, and mining slag or spoil. It may include HAZARDOUS WASTE, such as refuse from hospitals or industrial plants.

solifluction The slow downhill creep of soil and other loose material typical of areas where the ground is subjected to alternate freezing and thawing. Freezing expands the water between the particles, raising them slightly. When the ice thaws, the particles come to rest a little farther down the slope.

solonchak A soil with a high soluble salt concentration in the top 30 cm and no obvious horizons except for deposits of salts like gypsum and carbonates. It is formed from saline parent rock in areas of warm to hot climate with high evaporation rates and a pronounced dry season, such as parts of the Mediterranean and subtropics. Solonchaks require irrigation if they are to be cultivated. *See* soil.

solstice Either of the two moments in the year when the Sun in its orbit reaches its greatest distance north or south of the equator. It occurs on or about June 22 (summer solstice in the northern hemisphere and winter solstice in the southern hemisphere) and on or about December 22. The day of the summer solstice has the longest period of daylight and the shortest night of the year, and vice versa for the winter solstice.

somatic polymorphism The presence on the same individual of two or more forms of a particular organ, for example, leaves of different shape. *See* modular organism.

sonar *See* echo sounder.

soot Carbon dust contaminated with oily compounds derived from the incomplete combustion of wood or fossil fuels.

sorbent *See* sorption.

sorption The uptake of a gas or liquid by a solid by either ADSORPTION or ABSORPTION. The substance that takes up the gas or liquid is called the *sorbent*.

source–sink metapopulation *See* metapopulation.

Southern Oscillation (SO) A fluctuation of the intertropical atmospheric circulation in which air moves between the southeastern Pacific subtropical high and the Indonesian equatorial low, driven by the temperature difference between the two areas. When air is blown toward the western Pacific, it causes warm surface water to accumulate there, but promotes upwelling along the western coast of South America, bringing colder nutrient-rich surface water. At the opposite stage of the SO, warm western Pacific water is carried farther east, reducing the nutrient content of the surface waters and affecting marine ecosystems. These oscillations are also associated with periods of intense instability and storms. *See also* El Niño.

specialist A species with narrow food or HABITAT preferences. *Compare* generalist.

speciation The formation of one or more new SPECIES from an existing species. Speciation occurs when a POPULATION separates into isolated subpopulations that develop distinctive characteristics as a result of NATURAL SELECTION or random GENETIC DRIFT, and cannot then reproduce with the rest of the population, even if there are no geographical or other physical reasons to prevent them from doing so. Speciation that occurs as a result of such factors as genetic mutations and genetic drift, and changes in behavior, with no external barriers to reproduction is called *sympatric speciation*. Often there are more obvious barriers to interbreeding, such as geographical changes (e.g. in river courses) –

this is called *allopatric speciation*; habitats becoming uninhabitable (e.g. following urbanization), as changes in timing of flowering, or when populations at the margins of the main population experience different selection pressures as they colonize different environments (*parapatric speciation*). Another cause of rapid speciation in plants is POLYPLOIDY. *See also* adaptive radiation; reproductive barrier.

species A TAXON comprising one or more POPULATIONS, all the members of which are able to breed among themselves and produce fertile offspring. They are normally isolated reproductively from all other organisms, i.e. they cannot breed with any other organisms. Some species can interbreed with other related species, but the HYBRIDS are often wholly or partially sterile. For some plant species there may be an almost complete continuum between closely related species, as in certain orchids. This may also happen in disturbed areas or in regions where the range of two or more species or even genera overlap, producing *hybrid swarms* (*see* apomixis). Where species form apomictic clones that are very difficult to distinguish from each other, they are termed *aggregate species,* e.g. bramble (*Rubus fruticosus*). In general, species are regarded as distinct if they remain reproductively isolated for most of their geographical range and have recognizably distinct morphological characteristics. Within a species, there may be subgroups with distinct morphological or (especially in microorganisms) physiological characteristics; these groups are termed *subspecies* or *races*. Groups of similar species are classified together in genera. *See also* binomial system of nomenclature; genus; population; speciation.

species diversity The number of SPECIES in an area or community and their relative ABUNDANCE. There are various scales of diversity. *Alpha (local) diversity* is the number of species in a small area of fairly uniform habitat. *Gamma (regional) diversity* is the number of species in all the habitats in a region with which there are no significant barriers to dispersal. It is related to the local diversity and the number of different habitats in the region. The difference in species diversity between two habitats is termed the *beta diversity*, and is related to the whether the species are generalists or specialists and their degree of habitat specialization. The greater the habitat specialization and the greater the proportion of specialists the bigger the difference in species between habitats, and the greater the beta diversity. If all the species in both habitats are generalists, the two habitats are effectively one, and beta diversity = 1.

species pool The biota of a large area from which species may migrate to surrounding areas. For example, an island population may be augmented by migration from the nearby mainland. *See also* equilibrium theory.

species richness The number of species in a community compared with the number of individuals in the community.

species saturation A community that contains the maximum diversity of species possible under the prevailing conditions. In a stable environment, species diversity is limited by local interactions between species.

species turnover The change in species composition of a community or area that results from COLONIZATION, EXTINCTION, and recolonization events. Species turnover rates will be high in disturbed or newly available habitats (e.g. in the early stages of SUCCESSION), and low in long-established communities in stable environments.

spermatophyte Any seed-bearing plant.

Sphagnum A genus of mosses, often called peat mosses or bog mosses. A *Sphagnum* moss can hold over 20 times its own dry weight of water and these mosses are important BOG-formers, producing PEAT that can be harvested for fuel or, more recently, for horticultural use, a trade that has seriously reduced some bogs. In areas of high rainfall where rocks are fairly impermeable and soils are acidic, *Sphagnum*

bogs can cover larger areas, e.g. parts of northern Scotland and the Great Dismal Swamp in North Carolina and Virginia in the United States, which once covered 5700 sq km but is now reduced to 1940 sq km. *See also* hummock and hollow cycle.

spiders *See* Arthropoda.

sponges *See* Porifera.

sporangium (*pl.* sporangia) A reproductive body in which asexual SPORES are formed.

spore *See* sporangium.

sporophyte *See* alternation of generations.

spring A place where GROUNDWATER flows naturally out of the ground where the WATER TABLE intersects the ground surface.

spring overturn Mixing of the water of lakes in spring due to wind action, resulting in nutrients from the bottom sediments being mixed into the water column and oxygen from the surface being carried into deep water.

spring tide *See* tide.

stability 1. A situation in which air that is forced to rise tends to return to its previous level once the factor causing the rise is removed. This occurs if the rate at which the rising air cools with height due solely to its vertical movement (*see* adiabatic cooling and heating) is greater than the rate at which the surrounding air cools with height. The rising air becomes cooler and denser than the surrounding air, so it sinks down again. *See* atmosphere.
2. (stable equilibrium) The situation where the number of individuals in a population or the level of a resource quickly returns to its original value following a disturbance.

stabilization pond A large shallow pond in which WASTEWATER is allowed to decompose naturally by the action of microorganisms, algae, and sunlight, the waste being mixed and oxygenated by the wind.

stabilizing selection Selection that serves to maintain the genetic structure of a population as it is. It will normally be the dominant form of selection in a population in a stable environment with little competition. The range of variation in the population may also decrease as the most common forms are favored and the less well adapted forms decline in number with each generation, resulting in a restricted range of phenotypes. *See* natural selection.

stable equilibrium *See* stability.

stagnation Lack of movement in a mass of air or water. Stagnation leads to a build-up of POLLUTANTS, where present. Stagnant water tends to become deoxygenated, because there is no turbulence to mix air from the surface into deeper water.

standard deviation (*s*) In statistics, a measure of the dispersion of values assuming that they form approximately a normal distribution. It is calculated by squaring all the deviations from the mean, calculating their mean, then finding the square root. This gives a value *s*, which is the point of maximum slope either side of the center of the distribution curve.

standard error (SE) The estimated STANDARD DEVIATION of an estimate of a parameter. The standard error of the mean of a sample from a population with a normal distribution is given by:
$$SE = s/\sqrt{n}$$
where *s* is the sample standard deviation. The standard error indicates the reliability of the sample mean as an estimate of the true mean of the population and *n* is the sample size.

standard index of association A measure of the degree to which two species are associated with each other in their distribution:
$$d_{ik} = (O_{ik} - E_{ik}) / s_{ik}$$
where d_{ik} is the difference between the observed (O_{ik}) and the expected (E_{ik}) number

of locations in which species i and k would be expected to occur together by chance, and s_{ik} is the standard deviation of the expected number. *See also* association.

standing crop The total weight of all the living organisms present in an ECOSYSTEM at a given moment, usually expressed as dry weight per unit area.

starch *See* carbohydrates.

stationary source *See* point source.

steady state A state in which the input to a system (energy, resources, individuals) is at the same level as the output. In terms of a living COMMUNITY, the input comes from the energy and raw materials used in photosynthesis or other forms of autotrophism, and the output from respiration and decomposition. If input and output are in balance, there is no net loss or gain of energy or materials by a community or ecosystem. A CLIMAX COMMUNITY in equilibrium with the climate is considered to be in a steady state.

stem *See* shoot.

stenohaline Describing an organism that is extremely sensitive to changes in salinity, and is unable to tolerate much variation in osmotic pressure.

steppe Temperate GRASSLAND dominated by drought-resistant species of perennial grasses, found in regions of LOESS soil and extreme temperature range, often typically on CHERNOZEM soils, in a zone from Hungary eastward through the Ukraine and southern Russia to Central Asia and China. There are pronounced seasons, with hot summers and cold winters, with greater temperature extremes and lower rainfall in the east. The species composition varies with the climate. Typical grasses include feathergrasses (*Stipa*) and sheep's fescue (*Festuca ovina*), together with bunchgrass (*Schizachyrim*) and bluegrass (*Poa* spp.). Many of the herbs, such as *Tulipa*, *Allium*, and the grass *Poa bulbosa*, have underground PERENNATING OR-

GANS and die back at the start of summer. There are also many EPHEMERAL species that germinate as soon as rain falls, before the grasses take over. Large areas of former steppe are under cultivation for grain production.

sterile   1. Describing an organism that is unable to reproduce.
2. Describing a pure culture of microorganisms that is not contaminated by other species.

sterilization   1. The removal or destruction of all microorganisms, including spores, from an object.
2. The process of making an organism unable to reproduce.

Stevenson screen   A shelter for housing thermometers for use in meteorological measurements in conditions in which they give standard readings. It consists of a ventilated box with an air space between an inner an outer roof. It contains wet-and-dry-bulb thermometers for measuring humidity and temperature, which are screened from the sun's radiation,

stilt root   An enlarged form of *prop root*, seen in some mangroves and a few palms and other trees, which helps support plants in unstable soils. *See* aerial root.

stomata (*sing.* stoma)   One of a large number of pores in the epidermis of plants through which gaseous exchange occurs. *See* stomatal rhythm; transpiration.

stomatal rhythm   A daily rhythm of stomatal opening and closing, usually governed by a biological clock, and fine-tuned by environmental signals. The rhythm is maintained for some time in constant conditions, such as darkness, but eventually fades. The stomata of most plant species open in the morning and close in the evening to conserve moisture when conditions are no longer suitable for PHOTOSYNTHESIS. Some species living in hot dry climates open in the morning, then close for some hours around midday, opening again later in the afternoon, thus avoiding

water loss from TRANSPIRATION at a time when evaporation rates are at their highest. In plants with CRASSULACEAN ACID METABOLISM, the stomata remain closed by day and open at night, when carbon dioxide is fixed into organic acids.

storm sewer   A conduit that collects run-off from heavy rain and snowmelt, including street wash, and transports it back to the groundwater. It is not connected to the sewerage system that carries wastewater, although in some cases (e.g. some city runoff) it may direct the runoff into a treatment plant.

stormwater   Excess runoff due to storms and rapid snowmelt. *See* storm sewer.

strain   Any group of similar or identical individuals, such as a clone, mating strain, physiological race, or pure line. *See* clone; race.

stratification   1. A seed treatment that enables seeds that require VERNALIZATION to germinate the following spring: the seeds are placed between layers of moist sand or peat and exposed to low temperatures, usually by leaving them outside through the winter.
2. The existence of layers of water of different density in a LAKE or other body of water. The differences in density may be due to temperature or to salinity. In many lakes in summer there is a warm low-density layer (*epilimnion*) lying above a colder denser layer (the *hypolimnion*), the zone of rapid temperature change between the two layers being called the *thermocline*. In autumn, strong winds create turbulence, and the layers become mixed. While stratification persists, nutrients are continually lost from the epilimnion as planktonic and larger organisms die and their bodies sink below the thermocline. With the autumn 'turnover', nutrients are returned to the upper layers, ready to promote growth in spring (*see* fall bloom; spring overturn). *See also* dimictic.

stratigraphy   The study of sediments – their formation, composition, sequences, and consolidation into sedimentary rock. It includes deducing the history of a sedimentary rock from a study of the structure, organization, and relationships of the different sediment layers of which it is composed. *See* geological time scale.

stratopause   *See* atmosphere.

stratosphere *See* atmosphere; ozone layer.

stratospheric cooling   A cooling of the stratosphere (*see* atmosphere) due to GREENHOUSE GASES that cause warming in the troposphere but prevent infrared radiation from the Earth from reaching the stratosphere. *See* greenhouse effect.

stratum (*pl.* strata)   A horizontal or inclined layer of rock of similar composition, especially one that is part of a series of parallel layers arranged on top of each other.

stream   A relatively small body of water flowing in a channel.

street canyon   An urban area enclosed by buildings (e.g. a street) in which the high density of traffic leads to carbon monoxide and other pollutants reaching high concentrations.

stress tolerator   A plant or other organism that copes with a stressful habitat (e.g. dry, cold climate, or infertile soil) by growing only slowly, thus making low demands on its environment, living for a long time and reproducing only when it has sufficient resources.

strip cropping (strip farming)   On sloping land, the planting of long strips of land parallel to the contours with different crops to reduce soil erosion.

strip farming   *See* strip cropping.

strip mining   *See* open-pit mining.

stromatolite   A structure formed by microbial mats of CYANOBACTERIA that secrete layers of calcium carbonate, forming large cushion-like masses. They occur only in certain protected warm lagoons today. Fossilized stromatolites have been found in rocks dating back at least 3000 million years.

structural diversity   Within the area occupied by a COMMUNITY, the range of physical structures that may provide suitable habitats for species.

subclimax   *See* climax community.

subduction zone   *See* plate tectonics.

sublittoral   1. The marine zone extending from low tide to a depth of about 200 m, usually to the edge of the continental shelf. Light penetrates to the seabed, and the water is well oxygenated. Large algae (e.g. kelps) are found in shallower waters whereas certain red algae (Rhodophyta) may be found in deeper water. *Compare* benthic zone; littoral.
2. The zone in a lake or pond between the littoral and profundal zones, extending from the edge of the area occupied by rooted plants to a depth of about 6 to 10 m, where the water temperature declines. Its depth is limited by the *compensation level* – the depth at which the rate of photosynthesis is equaled by the rate of respiration, and below which plants cannot live (but some phytoplankton can). *Compare* littoral; profundal.

subpopulation (local population)   The individuals of a particular species that live in the same habitat patch. *Compare* metapopulation.

subsere   *See* sere.

subspecies   *See* species.

substitutable resource   A RESOURCE that is capable of being interchanged with another resource to satisfy the requirement of the consumer.

subtropical   Relating to the zone on either side of the Equator, between latitudes 23.5° and 34.0° in either hemisphere.

succession A progressive series of changes in vegetation and animal life of an area over time from initial colonization to the final stage, or *climax*. The climax is a dynamic equilibrium because, although the succession can progress no further under the environmental factors present at the time, the populations present change, e.g. trees die, creating gaps for other species to colonize. In addition, the climate is seldom completely stable – at best it is cyclical, with variations from year to year (*see* cyclic climax; sere). In some successions, such as those from wetlands to terrestrial communities, the early occupants change the environment, making it possible for later species to move in. In others, modification of the environment by early colonists has little or no effect on the subsequent performance of later species in the succession – this is called *successional tolerance*.

Where succession is driven only by process operating within the community, it is termed *autogenic*; where it is driven by external influences that alter environmental factors, it is a termed *allogenic*. A succession that takes place from a newly colonized habitat, starting with PIONEER SPECIES, is called a *primary succession*. One that follows a disturbance such as a fire, which destroys the original community, is called a *secondary succession*. An *arrested succession* is one that has been prevented from reaching its natural climax, for example by grazing or other disturbance. *See also* facilitation.

Where a succession of species involves mainly plants, it is termed an *autotrophic succession*; if involving mainly animal species, it is a *heterotrophic succession*.

successional tolerance *See* succession.

succulent A plant with swollen stems or leaves that have large parenchyma cells that store water. They normally have a water-resistant waxy cuticle. Succulents are characteristic of dry habitats such as deserts and many of these succulents are armed with spines or similar structures, e.g. cacti. They also occur in saline habitats such as salt marshes.

sugar *See* carbohydrates.

sulfur An essential element in living tissues, being contained in nearly all proteins. Plants take up sulfur from the soil as the sulfate ion SO_4^{2-}. The sulfides released by decay of organic matter are oxidized to sulfur by sulfur bacteria of the genera *Chromatium* and *Chlorobium*, and further oxidized to sulfates by bacteria of the genus *Thiobacillus*. There is thus a cycling of sulfur in nature.

sulfur bacteria Filamentous autotrophic chemosynthetic bacteria that derive energy by oxidizing sulfides to elemental sulfur and build up carbohydrates from carbon dioxide. They use sulfides instead of water as a source of electrons in photosynthesis, releasing sulfur instead of oxygen. An example is *Beggiatoa*. They are found mainly in sulfur-rich muds and springs, including HYDROTHERMAL VENTS. A few archaea, e.g. *Sulfolobus*, can oxidize elemental sulfur. As well as sulfides, some bacteria oxidize thiosulfates, polythionates, and sulfites. Sulfur bacteria play an important role in the cycling of sulfur in the ecosystem.

Sulfur-respiring anaerobic bacteria, such as *Desulfovibrio*, transfer electrons from carbon compounds to sulfur compounds. They are responsible for the release of hydrogen sulfide into anaerobic sediments. The mineral pyrite (iron sulfide) is thought to have been formed by the reaction of this hydrogen sulfide with iron compounds present in mud. *See also* photosynthetic bacteria.

sulfur cycle *See* sulfur; sulfur bacteria.

sulfur dioxide (SO_2) A colorless choking gas, readily soluble to form an unstable solution of sulfurous acid (H_2SO_3). Sulfur dioxide is released from many industrial processes and from the burning of FOSSIL FUELS. It is a serious air pollutant, causing respiratory irritation and triggering asthma in susceptible people. It is a major constituent of ACID RAIN.

sulfur dust A powdered form of sulfur

used to kill fungal pathogens on grapes and other crops.

Sun The star around which the Earth and other members of the Solar System orbit. Its average density is relatively low, because it is made up of at least 90% hydrogen and helium, but the density of its core is six times that of Earth's core, and the temperature here reaches 15 000 000K. The intense pressure and density in the core make possible reactions in which hydrogen atoms are converted into helium, generating the energy radiated by the Sun. Most of this energy is initially in the form of gamma rays, which are absorbed and re-emitted as heat and light by material farther from the core. Hot ionized particles stream out of the Sun as the solar wind, a spiraling stream of about 400 km per second, dragging magnetic fields with it and affecting the Earth's upper ATMOSPHERE.

sun plant A plant that can tolerate and thrive in high light intensities. Such plants have thicker leaves than SHADE PLANTS, have numerous stomata on the underside of the leaf, where transpiration rates are lower, and small air spaces so the leaf does not lose water too readily. *Compare* shade plant.

supercontinent *See* Gondwanaland; Pangaea.

superkingdom A taxonomic rank above the rank of kingdom but below the rank of DOMAIN. The prokaryotes and eukaryotes are sometimes considered as two superkingdoms. *See also* classification; taxonomy.

superorganism The concept of a COMMUNITY as a kind of giant organism, with member SPECIES closely bound together by their present interactions and common evolutionary history, the function of each promoting the wellbeing of the whole community.

supertramp species One of the earliest colonizers of a new habitat, highly adapted for COLONIZATION.

surface tension A property of the surface of a liquid, whereby it acts as if an elastic film were stretched across the surface. It is caused by the attractive forces between the particles within the liquid, and between the liquid and the gases, liquids or solids it is in contact with. Surface tension is responsible for the formation of water droplets. It slows the evaporation of water, helping plants to retain water. It is also partially responsible for *capillarity* – the movement of liquids through thin tubes – which affects the supply of water to plants from the soil and its transport through plant cell walls and the xylem.

surface-to-volume ratio The ratio of the surface area of an object, such as a cell, organ, or organism, to its volume. The higher the ratio, the easier it is for gases and dissolved minerals to diffuse from the outside to the interior.

surface water Water that is naturally open to the atmosphere – streams, rivers, springs, ponds, lakes, reservoirs, estuaries, and seas.

surplus yield model A simple model of the effects on a population of harvesting, in which the population is represented simply by its size or BIOMASS. It ignores population structure (*see* population).

survival of the fittest *See* fitness; natural selection.

suspended load The part of the total sediment load of a stream or river that is carried in suspension.

sustainability The degree to which an activity can be sustained with causing harm to the environment or the depletion of NONRENEWABLE RESOURCES.

sustainable development Economic development that takes into account the needs of living and future generations, but is based on the use of RENEWABLE RE-

SOURCES rather than the depletion of NON-RENEWABLE RESOURCES, and takes account of the environmental consequences of economic activity. It was defined by the World Commission on Environment and Development in 1987 as: development that meets the needs of the present without compromising the ability of future generations to meet their own needs.

swamp An area of vegetation dominated by trees that develops on ground that is normally waterlogged or covered by water all year round, such as the margin of a lake, a river floodplain, an area of water-retentive clay in an arid region, or along the shores of an estuary. Swamp vegetation represents the early stages of a HYDROSERE or HALOSERE. The presence of trees distinguishes it from a *marsh*. The vegetation typically consist of various reeds (e.g. *Phragmites*) and sedges (Cyperaceae), including papyrus (*Cyperus papyrus*) in parts of Africa, together with trees like gums (*Nyssa*), willows (*Salix*), alders (*Alnus*), and swamp cypress (*Taxodium distichum*). Coastal and estuarine swamps in warm climates are dominated by the MANGROVES. The swamp vegetation slows the flow of water and dead plant matter builds up, leading to a build-up of stagnant muds and bottom water lacking in oxygen. Decay is therefore slow, and PEAT may accumulate. *Compare* marsh.

switching The tendency for a predator to switch between prey species or categories according to their relative abundance

symbiogenesis *See* endosymbiont theory.

symbiosis (*pl.* symbioses) Any close association between two or more different organisms, as seen in parasitism, mutualism, and commensalism. Often one or both organisms is dependent on the other. The term is usually used more narrowly to mean mutualism, but many mutualistic associations may have once been parasitic or may become so at some stage in their life cycle, e.g. plants and their mycorrhizal fungi.

sympatric speciation *See* speciation.

synecology The study of all the interactions between the living organisms in a natural COMMUNITY and the effects of the nonliving components of the environment upon them and on their relationships with each other. *Compare* autecology.

systematic error *See* bias.

systematic random sampling *See* sampling.

systematics The area of biology that deals with the diversity of living organisms, their relationships to each other, and their classification. The term may be used synonymously with TAXONOMY.

systematic variation A variation in a measured variable that is not due to the parameter being measured, but to another linked factor that causes that parameter to vary in a similar way. For example, if the data showed an association between drinking coffee and coronary heart disease (i.e. that people who drink coffee are more likely to suffer from the disease) this may not be the true situation – in fact, people who drink coffee may be more likely to smoke, and smoking may be the true cause of the association. In this case, the link with smoking causes a systematic variation in the data attempting to link coffee-drinking with heart disease.

systemic Describing a chemical that is absorbed by a plant and distributed throughout its tissues. Systemic FUNGICIDES and INSECTICIDES make plant tissues toxic to fungi and insects, providing built-in defenses. Systemic weedkillers ensure that both shoots and roots are killed, and are particularly useful for plants that can regenerate from fragments of shoot or root, or which have rhizomes or runners.

systems ecology A branch of ECOLOGY concerned with the flow of energy and circulation of matter in an ECOSYSTEM.

T

tachytely A rate of evolution that is much faster than is usual for the taxonomic group concerned, e.g. species undergoing ADAPTIVE RADIATION.

taiga *See* biome; boreal forest.

take back A scheme in which householders are required to return WASTE to retailers, who pass it into a private recycling or waste disposal system rather the public one, or where the householder receives a refund of a deposit paid at the time of purchase.

talus Large blocks of rock that break away from cliff faces and form banks against the base of the cliff.

Tansley, Sir Arthur George (1871–1955) British plant ecologist. Tansley's thinking was greatly influenced by books by Eugenius WARMING and Andreas SCHIMPER. These – together with travels in Ceylon, Malaya, and Egypt – stimulated his interest in different vegetation types. In 1902 Tansley founded *The New Phytologist*, a journal designed to promote botanical communication and debate in Britain. In 1913 he founded and became the first president of the British Ecological Society and four years later founded and edited the *Journal of Ecology*. These activities, and his ecology courses at Cambridge University, played a large part in establishing the science of ecology.

tapeworms *See* Platyhelminthes.

taxon (*pl.* taxa) A group of any rank in taxonomy. Felidae (a family) and *Panthera* (a genus) are examples.

taxon cycle A cycle of expansion and contraction of the geographical range and population density of a species or higher TAXON.

taxonomic diversity DIVERSITY of taxonomic groups in a community or area. *See* biodiversity; species diversity.

taxonomy The area of SYSTEMATICS that covers the principles and procedures of classification, specifically the classification of variation in living organisms. *See* classification; systematics.

TBTO *See* tributyltin oxide.

tectonic plate A section of the Earth's crust that is capable of movement as a result of convection currents in the MANTLE. *See* continental drift; plate tectonics.

temperate deciduous forest A major BIOME found in the mid-latitudes (temperate zone) between the Tropics of Cancer and Capricorn and extending to the Arctic and Antarctic Circles. Temperate deciduous forests have moderate climates with abundant and fairly evenly distributed rainfall. There is a marked seasonal change. The trees are shorter than those of BOREAL FOREST and are dominated by broad-leaved trees (oak, beech, lime, birch, hazel), which shed their leaves in the fall, an adaptation to lack of available water in winter. Many of these forests contain several different species of trees and are termed *mixed deciduous forest*. Warm *temperate forests* occur in south China, for example, and *cool temperate rainforests* also exist, in New Zealand, for example. In temperate deciduous forests there is a shrub layer beneath the CANOPY and a rich herbaceous ground flora. The temperate is high enough to decompose the leaf litter, creating rich soils. In winter, many of the animals hibernate or become inactive and some birds migrate to warmer climates.

Broad-leaved evergreen forest is characteristic of Mediterranean-type climate with hot dry summers and mild wet winters. The leaves of the trees are adapted to water shortage and high evaporation in summer and usually have thick waxy cuticles (e.g. olive trees). The forest often merges into scrub woodland characteristic of GARRIGUE and MAQUIS.

temperate grassland *See* grassland; pampas; prairie; steppe.

temperate zone The region between the Tropic of Cancer and the Arctic circle or the Tropic of Capricorn and the Antarctic Circle, i.e. latitudes 23°27′ to 66°30′ N and 23°27′ to 66°30′ S.

temperature coefficient The ratio of the rate at which a process proceeds at one temperature and its rate at a temperature 10°C lower or higher. *See* Q_{10}.

temperature inversion *See* inversion.

temperature profile The relationship between temperature and elevation or depth.

temporal variation Variation in environmental factors through time, e.g. on an hourly, daily, or seasonal basis.

teratogen A physical or chemical agent that causes birth defects in a developing fetus by interfering with normal development.

termites (white ants) A group of social insects (order Isoptera) that contain symbiotic microorganisms in their guts, which help them digest CELLULOSE. Termites can be highly destructive, undermining house timbers and damaging commercial lumber.

terracing The construction of cutting terraces (flat platforms) or broad channels into slopes, sometimes banked up or protected by embankments, into order to trap runoff for the purposes of irrigation or to channel it to suitable outlets, thus preventing soil erosion.

terra rossa A type of CALCAREOUS SOIL that develops over limestone in parts of the Mediterranean region. It is a CLAY SOIL with low HUMUS and significant amounts of iron oxides, which give it a bright red color. *See* soil. *Compare* rendzina.

terrestrial biome *See* biome.

territorial competition *See* competition.

territoriality The establishment by one or more animals of an area from which others of the same or different species are excluded. Such an area is called a *territory*.

territory *See* territoriality.

Tertiary The larger and older period of the Cenozoic, being composed of the Paleocene, Eocene, Oligocene, Miocene, and Pliocene epochs (65–2 million years ago). Literally the 'third age', it is characterized by the rapid evolution and expansion of angiosperms and the emergence of mammals. The ferns underwent adaptive radiation, and the angiosperms underwent an even greater explosion of adaptive radiation, in which flowers became more specialized. This went hand in hand with the evolution of specialist insect pollinators, especially bees, butterflies, and moths. The climate was warm and wet at the start of the period, but became colder and drier later, and this was accompanied by the expansion in range of grasses. Many modern genera of conifers arose in the Tertiary, but other gymnosperms declined. *See also* geological time scale.

tetraethyl lead (lead tetraethyl) *See* lead.

tetraploid *See* haploid.

thallus (*pl.* thalli) A simple plant body

showing no differentiation into root, leaf, and stem and lacking a true vascular system. It may be uni- or multicellular, and is found in the algae, lichens, bryophytes, and the gametophyte generation of ferns, clubmosses, and horsetails.

thermal A spiraling column of upward-moving warm air that forms as the land warms faster than the air, so that warmed air next to the ground becomes less dense than the air above. Thermals are used by many birds to gain height.

thermal pollution POLLUTION that contributes excess heat to the environment, for example the cooling water discharged from power plants. Many aquatic and marine organisms can tolerate only a limited range of temperatures, and many are unable to cope with rapid changes in temperatures, because this seldom happens in natural bodies of water. As temperature rises, the oxygen concentration in the water decreases, yet metabolic rates also increase with temperature, requiring greater oxygen consumption.

thermocline *See* metalimnion.

thermohaline circulation Upwelling and downwelling movements of bodies or water due to gradients of temperature and salinity in the surface waters.

thermometer A simple instrument for measuring temperature, based on the expansion of certain liquids with temperature. It consists of a graduated sealed glass tube containing mercury or dyed alcohol. *See also* maximum and minimum thermometer.

thermophile A microorganism that requires high temperatures (around 60°C) for growth, e.g. certain bacteria (domain ARCHAEA) that grow in hot springs or compost and manure. *See* mesophilic; psychrophilic. *See also* extremophile; halophile; methanogen.

thermophilic digestion *See* digestion.

thermoregulation The maintenance of an optimum temperature range by an organism. Homoiotherms ('warm-blooded animals') have internal means of generating heat, and regulate temperature by both increasing heat generation and controlling heat loss. Poikilotherms ('cold-blooded animals') control their body heat by moving between warm and cool places, and by adopting a body stance and position in relation to the sun that helps them minimize passive heat gain or loss.

therophyte *See* Raunkiaer's life-form classification.

third-generation insecticides Synthetic insecticides, usually organic compounds, that aim not to harm nontargeted species. Some are synthetic versions or derivatives of naturally occurring insecticides.

thorn forest A transitional type of FOREST with vegetation resembling savanna and semidesert vegetation, grading into TROPICAL FOREST. It is found in Africa, Australia, and Central and South America.

threatened species *See* Endangered Species Act; Red Data Book.

Three Mile Island A nuclear power station near Harrisburg, Pennsylvania. Here, in 1979, a malfunction of the cooling system led to overheating and meltdown of the reactor core. At one point during the crisis radioactive gas was allowed into the atmosphere. Although there was no subsequent evidence of adverse affects on health, the incident contributed to a general mistrust of nuclear energy.

threshold The minimum stimulus intensity that will initiate a physiological or environmental response, or the lowest dose of a chemical or other form of pollutant at which a measurable effect occurs.

tidal wave *See* tsunami.

tide The periodic rise and fall in level of the oceans and other large bodies of water caused by the relative gravitational attractions of the Sun, Moon, and Earth. The effect of the Moon is much greater than that of the Sun, causing the cycle of rise and fall. The extent of the tidal rise and fall varies with the time of year as a consequence of changes in the relative positions of the Sun, Moon, and Earth, the Sun enhancing the effect of the Moon's tidal pull to form the high *spring tides* and, when it opposes the moon's pull, the *neap tides* (lowest tides of the year). Geographical variations in tides are related to variations in the contours of the seabed, and to the generally uneven distribution of water in some areas. Some areas have a single rise and fall in a tidal day (24 hours 50 minutes). Others have two periods of high and low water with a period of 12 hours 25minutes.

tillage 1. The process of cultivating the land.
2. Land under cultivation.

tiller A shoot that develops from an axillary or adventitious bud at the base of a stem, often in response to injury of the main stem. Tillering is characteristic of grasses, and is why they thrive under moderate grazing, which actually enhances the growth of the grasses by increasing the number of shoots produced. The tufted habit of many grasses is due to tillering. Tillering also occurs when a tree is coppiced. *See* coppice.

timber line (Waldgrenze) The altitudinal limit for the growth of trees of normal height that form a closed canopy. It lies at a lower level than the TREE LINE.

time delay (time lag) A delay in the response of a population, community, or ecosystem to changes in environmental factors.

time hypothesis The hypothesis that BIODIVERSITY is greater in areas that have experienced long periods of stability than areas of frequent disturbance.

tissue In a multicellular organism, a group of cells that is specialized for a particular function. Examples are connective tissue, muscular tissue, and nervous tissue. Several different tissues are often incorporated in the structure of each organ of the body, e.g. the heart or, in plants, a leaf.

tolerance 1. *See* succession.
2. The capacity of an organism to ingest certain chemicals or be exposed to certain physical agents without deleterious effect. The tolerance limit of a chemical or physical agent is the lowest level that is harmful to the organism.

tonnage The rate at which waste enters a LANDFILL, usually expressed as tons per month.

top-down control *See* food chain.

topography The surface relief of the land, or of an object – its shape and relative elevations.

top predator A predator that occurs at the top of the food chain or food web – usually one that eats other predators. For example, the lion is one of the top predators of the African savanna. *See also* keystone species.

topsoil 1. The uppermost layer of SOIL, usually including the organic layer in which most plants root, and which is turned over during cultivation.
2. The A horizon of a soil profile. *See* soil.

tornado A relatively small (about 100 m in diameter) funnel of rapidly rotating air (wind speeds may exceed 300 km h^{-1}) that forms around an intense low-pressure center. It is capable of sweeping up objects from the ground. Tornadoes are accompanied by violent down-drafts.

torpor A state of inactivity, low metabolic rate, and lowered body temperature developed in response to adverse environmental conditions, especially cold or heat. For example, many insects, reptiles, and temperate zone hummingbirds (whose small size and large surface-to volume ratio makes for rapid heat loss) enter a state of torpor overnight to conserve energy when temperatures fall. Torpor in swifts is thought to be linked to periods when flying insect prey is not available. Many poikilothermic vertebrates and a few homoiothermic ones, such as polar bears, enter true hibernation or winter torpor that may last for weeks or months. *Compare* hibernation.

torr   A unit of pressure that equals 133.3 pascals, or 1 mm Hg at 0°C.

total response of predator   The per capita mortality of a prey population in response to changing prey density. This includes the effects of the changing PREDATOR–PREY RATIO on the behavior of both predator and prey. For example, at low densities a predator may selectively kill weaker individuals, whereas at high densities the prey may indulge in high-risk strategies to compete for food. The shape of this response, when plotted on a graph, gives an indication of the type of predator–prey interactions occurring.

toxicity   The degree of danger posed by a chemical substance to living organisms or the environment.

toxic oil syndrome (eosinophilia myalgia syndrome)   An autoimmune disease caused by exposure to cooking oil contaminated with aniline derivatives.

trace element   *See* micronutrient.

trace fossil   *See* fossil.

trace metal   A metal element that is essential for growth but is required only in very small quantities.

tracheid   *See* xylem.

tracheophyte   Any plant with a differentiated vascular system; i.e. all plants except the liverworts, mosses, and hornworts.

trade winds   Persistent winds that blow toward the Equator in the latitude belt between 30°N and S (the horse latitudes) and the DOLDRUMS. The trade winds blow from the northeast in the northern hemisphere, and from the southeast in the southern hemisphere.

trait   *See* character.

transcription   The process in living cells whereby RNA is synthesized according to the template embodied in the base sequence of DNA, thereby converting the cell's genetic information into a coded message (messenger RNA, mRNA) for the assembly of proteins, or into the RNA components required for protein synthesis (ribosomal RNA and transfer RNA). The mRNA transcript leaves the nucleus to direct protein assembly on ribosomes in the cytoplasm, in the process called TRANSLATION.

transect   *See* sampling.

transgenic   Describing organisms, especially EUKARYOTES, containing foreign genetic material. Genetic engineering has created a wide range of transgenic animals, plants, and other organisms, for both experimental and commercial purposes. Examples include dairy cows that secrete drugs in their milk, herbicide-resistant crop plants, and plants that secrete pharmaceuticals. *See* recombinant DNA technology.

transhumance   The moving of herds to livestock up mountains to summer pasture and back to the shelter of the valleys in winter.

transient polymorphism   The occurrence of two or more different forms of a species (i.e. groups of individuals having distinctive combinations of alleles) or of particular alleles in a population during the period in which one or more forms are being replaced by others.

transition zone   *See* ecotone.

transit time   1. In a geochemical cycle, the average time a nutrient remains in a particular form.
2. The ratio of biomass to productivity.

translation The process whereby the genetic code (sequence of nitrogenous bases) of messenger RNA (mRNA) is deciphered by the machinery of a cell to make proteins. This takes place in the ribosomes, where amino acids corresponding to specific triplets of bases in the mRNA as assembled and linked together by peptide bonds. *See* RNA; transcription.

translocation 1. The moving of individuals of a species from a CAPTIVE BREEDING program or from a habitat where the species is plentiful to one where it is desired to reintroduce the species to boost its numbers.
2. The movement of mineral nutrients, recently synthesized food materials, and hormones through a plant.

transparency The degree to which a substance allows light to pass through it.

transpiration The loss of water vapor by EVAPORATION from the surface of a plant. It has been shown that most transpiration occurs through the STOMATA when they are open for gaseous exchange, but a small amount (typically about 5%) is lost directly from epidermal cells through the cuticle (*cuticular transpiration*) and a minute proportion through lenticels. A continuous flow of water, the *transpiration stream*, is thus maintained through the plant from the soil via root hairs, root cortex, xylem, and tissues such as leaf mesophyll served by xylem. Transpiration may be useful in maintaining a flow of solutes through the plant and in helping cool leaves through evaporation, but is often detrimental under conditions of water shortage, when WILTING may occur. It is favored by low humidity, high temperatures, and moving air. A major control is the degree of opening of the stomata. *Compare* guttation.

trap crops Plants that are planted between the plants of a main crop because they are more attractive to pests. They may suffice to draw the pest away from the main crop, or they may simply serve to concentrate the pests, which can then be sprayed without harming the main crop.

travel time 1. The time it takes a predator to move from one patch of high prey abundance to another. This is a factor in determining when to move on – if travel time is high, it may pay the predator to continue hunting in the same patch even if prey density is relatively low. *See also* giving-up time.
2. The time it takes a pollutant to travel from a point source to a given point.

tree A large persistent woody plant, usually with a single main stem (the trunk) that remains unbranched near the ground and doe not die back in winter or a dry season. Species of trees that sometimes have several trunks, such as beech (*Fagus sylvatica*) and oak (*Quercus*), have no ordinary branches leaving any trunk near the ground. *Compare* herb; shrub.

tree line (Baumgrenze) At high latitudes or high altitudes, the limit of tree growth, representing the taiga/TUNDRA and subalpine/alpine boundaries, respectively. The term also applies to lower tree limits such as frost hollows – a point beyond which trees cannot grow. Trees close to the tree line are often scattered and stunted or prostrate. *See* boreal forest. *Compare* timber line.

tree rings *See* annual rings.

Triassic The oldest period of the Mesozoic era, 250–215 million years ago. During the Triassic, the climate gradually changed from arid to more temperate. Conditions for preservation in the drier period were poor, so there are few fossils. There was a diversification of the gymnosperms, a decrease in the number and variety of cartilaginous fishes, and an increase in primitive amphibians and reptiles. *See also* geological time scale.

tributyltin oxide (TBTO) One of a group of organic tin-containing chemicals used as a BIOCIDES to prevent the fouling of ships' hulls, and as stabilizers for PVC manufacture. It is toxic to marine life, being released into the water by bacterial action or leached out by seawater. In shellfish such as oysters and whelks it causes females to develop secondary male characteristics, leading to sterility and consequent collapse of the population. Very low levels have been shown to have adverse effects on plankton. Its use is now banned in many countries.

triploid *See* polyploidy.

trophic Relating to the nutrition of organisms.

trophic cascade *See* food chain.

trophic level In complex natural communities, organisms whose food is obtained by the same number of energy-transfer steps are said to belong to the same trophic or energy level. The first and lowest trophic level contains the PRODUCERS, green plants that convert solar energy to food by photosynthesis. Herbivores occupy the second trophic level and are primary CONSUMERS: they eat the members of the first trophic level. At the third level carnivores eat the herbivores (the secondary consumer level), and at the fourth level secondary carnivores eat the primary carnivores (the tertiary consumer level). These are general categories, and many organisms feed on several trophic levels, for example omnivores eat both plants and animals. DECOMPOSERS or transformers occupy a separate trophic level, which consists of organisms such as fungi and bacteria that break down dead organic matter into nutrients usable by the producers. *See* food chain.

trophic structure The structure of a community expressed in terms of the energy flow through its various TROPHIC LEVELS.

trophospecies *See* food web.

tropical Located in or relating to the equatorial region of the Earth's surface between latitudes 23°27′ N and 23°27′ S.

tropical forest A major BIOME found between the Equator and the Tropics of Cancer and Capricorn. Tropical forests include RAINFOREST, RIVERINE FOREST, and MONSOON FOREST. *See also* thorn forest.

tropical grassland *See* savanna.

tropics Latitudes 23°27′ N (Tropic of Cancer) ad 23°27′ S (Tropic of Capricorn). At the summer solstice in the northern hemisphere the Sun is directly overhead at the Tropic of Cancer, and in the southern hemisphere at the summer solstice it is directly overhead at the Tropic of Capricorn.

tropopause *See* atmosphere.

troposphere *See* atmosphere.

true north The direction of the geographical North Pole, which lies at the northernmost end of the Earth's axis. It is not the same direction as the magnetic North Pole – the northernmost end of the Earth's magnetic field, which changes position with time.

tsunami (tidal wave) A giant wave caused by an undersea earthquake or landslide. As travels across the sea it becomes very large and when it reaches land may overrun and destroys coastal settlements.

t-test A statistical test used to compare the means of one or more normal distributions whose variance is not known but must be estimated from the data.

tundra A major BIOME located north of the BOREAL FOREST in the subarctic regions of North America, Europe, and Asia characterized by its lack of trees, permanently frozen subsoil (*permafrost*), high winds, and extremely low temperatures. Tundra occurs only in scattered localities on Antarctic islands, where it consists mainly of mosses and lichens. The vegetation in the northern tundra consists of low-growing grasses, lichens, and mosses. Tundra may also be classified as cold DESERT, the temperature rarely rising above 10°C and the topsoil is frozen for nine months of the year. Thus the *arctic vegetation* is subjected to both extreme PHYSIOLOGICAL DROUGHT (with water unavailable for most of the year) and to extreme cold. In the short summer, many meltwater pools contain insect larvae, which develop into adults as the temperature rises, providing food for migrating birds, e.g. waders, which return to the tundra to breed. Reindeer and other herbivores graze on the lichens and migrate south to the boreal forest when the winter becomes extreme, although they can locate lichens under the snow cover.

Similar conditions may prevail at high altitude in temperate and tropical regions and similar vegetation is found (*alpine tundra*). *See also* gley.

turbidity current A current caused by the slumping of sediment that has accumulated on the rim of the continental slope, especially at the head of a submarine canyon. It forms a huge slurry of mixed sediment and water that spreads out at great speed over the ocean floor, forming a layer of unusually coarse sand in deep water.

turnover 1. The proportion of a population that is lost or gained by deaths/emigration or births/immigration respectively over a given time.
2. The changing species composition of a COMMUNITY or habitat as species become extinct and their niches are occupied by new colonizers, which may be newly evolved local species or immigrants.
3. The ratio of productivity to BIOMASS: the proportion of its biomass that is taken into a community by primary productivity each year.

4. A measure of the nutrient flux in a BIOGEOCHEMICAL CYCLE: the rate at which a nutrient flows into or out of a nutrient pool divided by the quantity of nutrient in the pool.

type specimen An individual or sample or the whole or part of an organism that is used as an example of a particular taxonomic group. A *type specimen* is a designated specimen from which the initial criteria defining the characteristics of the group were taken. There are several categories of type specimen. A *holotype* is a single type specimen designated by the person who originally descried a species or subspecies, and used to verify the status of other specimens. It has priority over all subsequent specimens. If the holotype is unavailable, a second specimen, called a *neotype*, may be selected to substitute for it. If the original describer of the group did not designate a holotype, a specimen may be designated by someone else later – this is a *lectotype*. Sometimes a collection of specimens was used for the original description, but no single specimen was designated as a type specimen. Such a collection may then be used as a whole and is called a *syntype*. Specimens that were collected at the same locality and at the same time as the holotype, and used in the original designation of a new taxon, are called *paratypes*. Finally, specimens collected at the same time and from the same organism or local population as the holotype may be used as substitutes for the holotype, for example in different institutions or countries, or kept as safeguards against the destruction of the holotype. Such specimens are *isotypes*.

typhoon *See* depression.

UV

ultra-low sulfur diesel *See* diesel.

ultraviolet (UV radiation) Electromagnetic radiation (*see* electromagnetic energy) of wavelengths 4–400 nm, less than visible light but greater than x-rays. Lying just beyond the blue end of the visible spectrum, it contains more energy than ordinary light, and certain forms of UV radiation have been associated with eye cataracts, genetic damage, and skin cancer. These wavelengths are to some extent filtered out by the OZONE LAYER.

UNCED *See* Earth Summit.

underdispersion *See* dispersion.

understory The layer of vegetation between the ground cover or shrub layer and the canopy in a forest community. It consists of shade-tolerant species and the saplings of canopy and emergent species.

undulipodium (*pl.* undulipodia) Cilia and flagella.

UNEP *See* United Nations Environment Program.

UNFCCC *See* United Nations Framework Convention on Climate Change.

ungulate A hoofed mammal.

unicellular Describing organisms that exist as a single cell.

unitary organism *See* modular organism.

United Nations Conference on Environment and Development *See* Earth Summit.

United Nations Environment Program (UNEP) A program set up in 1972 to promote international cooperation on environmental issues, and provide scienific advice to United Nations organizations, with the aim of promoting the sustainable use of the world's resources. It plays a major role in organizing international conventions such as the MONTREAL PROTOCOL on Substances That Deplete the Ozone Layer, The Basel Convention on the Control of Transboundary Movements of Hazardous Wastes and Their Disposal, and the UN Convention on Biological Diversity, providing technical assistance, data, and assistance in implementing the resulting decisions and in monitoring implementation. Its Governing Council is elected by the UN General Assembly every four years. *See also* sustainable development.

United Nations Framework Convention on Climate Change (UNFCCC) An international treaty dating to 1992 that commits signatory countries to stabilize GREENHOUSE GAS emissions due to human activity to 'levels that would prevent dangerous anthropogenic interference with the climate system'. It requires countries to maintain national inventories of anthropogenic emissions of greenhouse gases, including those not covered by the MONTREAL PROTOCOL. *See also* Earth Summit.

unstable equilibrium A state in which a system is at present finely balanced, but small changes in the numbers of individuals in populations or metapopulations or in the concentration and availability of resources will lead to even larger perturbations, rather than the system returning eventually to its former state. *See also* equilibrium.

upwelling A upward movement of water, usually near coasts and driven by onshore wind, that brings deep water to the surface, along with nutrients.

uranium (U) A toxic silvery radioactive element occurring naturally in various minerals. It has three naturally occurring radioisotopes, of which the commonest (over 99%) is ^{239}U. ^{235}U is a major nuclear fuel and nuclear explosive, while ^{238}U is a source of fissionable ^{239}Pu (plutonium). *See also* fission; nuclear energy; isotopic dating.

urban Relating to towns and cities.

ur-**gene** *See* origin of life.

Urochordata *See* Chordata.

validation Determination of how well a model fits the process or system under investigation.

variable A characteristic or property that can posses a range of values. It may be a measured value of an environmental factor, such as wind speed or temperature, or it may represent the number of individuals in a population possessing a particular characteristic, e.g. a particular age or height. Variables are often used to plot graphs showing relationships or dependencies. In this situation, one variable may be fixed (e.g. a location along a transect) or under the control of the experimenter (e.g. time) – this is an *independent variable*. The other – the property being measured or the rate of reaction – is not controlled, and is a *dependent variable*.

variance (s^2) A statistical measure of the spread (DISPERSION) of the distribution of a quantitative variable – of its average deviation from the mean. It is the sum of the squares of the values of the variables divided by the number of samples. *Compare* standard deviation.

variation The extent to which the characteristics of the individuals of a species can vary. Variation can be caused by environmental and genetic factors. Environmental variation (phenotypic plasticity) results in differences in the appearance of individuals of a species because of differences in nutrition, disease, light intensity, etc. Genetic variation is caused by RECOMBINATION and occasionally MUTATION. These differences may be favored or discriminated against by NATURAL SELECTION. *See* qualitative variation; quantitative variation.

variety The taxonomic group below the subspecies level. The term is often loosely used to describe breeds of livestock or various cultivated forms of agricultural and horticultural species. Varieties are morphological variants, which may differ in color or growth habit. *See also* cultivar; race; species.

vascular plants Plants containing differentiated cells forming conducting tissue, which comprises the XYLEM and PHLOEM (water- and food-conducting tissues respectively). All plants except mosses (Bryophyta), liverworts (Hepatophyta), and hornworts (Anthocerophyta) are vascular plants. They are sometimes classed as a single division, the Tracheophyta (*see* tracheophyte).

vector (of parasite) **1.** An agent that carries a disease-causing organism to a healthy plant or animal, causing the latter to become infected.
2. An agent (*cloning vector*) used as a vehicle for introducing foreign DNA, for example a new gene, into host cells during GENETIC ENGINEERING.

vegetation texture The number, density, and arrangement of patches of plants or the species composition, density, and dispersion pattern of plants in a patch of habitat.

vegetative propagation (vegetative reproduction) Growth of any parts of a plant, other than flowers, that may get detached from the parent and become independent, although they are identical genetically. For example, bulbs, corms, stolons, and rhizomes.

veld Extensive open grazing land in southern Africa. *See* grassland.

ventilation The process by which an organism maintains a flow of air or water over its respiratory surfaces in the lungs, gills, or other respiratory organs. This increases the rate at which oxygen enters and carbon dioxide leaves the blood. Most active animals possess some form of ventilation mechanism, typically involving muscular movements (*respiratory movements*). For example, bony fishes can actively pump water over their gills by alternately expanding and contracting the mouth and gill cavities. In mammals, air is alternately drawn into and expelled from the lungs by muscular movements of the rib cage and diaphragm.

ventimeter *See* wind.

vermiculture The culture of worms that break down vegetable matter and turn it into compost. This may be done on a commercial scale in a worm farm, or on a domestic scale in special worm bins.

vermin Animals considered to be harmful to game or crops, for example mice, rats, etc., and parasitic or disease-carrying invertebrates such a fleas, lice, ticks, and parasitic worms.

vernalization The cold treatment of ungerminated or partially germinated seeds. Certain plants will germinate and flower only if exposed to low temperatures (1–2°C) at an early period of growth, i.e. they have a chilling requirement. Thus winter varieties of cereals will flower in summer only if sown the previous autumn. Spring-sown winter varieties remain vegetative throughout the season unless they have been vernalized. *Artificial vernalization* is an important technique in countries where severe winters can kill autumn-sown crops.

vertical life table *See* life table.

vertical mixing The exchange of water between surface layers and deeper layers in a body of water. This does not happen under conditions of STRATIFICATION, but is reestablished during the SPRING OVERTURN or fall overturn (*see* fall bloom). *Compare* stratification.

vessel *See* xylem.

Vienna Convention A convention held in Vienna, in 1985, in which nations agreed to take appropriate measure to protect the environment and human health against the adverse affects from human activities that modify or are likely to modify the OZONE LAYER. It sought to encourage research, cooperation between nations, and the exchange of information. The urgency for action was stressed when evidence of severe depletion of the ozone layer above Antarctica was published during 1985; in 1987 the MONTREAL PROTOCOL was drawn up.

virion *See* virus.

virus An extremely small infectious agent that causes a variety of diseases in plants and animals, such as smallpox, the common cold, and tobacco mosaic disease. Viruses can reproduce only in living tissues; outside the living cell they exist as inactive particles consisting of a core of DNA or RNA surrounded by a protein coat. Most plant viruses are single-stranded RNA viruses. The inert extracellular form of the virus, termed a *virion*, penetrates the host membrane and liberates the viral nucleic acid into the cell. Usually, the nucleic acid is translated by the host cell ribosomes to produce enzymes necessary for the reproduction of the virus and the formation of daughter virions. The virions are released by lysis (bursting) of the host cell. Plant viruses are transmitted by vectors such as aphids and nematodes. A virus that infects a bacterium is termed a *bacteriophage* (phage). Some viruses are associated with the formation of tumors.

visibility The distance that can be seen – the most distant object toward the horizon that can be seen with the naked eye. It is a measure of the transparency (or opacity) of the medium (air or water). *See also* turbidity current.

vitamins Organic chemical compounds that are essential in small quantities for metabolism but are not synthesized by animals, which therefore need to obtain them from food or microorganisms. The vitamins have no energy value; lack of vitamins results in the breakdown of normal bodily activities and produces disease symptoms. Such deficiency diseases can usually be remedied by including the necessary vitamins in the diet. Plants can synthesize vitamins from simple substances, but animals generally require them in their diet, though there are exceptions to this.

vitrification The encasing of something in a glassy material by heat and fusion. It is used to create fuel pellets for nuclear reactors, and also to immobilize and store nuclear wastes to minimize leaching into groundwater.

volatile organic compounds (VOCs) Organic compounds that readily evaporate at room temperature, especially those that take part in atmospheric photochemical reactions. *See also* photochemical smog.

volcanism (vulcanism; volcanic activity) The processes by which magma and associated gasses, or hot water and steam, rise into the Earth's crust, travel along lines of weakness, and are extruded onto the surface or ejected into the atmosphere. It encompasses the formation of volcanoes, fumaroles, and geysers on the Earth and other planets. *See also* plate tectonics; earthquake.

vortex *See* polar vortex.

vulnerable species *See* Endangered Species Act.

W

Waldgrenze   *See* timber line.

Wallacea   A zone of mixing between animal species from the Oriental (Southeast Asian) faunal region and the Australasian region, bounded to the west by *Huxley's line* (a modification of *Wallace's line* that runs east of Bali and Java and north between Borneo and Sulawesi, then east again, passing to the west of the Philippines, with the exception of Palawan, marking the easternmost spread of marsupials), and to the east by *Lydekker's line*, which follows the edge of Australia's continental shelf, marking the westernmost spread of Oriental fauna.

Wallace's line   *See* Wallacea.

warm event   *See* El Niño.

Warming, (Johannes) Eugenius Bülow (1841–1924) Danish botanist. In 1895 he published his book *Plantesamfund*. An enlarged and revised English translation was published in 1925 as *Oecology of Plants*. Warming is regarded as one of the founders of the subject of plant ecology.

warm temperate forest   *See* temperate deciduous forest.

warning coloration   *See* aposematism.

waste   Domestic or commercial refuse, including garbage, unwanted materials left over from industrial processes and construction work, spent nuclear fuel, and so on. Waste may be gaseous, liquid, or solid. *See also* commercial waste; hazardous waste; radioactive waste; wastewater.

waste disposal   Methods of dealing with WASTE. *See* activated sludge process; incineration; landfill; recycling.

wastewater   Water that as been used for a process and released because it is no longer required (e.g. industrial effluent, SEWAGE).

water   (H_2O) A colorless liquid that freezes at 0°C and, at atmospheric pressure, boils at 100°C. The maximum density of water occurs at 3.98°C, which is why ice floats on water. Without floating ice to insulate against further temperature fall and severe weather, aquatic communities in temperate and polar regions would perish. Water is a polar liquid, and the most powerful solvent known, partly because of its high dielectric constant and partly its ability to hydrate ions.

water balance index   *See* Langelier index.

water culture   *See* hydroponics.

water cycle (hydrological cycle; hydrologic cycle)   The cycling of water through the atmosphere, biosphere, lithosphere, and hydrosphere: water in the sea and lakes and rivers evaporates, forming clouds that deposit rain on the land, and the water runs off the land or percolates through soil and rocks to the water table, eventually returning to the sea, lakes, and rivers.

water-flow measurements   The simplest methods for measuring water flow measure the time taken for a float to travel between two fixed points. The *Lagrangian measurements* measure only the flow of the surface waters. For measurement at depth, current meters are used – these are called

Eulerian measurements. A *current meter* has a propeller that is driven by the water movement, its revolutions per unit time increasing with the rate of water flow.

Water flow is measured in units of discharge (Q), usually m^3s^{-1}. These are calculated from the *flow velocity* ($m\ s^{-1}$) and the cross-sectional area of the riverbed, pipe, or channel (m^2). The rate at which sediment is transported in a river (the *instantaneous suspended sediment discharge*) is calculated by: water discharge (m^3s^{-1}) times suspended sediment concentration ($mg\ l^{-1}$). Regular measurements of depth and flow may be automated at *gauging stations* anchored in the water at intervals along a river.

water hyacinth   A floating aquatic plant (*Eichhornia crassipes*) with showy bluish-purple flowers, native to tropical America. Water hyacinth is a serious pest in the southern United States and in many countries to which it has been introduced, forming dense masses and choking rivers, ponds, and waterways, suffocating fish and blocking formerly navigable channels.

water potential   *See* osmosis.

water sampler   A manual or automatic device for collecting samples of water, consisting of an open tube of fixed volume that is lowered to the desired depth, then a releasing mechanism triggers two tightly closing rubber stoppers to seal its ends.

watershed   *See* divide.

water table   The upper surface of the GROUNDWATER, below which the soil and rocks are saturated with water. The topography of the water table usually mirrors that of the ground surface, but with less exaggerated peaks and troughs. In places the water table may be above the surface of the ground, forming lakes, rivers, or springs.

water vapor   The gaseous form of WATER, especially when below its boiling point and diffused in the atmosphere. It is the most abundant GREENHOUSE GAS in the atmosphere, and helps regulate global temperature, weather, and climate through cloud formation.

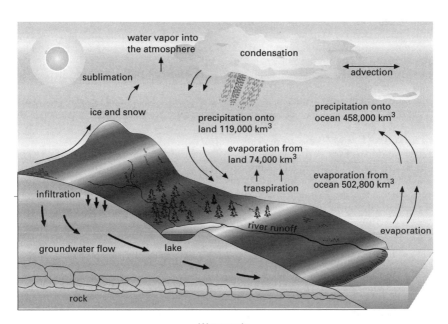

Water cycle

wave power The generation of electricity using the kinetic energy of waves to drive turbines or rocking floats.

WCMC *See* World Conservation Monitoring Center.

WCU *See* World Conservation Union.

weather The state of the ATMOSPHERE over a short period of time – air movement (wind), air pressure, humidity, cloud cover, the nature of precipitation, air temperature, and periodic disturbances such as storms or hurricanes. Weather occurs mainly in the troposphere, but is to a small extent influenced by events in the upper atmosphere. It is also affected by topography, especially by large bodies of water and mountains.

weathering The physical and chemical breakdown of rock and its component minerals. It occurs both on bare rock surfaces and at the underground boundary between rock and soil.

Chemical weathering breaks down the minerals and cement in rocks, so loosening their structure and making them more liable to crumble. Chemicals present in the atmosphere (e.g. oxygen) and in rain (e.g. carbonic acid and ACID RAIN) act on exposed rocks, and solutions of chemicals percolating through rocks and sediments may selectively dissolve out minerals and redeposit them in other sediments.

Physical weathering includes the effects of differential expansion and contraction of the surface layers of rocks when exposed to the sun or to frost. Water in the interstices of rocks, and in cracks and faults, expands when it freezes, forcing blocks of rock apart, leading to the formation of screes and landslides. Runoff is an important lubricating factor in landslides, helping loosened blocks of rocks fall away, exposing new surfaces to weathering.

Where rocks lie under vegetation, as on lichen-covered rocks or at the soil–bedrock interface, *biological weathering* is important, plants secreting organic acids that eat into the rocks. Plant roots can exert extraordinary pressures as they grow, especially where a tree takes root in a rock crevice. They can force apart walls and push up sidewalks.

weather station A collection of instruments for recording and reporting meteorological observations at a single site. The simplest measure atmospheric pressure, temperature (or maximum and minimum temperatures), precipitation, and wind speed and direction.

weed Any plant growing where it is not wanted. Many weed species are adapted to colonize disturbed ground, such as a newly plowed field, and have rapid growth, good seed dispersal, and are frequently self-pollinated. Others have persistent rhizomes or roots that can regenerate from fragments, e.g. bindweed (*Convolvulus*).

weedkiller *See* herbicide.

weighted Describing data that have been adjusted for the fact that some items are less reliable or to reduce the effect of a particular bias.

Westerlies The prevailing winds in the middle latitudes that blow from a westerly direction as a result of the Earth's rotation *See* Coriolis force. *Compare* trade winds.

wet-and-dry-bulb thermometer *See* humidity.

wetfall Nutrients that wash into the ecosystem in rain, snow, and fog.

wetland An area that is waterlogged for most of the year with surface or groundwater and supports vegetation adapted for such conditions. *See* bog; fen; estuary; marsh; salt marsh; swamp.

whirling psychrometer *See* humidity.

white ant *See* termites.

white goods Large (usually white) electrical appliances such as refrigerators, stoves, washing machines, washing-up machines, and air conditioners, for which special waste disposal and recycling facilities are often made available.

wildfire A destructive and rapidly spreading natural conflagration, especially in a wilderness area. Such fires are often started by lightning and are seasonal occurrences in such habitats as grasslands and dry scrub, and in certain forests whose trees give off flammable aromatic oils.

wildlife management The management of land for the benefit of wildlife, including, where necessary, the culling of populations of animals to ensure the continued good health of the population or to prevent overcrowding leading to disease, starvation, or destruction of the habitat.

wildlife refuge An area designated for the protection of wild animals, within which hunting and fishing are strictly controlled or prohibited.

Wilson, Edward Osborne (1929–) American entomologist, ecologist, and sociobiologist. Wilson collaborated with Robert MACARTHUR in developing a theory on the equilibrium of island populations from which emerged their *Theory of Island Biogeography* (1967). To test such ideas Wilson conducted a number of remarkable experiments with Daniel Simberloff in the Florida Keys. They selected six small mangrove clumps and made a survey of the number of insect species present. They then fumigated the islands to eliminate all the 75 insect species found. Careful monitoring over the succeeding months revealed that the islands had been recolonized by the same number of species, thus confirming the prediction that a dynamic equilibrium number of species exists for any island. Later Wilson caused some controversy with his book *Sociobiology* (1975); he has also written a work on ecology, *The Diversity of Life* (1992), and a revealing autobiography, *Naturalist* (1994).

wilting The loss of turgor in a plant due to lack of water.

wind A directional movement of air.

Wind speed can be estimated by observing the effects of the wind on objects. The BEAUFORT SCALE of wind speed, used in weather and shipping forecasts, is based on this. Wind speed may be measured by a hand-held instrument called a *ventimeter*, which consists of a calibrated tube over which wind passes, causing a reduction in the pressure inside the tube that induces a pointer to rise in proportion to wind speed. *Anemometers* are fixed or hand-held wind speed measurers. They are used in WEATHER STATIONS, mounted in locations where wind is not obstructed by buildings or other objects. An anemometer consists of three revolving cups connected to a meter that counts the number of rotations per unit time and converts this to wind speed. The strength and direction of the wind varies at different heights above the Earth's surface. The variation in wind speed with elevation is called the *wind profile*.

wind chill The cooling effect of the wind in disrupting the boundary layer between the land and the atmosphere and carrying away heat that has moved into the air by convection. It makes the air feel much colder than it really is, an effect (the *chill factor*) that increases with wind speed.

wind power The generation of electricity using propeller-shaped wind vanes mounted on tall supports (windmills) to drive turbines that generate electricity. Groups of windmills are usually sited together and linked to a common generating center – such clusters are called *wind farms*.

wind profile *See* wind.

Winkler titration A laboratory method that uses an iodine titration to determine the dissolved oxygen content of a sample of water.

wood The hard fibrous structure found in woody PERENNIALS such as trees and shrubs. It is normally formed from the secondary XYLEM and thus found in plants that show *secondary thickening* (the addition of extra layers of tissues, including vascular tissues, to allow increase in girth as the plant grows taller), mainly conifers and dicotyledons. Water and nutrients are transported only in the outermost youngest wood, termed the *sapwood*. The nonfunctional compacted wood of previous seasons' growth is called the *heartwood* and it is this that is important commercially. Wood is classified as *hardwood* or *softwood* depending on whether it is derived from dicotyledons, e.g. oak (*Quercus*), or conifers, e.g. pine (*Pinus*). Hardwood is generally harder than softwood but the distinction is actually based on whether or not the wood contains fibers and vessels in addition to tracheids and parenchyma. *See also* annual rings.

woodland A plant COMMUNITY similar to a FOREST but whose large trees are more widely spaced, and their crowns do not form a closed CANOPY. The ground flora may consist of grass, herbs, heath, or scrub. While there may be scattered understory species, the understory does not form such a dense and well-defined layer as in a true forest.

woody perennial *See* perennial.

World Conservation Monitoring Center (WCMC) An information service set up by the World Conservation Union, the Worldwide Fund for Nature, and the United Nations Environment Program to compile information on conservation and the sustainable use of the world's natural resources. The WCMC collects data on threatened and endangered species and habitats, National Parks and nature reserves, international agreements, and conservation and environment programs. It also assists in the development of other in-

formation sources and the training of staff to support them. *See* Red Data Book.

World Conservation Union (WCU) Formerly the International Union for the Conservation of Nature (IUCN), founded in 1948, an organization that brings together states, government agencies, and a range of nongovernmental organizations to promote conservation and the sustainable use of natural resources. Expert volunteers collect information on particular species and biodiversity conservation projects, providing a base of expertise that enables the WCU to advise countries on national conservation strategies, assisting them in devising strategy and management plans, and providing technical support. The WCU has an expanding network of regional and country offices, especially in developing countries.

World Health Organization (WHO) An agency of the United Nations that promotes international cooperation to improve global health, providing guidance on and finance for the control of epidemic and endemic diseases, quarantine measures, and public health, sending teams into the field to train medical and health workers, and disseminating information on research relating to health and disease.

Worldwide Fund for Nature The largest independent conservation organization in the world, comprising a global network of National Organizations, Associates, and Program Offices. It provides conservation services based on global policy, fieldwork, and scientific information. Its goal is to protect genetic, species and ecosystem diversity, ensure the sustainable use of resources, and promote actions to reduce pollution.

XYZ

xenobiota Any organisms displaced from their normal habitat.

xenon (Xe) A colorless odorless inert gas that occurs in trace amounts in air (*see* atmosphere).

xeric Relating to or describing an environment that is dry. *Compare* hydric; mesic.

xeromorphic Structurally adapted to withstand dry conditions. *See* xerophyte.

xerophyte Any plant adapted to growing in dry conditions or in a physiologically dry habitat, such as an acid bog, sandy desert (*see* dune), or a salt marsh, or an exposed very windy situation. Xerophytes store available water, reducing water loss, or possess a deep root system. Succulents, such as cacti and agaves (family Agavaceae), have thick fleshy stems or leaves that store water. Features associated with reducing water loss include: shedding or dieback of leaves, e.g. ocotillo (*Fouquieria splendens*); waxy leaf coatings coupled with closure or plugging of stomata, e.g. *Kalanchoe*; sunken or protected stomata, e.g. marram grass (*Ammophila arenaria*); folding, rolling, or repositioning of leaves to reduce sunlight absorption, e.g. marram grass; a dense covering of white reflective spines, e.g. cacti; and the development of a dense hairy leaf covering, e.g. *Espeletia*. *Compare* hydrophyte; mesophyte. *See also* halophyte; physiological drought.

xerosere Any plant community in a succession that starts in dry conditions. The effect of increasing vegetation on the environment can lead toward more mesophytic conditions. The seeds of XEROPHYTES estab-

lish a pioneer community called a PSAMMOSERE, dominated by plants that can tolerate abrasion by blown sand, high winds, temperature extremes by day and night, and in coastal areas, salt spray. Various xerophytic grasses such as marram grass (*Ammophila arenaria*) grow and trap sand particles, raising the level of the DUNE. Their rotting remains contribute to soil formation. *See* sere; succession.

xylem The water-conducting tissue in vascular plants. Xylem consists of dead hollow cells (the *tracheids* and *vessels*), which are the conducting elements. It also contains additional supporting tissue in the form of fibers and sclereids and some living packing tissue (parenchyma). The secondary cell walls of xylem vessels and tracheids eventually become thickened with lignin to give greater support. *Compare* phloem.

yield The total amount of BIOMASS produced by an ecosystem, community, or population in a given time (usually a year). It may be calculated by multiplying the density of a population by the mean weight per individual.

zero net growth isocline (ZNGI) An ISOCLINE along which the rate of population growth is zero.

zinc A micronutrient that is toxic in large quantities and can pollute the soil, allowing only certain tolerant plants to grow.

ZNGI *See* zero net growth isocline.

zonation The distribution of species and habitats along environmental gradients.

zone of depletion An area around an individual plant from which its roots draw nutrients and water. Where the zones of depletion of adjacent plants overlap, there may be competition for resources.

zoogeography The study of the geographical distribution of animal species. Such study shows that the earth can be divided into distinct geographical regions, each having its own unique collection of animal species (*see* AUSTRALASIA, ETHIOPIAN, NEARCTIC, NEOTROPICAL, ORIENTAL, PALAEARCTIC). For example, the continents of the southern hemisphere – Australia, Africa (south of the Sahara), and South America – each have a characteristic fauna not found elsewhere. Anteaters, sloths, and armadillos are native to South America; marsupial and monotreme mammals are characteristic of Australia; while Africa shows a greater diversity of fauna than any other region.

zooid An individual module of a modular organism such as a colonial coral, hydroid, or bryozoan.

Zoological Code of Nomenclature *See* International Codes of Nomenclature.

zoology The study of animals, including their classification, structure, morphology, physiology, interrelationships, behavior, population dynamics, and evolution.

zooplankton *See* plankton.

zooxanthellae *See* corals; symbiosis.

zygote The diploid cell resulting from the fusion of two haploid gametes. *See* sexual reproduction.

APPENDIXES

Appendix I

SI Units

BASE AND DIMENSIONLESS SI UNITS

Physical quantity	Name of SI unit	Symbol for SI unit
length	meter	m
mass	kilogram(me)	kg
time	second	s
electric current	ampere	A
thermodynamic temperature	kelvin	K
luminous intensity	candela	cd
amount of substance	mole	mol
*plane angle	radian	rad
*solid angle	steradian	sr
*supplementary units		

DERIVED SI UNITS WITH SPECIAL NAMES

Physical quantity	Name of SI unit	Symbol for SI unit
frequency	hertz	Hz
energy	joule	J
force	newton	N
power	watt	W
pressure	pascal	Pa
electric charge	coulomb	C
electric potential difference	volt	V
electric resistance	ohm	Ω
electric conductance	siemens	S
electric capacitance	farad	F
magnetic flux	weber	Wb
inductance	henry	H
magnetic flux density	tesla	T
luminous flux	lumen	lm
illuminance (illumination)	lux	lx
absorbed dose	gray	Gy
activity	becquerel	Bq
dose equivalent	sievert	Sv

DECIMAL MULTIPLES AND SUBMULTIPLES USED WITH SI UNITS

Submultiple	Prefix	Symbol	Multiple	Prefix	Symbol
10^{-1}	deci-	d	10^{1}	deca-	da
10^{-2}	centi-	c	10^{2}	hecto-	h
10^{-3}	milli-	m	10^{3}	kilo-	k
10^{-6}	micro-	μ	10^{6}	mega-	M
10^{-9}	nano-	n	10^{9}	giga-	G
10^{-12}	pico-	p	10^{12}	tera-	T
10^{-15}	femto-	f	10^{15}	peta-	P
10^{-18}	atto-	a	10^{18}	exa-	E
10^{-21}	zepto-	z	10^{21}	zetta-	Z
10^{-24}	yocto-	y	10^{24}	yotta-	Y

Appendix II
Webpages

The following are university department websites:

University of California, Irvine, Department of Ecology and Evolutionary Biology
ecoevo.bio.uci.edu
University of Connecticut, Department of Ecology and Evolutionary Biology
www.eeb.uconn.edu
Harvard University, Department of Organismic and Evolutionary Biology
www.oeb.harvard.edu
Princeton University, Department of Ecology and Evolutionary Biology
www.eeb.princeton.edu
University of Tennessee, Department of Ecology and Evolutionary Biology
eeb.bio.utk.edu
Yale University, Department of Ecology and Evolutionary Biology
www.eeb.yale.edu

There are many museum sites on the Web:

American Museum of Natural History   www.amnh.org
Harvard Museum of Natural History   www.hmnh.harvard.edu
Smithsonian Institution National Museum of Natural History
www.mnh.si.edu
The Natural History Museum, London   www.nhm.ac.uk
Yale Peabody Museum of Natural History   www.peabody.yale.edu

Useful ecology sites are:

The British Ecological Society   www.britishecologicalsociety.org
The Ecological Society of America   www.esa.org
The National Center for Ecological Analysis and Synthesis
biology.usgs.gov

Websites dealing with the environment include:

The US Environmental Protection Agency   www.epa.gov/enviroed
Environmental Sciences Division, Oak Ridge National Laboratory
www.esd.ornl.gov
Smithsonian Environmental Research Center www.serc.si.edu

A general site for biology resources is:

The US Geological Survey   biology.usgs.gov

Bibliography

Begon, M.; Harper, J. L.; & Townsend, C. R. *Ecology: Individuals, Populations and Communities.* Oxford, U.K.: Blackwell Science, 1996

Evans, G. M. & Furlong, J. C. *Environmental Biotechnology: Theory and Application.* New York: Wiley, 2002

Glasson, J.; Therivel, R.; & Chadwick, A. *Introduction to Environmental Impact Assessment.* 2nd ed. London: UCL Press, 2001

Gotelli, N. J. *A Primer of Ecology.* 3rd ed. Sunderland, Mass.: Sinauer Associates, 2000

Harrison, R. M. *Pollution: Causes, Effects and Control* London: Royal Society of Chemistry, 2001

Pepper, D. *Modern Environmentalism: A Critical Assessment.* 2nd ed. London: Taylor & Francis , 1999

Smith, A. *The Weather* London: Arrow Books, 2000

Southwood, T. R. E. & Henderson, P. A. *Ecological Methods.* 3rd ed. Oxford, U.K.: Blackwell Science, 2000

Van Loon, G. W. & Duffy, S. *Environmental Chemistry.* Oxford, U.K.: Oxford University Press, 2000